T0074828

An Introduction to Functional Analysis

This accessible text covers key results in functional analysis that are essential for further study in the calculus of variations, analysis, dynamical systems, and the theory of partial differential equations. The treatment of Hilbert spaces covers the topics required to prove the Hilbert–Schmidt Theorem, including orthonormal bases, the Riesz Representation Theorem, and the basics of spectral theory. The material on Banach spaces and their duals includes the Hahn–Banach Theorem, the Krein–Milman Theorem, and results based on the Baire Category Theorem, before culminating in a proof of sequential weak compactness in reflexive spaces. Arguments are presented in detail, and more than 200 fully-worked exercises are included to provide practice applying techniques and ideas beyond the major theorems. Familiarity with the basic theory of vector spaces and point-set topology is assumed, but knowledge of measure theory is not required, making this book ideal for upper undergraduate-level and beginning graduate-level courses.

JAMES ROBINSON is a professor in the Mathematics Institute at the University of Warwick. He has been the recipient of a Royal Society University Research Fellowship and an EPSRC Leadership Fellowship. He has written six books in addition to his many publications in infinite-dimensional dynamical systems, dimension theory, and partial differential equations.

An Introduction to Functional Analysis

JAMES C. ROBINSON
University of Warwick

CAMBRIDGE UNIVERSITY PRESS

CAMBRIDGE
UNIVERSITY PRESS

University Printing House, Cambridge CB2 8BS, United Kingdom

One Liberty Plaza, 20th Floor, New York, NY 10006, USA

477 Williamstown Road, Port Melbourne, VIC 3207, Australia

314–321, 3rd Floor, Plot 3, Splendor Forum, Jasola District Centre,
New Delhi – 110025, India

79 Anson Road, #06–04/06, Singapore 079906

Cambridge University Press is part of the University of Cambridge.

It furthers the University's mission by disseminating knowledge in the pursuit of
education, learning, and research at the highest international levels of excellence.

www.cambridge.org
Information on this title: www.cambridge.org/9780521899642
DOI: 10.1017/9781139030267

© James C. Robinson 2020

First published 2020

Printed in the United Kingdom by TJ International Ltd. Padstow Cornwall

A catalogue record for this publication is available from the British Library.

ISBN 978-0-521-89964-2 Hardback
ISBN 978-0-521-72839-3 Paperback

Cambridge University Press has no responsibility for the persistence or accuracy of
URLs for external or third-party internet websites referred to in this publication
and does not guarantee that any content on such websites is, or will remain,
accurate or appropriate.

To Mum & Dad

Contents

Preface

This book is intended to cover the core functional analysis syllabus and, in particular, presents many of the results that are needed in partial differential equations, the calculus of variations, or dynamical systems. The material is developed far enough that the next step would be application to one of these areas or further pursuit of 'functional analysis' itself at a significantly more advanced level.

The content is based on the two functional analysis modules taught at the University of Warwick to our third-year undergraduates. As such, it should be straightforward to use this book (with some judicious pruning) as the basis of a two-term course, with Part III (Hilbert spaces) taught in the first term and Part IV (Banach spaces) in the second term. Part II contains foundational material (a general theory of normed spaces and a collection of example spaces) that is needed for both Parts III and IV; some of this material could find a home in either term, according to taste. A one-term standalone module on Banach spaces could be based on Part II; Chapters 11, 14, and 15 from Part III; and Part IV.

Familiarity is assumed with the theory of finite-dimensional vector spaces and basic point-set topology (metric spaces, open and closed sets, compactness, and completeness), which is revised, at a fairly brisk pace and with some proofs omitted, in the first two chapters. No knowledge of measure theory or Lebesgue integration is required: the Lebesgue spaces are introduced as completions of the space of continuous functions in Chapter 7, with the standard construction of the Lebesgue integral outlined in Appendix B. The canonical examples of non-Hilbert Banach spaces used in Part IV are the sequence spaces ℓ^p rather than the Lebesgue spaces L^p; I hope that this will make the book accessible to a wider audience. In the same spirit I have tried to spell out all the arguments in detail; there are no[1] four-line proofs that when written with all the details expand to fill the same number of pages.

For the most part the approach adopted here is to cover the simpler case of Hilbert spaces in Part III before turning to Banach spaces, for which the theory becomes more abstract, in Part IV. There is an argument that it is more efficient to prove results in Banach spaces before specialising to Hilbert spaces, but my suspicion is that this is a product of familiarity and experience: in the same way one might argue that it is more economical to teach analysis in metric spaces before specialising to the particular case of real sequences and real-valued functions. That said, some basic concepts and results are not significantly simpler in Hilbert spaces, so portions of Parts II and III deal with Banach rather than Hilbert spaces.

By way of a very brief overview of the contents of the book, it is perhaps useful to describe the end points of Parts III and IV. Part III works towards the Hilbert–Schmidt Theorem that decomposes a self-adjoint compact operator on a Hilbert space in terms of its eigenvalues and eigenfunctions, and then applies this to the example of the Sturm–Liouville eigenvalue problem. It therefore covers orthonormal bases, orthogonal projections, the Riesz Representation Theorem, and the basics of spectral theory. Part IV culminates with the result that the closed unit ball in a reflexive Banach space is weakly sequentially compact. So this part covers dual spaces in more detail, the Hahn–Banach Theorem and applications to convex sets, results for linear operators based on the Baire Category Theorem, reflexivity, and weak and weak-∗ convergence.

Almost every chapter ends with a collection of exercises, and full solutions to these are given at the end of the book.

There are three appendices. The first shows the equivalence of Zorn's Lemma and the Axiom of Choice; the second provides a quick overview of the construction of the Lebesgue integral and proves properties of the Lebesgue spaces that rely on measure-theoretic techniques; and the third proves the Banach–Alaoglu Theorem on weak-∗ compactness of the closed unit ball in an arbitrary Banach space, a topological result that lies outside the scope of the main part of the book.

I am indebted to those at Warwick who taught the Functional Analysis courses before me, both in the selection of the material and the general approach. Although I have adapted both over the years, the skeleton of this book was provided by Robert MacKay and Keith Ball, to whom I am very

---

[1]   Actually, there is one. An abridged version of the proof that $(L^p)^* \equiv L^q$ appears in Chapter 18 and takes about half a page. The detailed proof, which requires some non-trivial measure theory, takes up two pages Appendix B.

grateful. Those who have subsequently taught the same material, Richard Sharp and Vassili Gelfreich, have also been extremely helpful.

Writing a textbook encourages a magpie approach to results, proofs, and examples. I have been extremely fortunate that there are already a large number of texts on functional analysis, and I have tried to take advantage of the many insights and the imaginative problems that they contain. Just as there are standard results and standard proofs, there are many standard exercises, but I have credited those that I have adopted that seemed particularly imaginative or unusual. In addition, there is a long list of references at the back of the book, and each of these has contributed something to this text. I would particularly like to acknowledge the book by Rynne and Youngson (2008) and the older texts by Kreyszig (1978) and Pryce (1973) as consistent sources of inspiration. The books by Giles (2000) and Lax (2002) contain many interesting examples and exercises.

I have not tried to trace the history of the many now 'classical' results that occur throughout the book. For those who are interested in this aspect of the subject, Giles (2000) has an appendix that gives a nice overview of the historical background, and historical comments are woven throughout the text by Lax (2002). Banach's 1932 monograph contains a significant proportion of the results in Part IV.

Many staff at Cambridge University Press have been involved with this project over the years: Clare Dennison, Sam Harrison, Amy He, Kaitlin Leach, Peter Thompson, and David Tranah. Given such a long list of names, it goes without saying that I would like to thank them all for their patience and support (and apologise to anybody I have missed). I would particularly like to thank Kaitlin for ultimately holding me to a deadline that meant I finally finished the book.

Lastly, I am extremely grateful to Wojciech Ożański, who read a draft version of this book and provided me with many corrections, suggestions, and insightful comments.

PART I

Preliminaries

1
Vector Spaces and Bases

Much of the theory of 'functional analysis' that we will consider in this book is an infinite-dimensional version of results familiar for linear operators between finite-dimensional vector spaces. We therefore start by recalling some of the basic theory of linear algebra, beginning with the formal definition of a vector space. We then discuss linear maps between vector spaces, and end by proving that every vector space has a basis using Zorn's Lemma. Proofs of basic results from linear algebra can be found in Friedberg et al. (2004) or in Chapter 4 of Naylor and Sell (1982), for example.

1.1 Definition of a Vector Space

The linear spaces that occur naturally in functional analysis are vector spaces defined over $\mathbb{R}$ or $\mathbb{C}$; we will refer to real or complex vector spaces respectively, but generally we will omit the word 'real' or 'complex' unless we need to make an explicit distinction between the two cases.

Throughout the book we use the symbol $\mathbb{K}$ to denote either $\mathbb{R}$ or $\mathbb{C}$.

Definition 1.1 A *vector space V over* $\mathbb{K}$ is a set V along with notions of addition in V and multiplication by scalars, i.e.

$$x + y \in V \quad \text{for} \quad x, y \in V \qquad \text{and} \qquad \lambda x \in V \quad \text{for} \quad \lambda \in \mathbb{K}, \ x \in V,$$
(1.1)

such that

(i) additive and multiplicative identities exist: there exists a zero element $0 \in V$ such that $x + 0 = x$ for all $x \in V$; and $1 \in \mathbb{K}$ is the identity for scalar multiplication, $1x = x$ for all $x \in V$;

(ii) there are additive inverses: for every $x \in V$ there exists an element $-x \in V$ such that $x + (-x) = 0$;

3

(iii) addition is commutative and associative,

$$x + y = y + x \qquad \text{and} \qquad x + (y + z) = (x + y) + z,$$

for all $x, y, z \in V$; and

(iv) multiplication is associative,

$$\alpha(\beta x) = (\alpha\beta)x \qquad \text{for all} \qquad \alpha, \beta \in \mathbb{K}, \ x \in V,$$

and distributive,

$$\alpha(x + y) = \alpha x + \alpha y \qquad \text{and} \qquad (\alpha + \beta)x = \alpha x + \beta x$$

for all $\alpha, \beta \in \mathbb{K}, x, y \in V$.

In checking that a particular collection V is a vector space over $\mathbb{K}$, properties (i)–(iv) are often immediate; one usually has to check only that V is closed under addition and scalar multiplication (i.e. that (1.1) holds).

1.2 Examples of Vector Spaces

Of course, $\mathbb{R}^n$ is a real vector space over $\mathbb{R}$; but is not a vector space over $\mathbb{C}$, since $i x \notin \mathbb{R}^n$ for any[1] $x \in \mathbb{R}^n$. In contrast, $\mathbb{C}^n$ can be a vector space over both $\mathbb{R}$ and $\mathbb{C}$; the space $\mathbb{C}^n$ over $\mathbb{R}$ is (according to the terminology introduced above) a 'real vector space'. This example is a useful illustration that the real/complex label refers to the field $\mathbb{K}$, i.e. the allowable scalar multiples, rather than to the elements of the space itself.

Given any two vector spaces V_1 and V_2 over $\mathbb{K}$, the product space $V_1 \times V_2$ consisting of all pairs (v_1, v_2) with $v_1 \in V_1$ and $v_2 \in V_2$ is another vector space if we define

$$(v_1, v_2) + (u_1, u_2) := (v_1 + u_1, v_2 + u_2) \qquad \text{and} \qquad \alpha(v_1, v_2) := (\alpha v_1, \alpha v_2),$$

for $v_1, u_1 \in V_1, v_2, u_2 \in V_2, \alpha \in \mathbb{K}$.

We now introduce some less trivial examples.

Example 1.2 The space $\mathcal{F}(U, V)$ of all functions $f : U \to V$, where U and V are both vector spaces over the same field $\mathbb{K}$, is itself a vector space, if we use the obvious definitions of what addition and scalar multiplication should mean for functions. We give these definitions here for the one and only time:

---

[1] Throughout this book we will use a bold x for elements of $\mathbb{R}^n$ (also of $\mathbb{C}^n$), with x given in components by $x = (x_1, \ldots, x_n)$.

for $f, g \in \mathcal{F}(U, V)$ and $\alpha \in \mathbb{K}$, we denote by $f + g$ the function from U to V whose values are given by

$$(f + g)(x) = f(x) + g(x), \qquad x \in U,$$

('pointwise addition') and by αf the function whose values are

$$(\alpha f)(x) = \alpha \, f(x), \qquad x \in U$$

('pointwise multiplication').

Example 1.3 The space $C([a, b]; \mathbb{K})$ of all $\mathbb{K}$-valued continuous functions on the interval $[a, b]$ is a vector space. We will often write $C([a, b])$ for $C([a, b]; \mathbb{R})$.

Proof The sum of two continuous functions is again continuous, as is any scalar multiple of a continuous function. $\qquad \square$

Example 1.4 The space $\mathcal{P}(I)$ of all real polynomials on any interval $I \subset \mathbb{R}$,

$$\mathcal{P}(I) = \left\{ p : I \to \mathbb{R} : \; p(x) = \sum_{j=0}^{n} a_j x^j, \; n = 0, 1, 2, \ldots, \; a_j \in \mathbb{R} \right\}$$

is a vector space.

The next example introduces a family of spaces that will prove to be particularly important.

Example 1.5 For $1 \le p < \infty$ the space $\ell^p(\mathbb{K})$ consists of all pth power summable sequences $x = (x_j)_{j=1}^{\infty}$ with elements in $\mathbb{K}$, i.e.

$$\ell^p(\mathbb{K}) = \left\{ x = (x_j)_{j=1}^{\infty} : \; x_j \in \mathbb{K}, \; \sum_{j=1}^{\infty} |x_j|^p < \infty \right\}.$$

For $p = \infty$, $\ell^{\infty}(\mathbb{K})$ is the space of all bounded sequences in $\mathbb{K}$. Sometimes we will simply write ℓ^p for $\ell^p(\mathbb{K})$. Note that, as with $\mathbb{K}^n$, we will use a bold x to denote a particular sequence in ℓ^p.

For $x, y \in \ell^p(\mathbb{K})$ we set

$$x + y := (x_1 + y_1, x_2 + y_2, \ldots),$$

and for $\alpha \in \mathbb{K}$, $x \in \ell^p$, we define

$$\alpha x := (\alpha x_1, \alpha x_2, \ldots).$$

With these definitions $\ell^p(\mathbb{K})$ is a vector space.

Proof The only thing that is not immediate is whether $x + y \in \ell^p(\mathbb{K})$ if $x, y \in \ell^p(\mathbb{K})$. This is clear when $p = \infty$, since

$$\sup_{j \in \mathbb{N}} |x_j + y_j| \leq \sup_{j \in \mathbb{N}} |x_j| + \sup_{j \in \mathbb{N}} |y_j| < \infty.$$

For $1 \leq p < \infty$ this follows using the inequality

$$(a + b)^p \leq [2 \max(a, b)]^p \leq 2^p (a^p + b^p), \qquad \text{for } a, b \geq 0; \qquad (1.2)$$

for every $n \in \mathbb{N}$ we have

$$\sum_{j=1}^{n} |x_j + y_j|^p \leq \sum_{j=1}^{n} 2^p (|x_j|^p + |y_j|^p) \leq 2^p \sum_{j=1}^{\infty} |x_j|^p + 2^p \sum_{j=1}^{\infty} |y_j|^p < \infty$$

and so $\sum_{j=1}^{\infty} |x_j + y_j|^p < \infty$ as required. $\qquad\qquad\qquad\square$

(The factor 2^p in (1.2) can be improved to 2^{p-1}; see Exercise 1.1.)

1.3 Linear Subspaces

If V is a vector space (over $\mathbb{K}$) then any subset $U \subset V$ is a *subspace* of V if U is again a vector space, i.e. if it is closed under addition and scalar multiplication, i.e. $u_1 + u_2 \in U$ for every $u_1, u_2 \in U$ and $\lambda u \in U$ for every $\lambda \in \mathbb{K}, u \in U$.

Example 1.6 For any $y \in \mathbb{R}^n$, the set

$$\{x \in \mathbb{R}^n : x \cdot y = 0\}$$

is a subspace of $\mathbb{R}^n$.

Example 1.7 The set

$$X = \left\{ f \in C([-1, 1]) : \int_{-1}^{0} f(x)\, dx = 0, \int_{0}^{1} f(x),\, dx = 0 \right\}$$

is a subspace of $C([-1, 1])$.

Example 1.8 The space $c_0(\mathbb{K})$ of all null sequences, i.e. of all sequences $x = (x_j)_{j=1}^{\infty}$ such that $x_j \to 0$ as $j \to \infty$, is a subspace of $\ell^\infty(\mathbb{K})$, and for every $1 \leq p < \infty$ the space $\ell^p(\mathbb{K})$ is a subspace of $c_0(\mathbb{K})$.

The space $c_{00}(\mathbb{K})$ of all sequences with only a finite number of non-zero terms is a subspace of $c_0(\mathbb{K})$ and of $\ell^p(\mathbb{K})$ for every $1 \leq p \leq \infty$.

Proof For the inclusion properties of $c_0(\mathbb{K})$, note that any convergent sequence (in particular any null sequence) is bounded, which shows that $c_0(\mathbb{K}) \subset \ell^\infty(\mathbb{K})$. If $x \in \ell^p$, $1 \le p < \infty$, then $\sum_{j=1}^\infty |x_j|^p < \infty$, which implies that $|x_j|^p \to 0$ as $j \to \infty$, so $x \in c_0(\mathbb{K})$. The properties of $c_{00}(\mathbb{K})$ are immediate. $\qquad\square$

1.4 Spanning Sets, Linear Independence, and Bases

We now recall the definition of a vector-space basis, which will also allow us to define the dimension of a vector space.

Definition 1.9 The *linear span* of a subset E of a vector space V is the collection of all finite linear combinations of elements of E:

$$\mathrm{Span}(E) = \left\{ v \in V : v = \sum_{j=1}^n \alpha_j e_j, \text{ for some } n \in \mathbb{N}, \ \alpha_j \in \mathbb{K}, \ e_j \in E \right\}.$$

We say that E *spans* V if $V = \mathrm{Span}(E)$.

If E spans V this means that we can write any $v \in V$ in the form

$$v = \sum_{j=1}^n \alpha_j e_j,$$

i.e v can be expressed as a finite linear combination of elements of E. (Once we have a way to discuss convergence we will also be able to consider 'infinite linear combinations', but these are not available when we can only use the vector-space axioms.)

Definition 1.10 A set $E \subset V$ is *linearly independent* if any finite collection of elements of E is linearly independent, i.e.

$$\sum_{j=1}^n \alpha_j e_j = 0 \quad \Rightarrow \quad \alpha_1 = \cdots = \alpha_n = 0$$

for any choice of $n \in \mathbb{N}$, $\alpha_j \in \mathbb{K}$, and $e_j \in E$.

To distinguish the standard definition of a basis for a vector space from the notion of a 'Schauder basis', which we will meet later, we refer to such a basis as a 'Hamel basis'.

Definition 1.11 A *Hamel basis* for a vector space V is any linearly independent spanning set.

Expansions in terms of basis elements are unique (for a proof see Exercise 1.3).

Lemma 1.12 *If E is a Hamel basis for V, then any element of V can be written uniquely in the form*

$$v = \sum_{j=1}^{n} \alpha_j e_j$$

for some $n \in \mathbb{N}$, $\alpha_j \in \mathbb{K}$, and $e_j \in E$.

Any Hamel basis E of V must be a maximal linearly independent set, i.e. E is linearly independent and $E \cup \{v\}$ is not linearly independent for any $v \in V \setminus E$. We now show that this can be reversed.

Lemma 1.13 *If $E \subset V$ is maximal linearly independent set, then E is a Hamel basis for V.*

Proof To show that E is a Hamel basis we only need to show that it spans V, since it is linearly independent by assumption.

If E does not span V, then there exists some $v \in V$ that cannot be written as any finite linear combination of the elements of E. To obtain a contradiction, we show that in this case $E \cup \{v\}$ must be a linearly independent set. Choose $n \in \mathbb{N}$ and $\{e_j\}_{j=1}^{n} \in E$, and suppose that

$$\sum_{j=1}^{n} \alpha_j e_j + \alpha_{n+1} v = 0.$$

Since v cannot be written as a sum of any finite collection of the $\{e_j\}$, we must have $\alpha_{n+1} = 0$, which leaves $\sum_{j=1}^{n} \alpha_j e_j = 0$. However, since E is linearly independent and $\{e_j\}_{j=1}^{n}$ is a finite subset of E it follows that $\alpha_j = 0$ for all $j = 1, \ldots, n$. Since we already have $\alpha_{n+1} = 0$, it follows that $E \cup \{v\}$ is linearly independent, contradicting the fact that E is a maximal linearly independent set. So E spans V, as claimed. $\qquad\square$

If V has a basis consisting of a finite number of elements, then every basis of V contains the same number of elements (for a proof see Exercise 1.4).

Lemma 1.14 *If V has a basis consisting of n elements, then every basis for V has n elements.*

This result allows us to make the following definition of the dimension of a vector space.

Definition 1.15 If V has a basis consisting of a finite number of elements, then V is *finite-dimensional* and the *dimension* of V is the number of elements in this basis. If V has no finite basis, then V is *infinite-dimensional*.

Since a basis is a maximal linearly independent set (Lemma 1.13), it follows that a space is infinite-dimensional if and only if for every $n \in \mathbb{N}$ one can find a set of n linearly independent elements of V.

Example 1.16 For every $1 \leq p \leq \infty$ the space $\ell^p(\mathbb{K})$ is infinite-dimensional.

Proof Let us define for each $j \in \mathbb{N}$ the sequence

$$e^{(j)} = (0, 0, \ldots, 1, 0, \ldots), \qquad (1.3)$$

which consists entirely of zeros apart from having 1 as its jth term. We can also write

$$e_i^{(j)} = \delta_{ij} := \begin{cases} 1 & i = j \\ 0 & i \neq j, \end{cases} \qquad (1.4)$$

where δ_{ij} is the Kronecker delta. These are all elements of $\ell^p(\mathbb{K})$ for every $p \in [1, \infty]$, and will frequently prove useful in what follows.

For any $n \in \mathbb{N}$ the n elements $\{e^{(j)}\}_{j=1}^n$ are linearly independent, since

$$\sum_{j=1}^n \alpha_j e^{(j)} = (\alpha_1, \alpha_2, \ldots, \alpha_n, 0, 0, 0, \ldots) = \mathbf{0}$$

implies that $\alpha_1 = \alpha_2 = \cdots = \alpha_n = 0$. It follows that $\ell^p(\mathbb{K})$ is an infinite-dimensional vector space. $\square$

Example 1.17 The vector space $C([0, 1]; \mathbb{K})$ is infinite-dimensional.

Proof For any $n \in \mathbb{N}$ the functions $\{1, x, x^2, \ldots, x^n\}$ are linearly independent: if

$$f(x) := \sum_{j=0}^n \alpha_j x^j = 0 \qquad \text{for every } x \in [0, 1],$$

then $\alpha_j = 0$ for every j. To see this, first set $x = 0$, which shows that $\alpha_0 = 0$, then differentiate once to obtain

$$f'(x) = \sum_{j=1}^{n} \alpha_j j x^{j-1} = 0$$

and set $x = 0$ to show that $\alpha_1 = 0$. Continue differentiating repeatedly, each time setting $x = 0$ to show that $\alpha_j = 0$ for all $j = 0, \ldots, n$. $\qquad\square$

1.5 Linear Maps between Vector Spaces and Their Inverses

Vector spaces have a linear structure, i.e. we can add elements and multiply by scalars. When we consider maps from one vector space to another, it is natural to consider maps that respect this linear structure.

Definition 1.18 If X and Y are vector spaces over $\mathbb{K}$, then a map $T : X \to Y$ is *linear* if

$$T(x + x') = T(x) + T(x') \quad \text{and} \quad T(\alpha x) = \alpha T(x), \qquad \alpha \in \mathbb{K},\ x, x' \in X.$$

(This is the same as requiring that $T(\alpha x + \beta x') = \alpha T(x) + \beta T(x')$ for any $\alpha, \beta \in \mathbb{K}, x, x' \in U$.)

We often omit the brackets around the argument, and write Tx for $T(x)$ when T is linear.

Note that the definition of what it means to be linear involves the field $\mathbb{K}$. So, for example, if we take $X = Y = \mathbb{C}$ and let $T(z) = \bar{z}$ (the complex conjugate of z), this map is linear if we take $\mathbb{K} = \mathbb{R}$, but not if we take $\mathbb{K} = \mathbb{C}$. We always have

$$T(z + w) = \overline{z + w} = \bar{z} + \bar{w} = T(z) + T(w), \qquad z, w \in \mathbb{C},$$

but the linearity property for scalar multiples only holds if $\alpha \in \mathbb{R}$, since

$$T(\alpha z) = \overline{\alpha z} = \bar{\alpha}\,\bar{z}$$

and this is equal to $\alpha \bar{z} = \alpha T(z)$ if and only if $\alpha \in \mathbb{R}$.

This kind of 'conjugate-linear' behaviour is common enough that it is worth making a formal definition.

Definition 1.19 If X and Y are vector spaces over $\mathbb{C}$, then a map $T : X \to Y$ is *conjugate-linear* if

$$T(x + x') = Tx + Tx' \quad \text{and} \quad T(\alpha x) = \overline{\alpha}\, Tx, \qquad \alpha \in \mathbb{C}, \ x, x' \in X.$$

(Such maps are sometimes called anti-linear.)

The space of all linear maps from X into Y we write as $L(X, Y)$, and when $Y = X$ we abbreviate this to $L(X)$. This is a vector space: for $T_1, T_2 \in L(X, Y)$ and $\alpha \in \mathbb{K}$ we define $T_1 + T_2$ and αT_1 by setting

$$(T_1 + T_2)(x) = T_1 x + T_2 x \qquad \text{and} \qquad (\alpha T_1)(x) = \alpha T_1 x, \qquad x \in X.$$

With these definitions a linear combination of two linear maps is again a linear map:

$$T_1, T_2 \in L(X, Y) \quad \Rightarrow \quad \alpha T_1 + \beta T_2 \in L(X, Y), \qquad \alpha, \beta \in \mathbb{K}.$$

Similarly the composition of compatible linear maps is again linear,

$$T \in L(X, Y), \ S \in L(Y, Z) \quad \Rightarrow \quad S \circ T \in L(X, Z)$$

since

$$(S \circ T)(\alpha x + \beta x') = S(\alpha Tx + \beta Tx') = \alpha(S \circ T)x + \beta(S \circ T)x'.$$

Definition 1.20 If $T \in L(X, Y)$, then we define its *kernel* as

$$\mathrm{Ker}(T) := \{x \in X : Tx = 0\}$$

and its *range* (or image) as

$$\mathrm{Range}(T) := \{y \in Y : y = Tx \text{ for some } x \in X\}.$$

These are both vector spaces (see Exercise 1.5).

One particularly simple (but important) example of a linear map is the *identity map* $I_X : X \to X$ given by $I_X(x) = x$.

Recall that a map $T : X \to Y$ is *injective* (or *one-to-one*) if

$$Tx = Tx' \quad \Rightarrow \quad x = x'.$$

To check if a linear map $T : X \to Y$ is injective, it is enough to show that its kernel is trivial, i.e. that $\mathrm{Ker}(T) = \{0\}$.

Lemma 1.21 *A map $T \in L(X, Y)$ is injective if and only if* $\mathrm{Ker}(T) = \{0\}$.

Proof We prove the equivalent statement that T is not injective if and only if $\mathrm{Ker}(T) \neq \{0\}$.

If T is not injective, then there exist $x_1, x_2 \in X$ with $x_1 \neq x_2$ such that $Tx_1 = Tx_2$, i.e. $T(x_1 - x_2) = 0$, and so $x_1 - x_2 \in \mathrm{Ker}(T)$ and therefore $\mathrm{Ker}(T) \neq \{0\}$. On the contrary, if $z \in \mathrm{Ker}(T)$ with $z \neq 0$, then for any $x_1 \in X$ we have $T(x_1 + z) = Tx_1$ and T is not injective. $\qquad\square$

A map $T \colon X \to Y$ is *surjective* (or *onto*) if for every $y \in Y$ there exists $x \in X$ such that $Tx = y$.

Lemma 1.22 *If X is a finite-dimensional vector space and $T \in L(X)$, then T is injective if and only if T is surjective.*

Proof The Rank–Nullity Theorem (e.g. Theorem 2.3 in Friedberg et al. (2014) or Theorem 4.7.7 in Naylor and Sell (1982)) guarantees that

$$\dim(\mathrm{Ker}(T)) + \dim(\mathrm{Range}(T)) = \dim(X)$$

(the 'nullity' is the dimension of $\mathrm{Ker}(T)$ and the 'rank' is the dimension of $\mathrm{Range}(T)$). By Lemma 1.21, T is injective when $\dim(\mathrm{Ker}(T)) = 0$, which then implies that $\dim(\mathrm{Range}(T)) = \dim(X)$ so that T is onto; similarly, if T is onto, then $\dim(\mathrm{Range}(T)) = \dim(X)$ which implies that $\dim(\mathrm{Ker}(T)) = 0$, and so T is also injective. $\qquad\square$

A map is *bijective* or *a bijection* if it is both injective and surjective. When T is a bijection we can define its inverse.

Definition 1.23 A map $T \colon X \to Y$ has an *inverse* $T^{-1} \colon Y \to X$ if T is a bijection, and in this case for each $y \in Y$ we define $T^{-1}y$ to be the unique $x \in X$ such that $Tx = y$.

Note that if $\mathrm{Ker}(T) = \{0\}$, then the linear map $T \colon X \to \mathrm{Range}(T)$ always has an inverse; if X is infinite-dimensional T may not map X onto Y, but it always maps X onto $\mathrm{Range}(T)$, by definition.

The following lemma shows that when $T \in L(X, Y)$ has an inverse, the map $T^{-1} \colon Y \to X$ is also linear.

Lemma 1.24 *A linear map $T \in L(X, Y)$ has an inverse if and only if there exists $S \in L(Y, X)$ such that*

$$ST = I_X \qquad and \qquad TS = I_Y, \tag{1.5}$$

and then $T^{-1} = S$.

Proof Suppose that $T\colon X \to Y$ is a bijection, so that it has an inverse $T^{-1}\colon Y \to X$; from the definition it follows that $TT^{-1} = I_Y$ and $T^{-1}T = I_X$. It remains to check that $T^{-1}\colon Y \to X$ is linear; this follows from the injectivity of T, since

$$T[T^{-1}(\alpha y + \beta z)] = \alpha y + \beta z = T[\alpha T^{-1} y + \beta T^{-1} z]$$

therefore implies that

$$T^{-1}(\alpha y + \beta z) = \alpha T^{-1} y + \beta T^{-1} z.$$

For the converse, we note that $TS = I_Y$ implies that $T\colon X \to Y$ is onto, since $T(Sy) = y$, and that $ST = I_X$ implies that $T\colon X \to Y$ is one-to-one, since

$$Tx = Ty \quad \Rightarrow \quad S(Tx) = S(Ty) \quad \Rightarrow \quad x = y.$$

It follows that if (1.5) holds, then T has an inverse T^{-1}, and applying T^{-1} to both sides of $TS = I_Y$ shows that $S = T^{-1}$. □

Note that if $T \in L(X, Y)$ and $S \in L(Y, Z)$ are both invertible, then so is $ST \in L(X, Z)$, with

$$(ST)^{-1} = T^{-1}S^{-1}; \tag{1.6}$$

since ST is a bijection it has an inverse $(ST)^{-1}$ such that $ST(ST)^{-1} = I_Z$; multiplying first by S^{-1} and then by T^{-1} yields (1.6).

1.6 Existence of Bases and Zorn's Lemma

We end this chapter by showing that every vector space has a Hamel basis. To prove this, we will use Zorn's Lemma, which is a very powerful result that will allow us to prove various existence results throughout this book. To state this 'lemma' (which is in fact equivalent to the Axiom of Choice, as shown in Appendix A) we need to introduce some auxiliary concepts.

Definition 1.25 A *partial order* on a set $\mathcal{P}$ is a binary relation $\preceq$ on $\mathcal{P}$ such that for $a, b, c \in \mathcal{P}$

(i) $a \preceq a$;
(ii) $a \preceq b$ and $b \preceq a$ implies that $a = b$; and
(iii) $a \preceq b$ and $b \preceq c$ implies that $a \preceq c$.

The order is 'partial' because two arbitrary elements of $\mathcal{P}$ need not be ordered: consider for example, the case when $\mathcal{P}$ consists of all subsets of $\mathbb{R}$ and $X \preceq Y$ if $X \subseteq Y$; one cannot order $[0, 1]$ and $[1, 2]$.

Definition 1.26 Two elements $a, b \in \mathcal{P}$ are *comparable* if $a \preceq b$ or $b \preceq a$ (or both if $a = b$). A subset $\mathcal{C}$ of $\mathcal{P}$ is called a *chain* if any pair of elements of $\mathcal{C}$ are comparable.

An element $b \in \mathcal{P}$ is an *upper bound* for a subset $\mathcal{S}$ of $\mathcal{P}$ if $s \preceq b$ for all $s \in \mathcal{S}$. An element m of $\mathcal{P}$ is *maximal* if $m \preceq a$ for some $a \in \mathcal{P}$ implies that $a = m$.

Note that among any finite collection of elements in a chain $\mathcal{C}$ there is always a maximal and a minimal element: if $c_1, \ldots, c_n \in \mathcal{C}$, then there are indices $j, k \in \{1, \ldots, n\}$ such that

$$c_j \preceq c_i \preceq c_k \qquad i = 1, \ldots, n; \tag{1.7}$$

this can easily be proved by induction on n; see Exercise 1.7.

Theorem 1.27 (Zorn's Lemma) *If $\mathcal{P}$ is a non-empty partially ordered set in which every chain has an upper bound, then $\mathcal{P}$ has at least one maximal element.*

It is easy to find examples in which there is more than one maximal element. For example, let $\mathcal{P}$ consist of all points in the two disjoint intervals $I_1 = [0, 1]$ and $I_2 = [2, 3]$, and say that $a \preceq b$ if a and b are contained in the same interval and $a \leq b$. Then every chain in $\mathcal{P}$ has an upper bound, and $\mathcal{P}$ contains two maximal elements, 1 and 3.

Theorem 1.28 *Every vector space has a Hamel basis.*

Proof If V is finite-dimensional, then V has a finite-dimensional basis, by definition.

So we assume that V is infinite-dimensional. Let $\mathcal{P}$ be the collection of all linearly independent subsets of V. We define a partial order on $\mathcal{P}$ by declaring that $E_1 \preceq E_2$ if $E_1 \subseteq E_2$. If $\mathcal{C}$ is a chain in $\mathcal{P}$, then set

$$E^* = \bigcup_{E \in \mathcal{C}} E.$$

Note that E^* is linearly independent, since by (1.7) any finite collection of elements of E^* must be contained in one $E \in \mathcal{C}$ (which is linearly independent). Clearly $E \preceq E^*$ for all $E \in \mathcal{C}$, so E^* is an upper bound for $\mathcal{C}$.

It follows from Zorn's Lemma that $\mathcal{P}$ has a maximal element, i.e. a maximal linearly independent set, and by Lemma 1.13 this is a Hamel basis for V. $\quad\square$

As an example of a Hamel basis for an infinite-dimensional vector space, it is easy to see that the countable set $\{e^{(j)}\}_{j=1}^{\infty}$ (as defined in (1.3)) is a Hamel basis for the space c_{00} from Example 1.8. However, this is a somewhat artificial example. We will see later (Exercises 5.7 and 22.1) that no Banach space (the particular class of vector spaces that will be our main subject in most of the rest of this book) can have a countable Hamel basis.

Exercises

1.1   Show that if $p \geq 1$ and $a, b \geq 0$, then

$$(a + b)^p \leq 2^{p-1}(a^p + b^p).$$

[Hint: find the maximum of the function $f(x) = (1 + x)^p/(1 + x^p)$.]

1.2   For $1 \leq p < \infty$, show that the set $\tilde{L}^p(0, 1)$ of all continuous real-valued functions on $(0, 1)$ for which

$$\int_0^1 |f(x)|^p \, \mathrm{d}x < \infty$$

is a vector space (with the obvious pointwise definitions of addition and scalar multiplication).

1.3   Show that if E is a basis for a vector space V, then every non-zero $v \in V$ can be written *uniquely* in the form $v = \sum_{j=1}^{n} \alpha_j e_j$, for some $n \in \mathbb{N}$, $e_j \in E$, and non-zero coefficients $\alpha_j \in \mathbb{K}$.

1.4   Show that if V has a basis consisting of n elements, then every basis for V has n elements.

1.5   If $T \in L(X, Y)$ show that $\mathrm{Ker}(T)$ and $\mathrm{Im}(T)$ are both vector spaces.

1.6   If X is a vector space over $\mathbb{K}$ and U is a subspace of X define an equivalence relation on X by

$$x \sim y \qquad \Leftrightarrow \qquad x - y \in U.$$

The quotient space X/U is the set of all equivalence classes

$$[x] = x + U := \{x + u : u \in U\}$$

for $x \in X$. Show that this is a vector space over $\mathbb{K}$ if we define

$$[x] + [y] := [x + y] \qquad \lambda[x] := [\lambda x], \qquad x, y \in X, \ \lambda \in K,$$

and deduce that the quotient map $Q \colon X \to X/U$ given by $x \mapsto [x]$ is linear.

1.7 Show that among any finite collection of elements in a chain $\mathcal{C}$ there is always a maximal and a minimal element: if $c_1, \ldots, c_n \in \mathcal{C}$, then there exist $j, k \in \{1, \ldots, n\}$ such that

$$c_j \preceq c_i \preceq c_k \qquad i = 1, \ldots, n.$$

(Use induction on n.)

1.8 Let Z be a linearly independent subset of a vector space V. Use Zorn's Lemma to show that V has a Hamel basis that contains Z.

2

Metric Spaces

Most of the results in this book concern normed spaces; but these are particular examples of metric spaces, and there are some 'standard results' that are
no harder to prove in the more general context of metric spaces. In this chapter we therefore recall the definition of a metric space, along with definitions
of convergence, continuity, separability, and compactness. The treatment in
this chapter is intentionally brisk, but proofs are included. For a more didactic
treatment see Sutherland (1975), for example.

2.1 Metric Spaces

A metric on a set X is a generalisation of the 'distance between two points'
familiar in Euclidean spaces.

Definition 2.1 A metric d on a set X is a map $d: X \times X \to [0, \infty)$ that
satisfies

(i) $d(x, y) = 0$ if and only if $x = y$;
(ii) $d(x, y) = d(y, x)$ for every $x, y \in X$; and
(iii) $d(x, z) \leq d(x, y) + d(y, z)$ for $x, y, z \in X$ ('the triangle inequality').

Even on a familiar space there can be many possible metrics.

Example 2.2 Take $X = \mathbb{K}^n$ with any one of the metrics

$$
d_{\ell^p}(\boldsymbol{x}, \boldsymbol{y}) = \begin{cases} \left(\sum_{j=1}^{n} |x_j - y_j|^p\right)^{1/p} & 1 \leq p < \infty, \\ \max_{j=1,\dots,n} |x_j - y_j| & p = \infty. \end{cases}
$$

The 'standard metric' on $\mathbb{K}^n$ is

$$d_{\ell^2}(\boldsymbol{x}, \boldsymbol{y}) = \left(\sum_{j=1}^{n} |x_j - y_j|^2 \right)^{1/2} ;$$

this is the metric we use on $\mathbb{K}^n$ (or subsets of $\mathbb{K}^n$) if none is specified.

Proof Property (i) is trivial, since $d_{\ell^p}(\boldsymbol{x}, \boldsymbol{y}) = 0$ implies that $x_j = y_j$ for each j, and property (ii) is immediate.

We show here that d_{ℓ^p} satisfies (iii) only for $p = 1, 2, \infty$, the most common cases. The proof for general p is given in Lemma 3.6.

For $p = 1$

$$d_{\ell^1}(\boldsymbol{x}, \boldsymbol{z}) = \sum_{j=1}^{n} |x_j - z_j| \leq \sum_{j=1}^{n} |x_j - y_j| + |y_j - z_j|$$

$$= \sum_{j=1}^{n} |x_j - y_j| + \sum_{j=1}^{n} |y_j - z_j| = d_{\ell^1}(\boldsymbol{x}, \boldsymbol{y}) + d_{\ell^1}(\boldsymbol{y}, \boldsymbol{z}),$$

using the triangle inequality in $\mathbb{K}$. For $p = \infty$ we have similarly

$$d_{\ell^\infty}(\boldsymbol{x}, \boldsymbol{z}) = \max_{j=1,\ldots,n} |x_j - z_j| \leq \max_{j=1,\ldots,n} |x_j - y_j| + |y_j - z_j|$$

$$\leq \max_{j=1,\ldots,n} |x_j - y_j| + \max_{j=1,\ldots,n} |y_j - z_j|$$

$$= d_{\ell^\infty}(\boldsymbol{x}, \boldsymbol{y}) + d_{\ell^\infty}(\boldsymbol{y}, \boldsymbol{z}).$$

For $p = 2$, writing $\xi_j = |x_j - y_j|$ and $\eta_j = |y_j - z_j|$,

$$d_{\ell^2}(\boldsymbol{x}, \boldsymbol{z})^2 = \sum_{j=1}^{n} |x_j - z_j|^2 \leq \sum_{j=1}^{n} \left[|x_j - y_j| + |y_j - z_j| \right]^2$$

$$= \sum_{j=1}^{n} \xi_j^2 + 2\xi_j \eta_j + \eta_j^2 \qquad (2.1)$$

$$\leq \left(\sum_{j=1}^{n} \xi_j^2 \right) + 2 \left(\sum_{j=1}^{n} \xi_j^2 \right)^{1/2} \left(\sum_{j=1}^{n} \eta_j^2 \right)^{1/2} + \left(\sum_{j=1}^{n} \eta_j^2 \right) \quad (2.2)$$

$$= \left[\left(\sum_{j=1}^{n} \xi_j^2 \right)^{1/2} + \left(\sum_{j=1}^{n} \eta_j^2 \right)^{1/2} \right]^2$$

$$= [d_{\ell^2}(\boldsymbol{x}, \boldsymbol{y}) + d_{\ell^2}(\boldsymbol{y}, \boldsymbol{z})]^2,$$

where to go from (2.1) to (2.2) we used the Cauchy–Schwarz inequality

$$\left(\sum_{j=1}^{n} \xi_j \eta_j\right)^2 \le \left(\sum_{j=1}^{n} \xi_j^2\right)\left(\sum_{j=1}^{n} \eta_j^2\right);$$

see Exercise 2.1 (and Lemma 8.5 in a more general context). $\qquad\square$

Note that the space X in the definition of a metric need not be a vector space. The following example provides a metric on any set X; it is very useful for counterexamples.

Example 2.3 The *discrete metric* on any set X is defined by setting

$$d(x, y) = \begin{cases} 0 & x = y, \\ 1 & x \ne y. \end{cases}$$

If A is a subset of X and d is a metric on X, then $(A, d|_{A \times A})$ is another metric space, where by $d|_{A \times A}$ we denote the restriction of d to $A \times A$, i.e.

$$d|_{A \times A}(a, b) = d(a, b) \qquad a, b \in A; \tag{2.3}$$

we usually drop the $|_{A \times A}$ since this is almost always clear from the context.

If we have two metric spaces (X_1, d_1) and (X_2, d_2), then we can choose many possible metrics on the product space $X_1 \times X_2$. The most useful choices are

$$\varrho_1\big((x_1, x_2), (y_1, y_2)\big) := d_1(x_1, y_1) + d_2(x_2, y_2) \tag{2.4}$$

and

$$\varrho_2\big((x_1, x_2), (y_1, y_2)\big) := \left[d_1(x_1, y_1)^2 + d_2(x_2, y_2)^2\right]^{1/2}. \tag{2.5}$$

These have obvious generalisations to the product of any finite number of metric spaces. While the expression in (2.4) is simpler and easier to work with, the definition in (2.5) ensures that the metric on $\mathbb{K}^n$ that comes from viewing it as the n-fold product $\mathbb{K} \times \mathbb{K} \times \cdots \times \mathbb{K}$ agrees with the usual Euclidean distance. Exercise 2.2 provides a larger family ϱ_p of product metrics.

2.2 Open and Closed Sets

The notion of an open set is fundamental in the study of metric spaces, and forms the basis of the theory of topological spaces (see Appendix C). We begin with the definition of an open ball.

Definition 2.4 If $r > 0$ and $a \in X$ we define the open ball of radius r centred at a as

$$B_X(a, r) := \{x \in X : d(x, a) < r\}.$$

If the space X is clear from the context (as in some of the following definitions), then we will omit the X subscript.

Definition 2.5 A subset A of a metric space (X, d) is *open* if for every $x \in A$ there exists $r > 0$ such that $B(x, r) \subseteq A$. A subset A of (X, d) is *closed* if $X \setminus A$ is open.

Note that the whole space X and the empty set $\varnothing$ are always open, so at the same time X and $\varnothing$ are also always closed. The open ball $B(x, r)$ is open for any $x \in X$ and any $r > 0$ (see Exercise 2.6) and any open subset of X can be written as the union of open balls (see Exercise 2.7).

Note that in any set X with the discrete metric, any subset A of X is open (since if $x \in A$, then $B(x, 1/2) = \{x\} \subseteq A$) and any subset is closed (since $X \setminus A$ is open).

Lemma 2.6 *Any finite intersection of open sets is open, and any union of open sets is open. Any finite union of closed sets is closed, and any intersection of closed sets is closed.*

Proof We prove the result for open sets; for the corresponding results for closed sets (which follow by taking complements) see Exercise 2.5.

Let $U = \cup_{\alpha \in \mathbb{A}} U_\alpha$, where $\mathbb{A}$ is any index set; if $x \in U$, then $x \in U_\alpha$ for some $\alpha \in \mathbb{A}$, and then there exists $r > 0$ such that $B(x, r) \subseteq U_\alpha \subseteq U$, so U is open.

If $U = \cap_{j=1}^n U_j$ and $x \in U$, then for each j we have $x \in U_j$, and so $B(x, r_j) \subseteq U_j$ for some $r_j > 0$. Taking $r = \min_j r_j$ it follows that

$$B(x, r) \subseteq \cap_{j=1}^n U_j = U. \qquad \square$$

In many arguments in this book it will be useful to have a less 'topological' definition of a closed set, based on the limits of sequences. We first define what it means for a sequence to converge in a metric space.

Throughout this book we will use the notation $(x_n)_{n=1}^\infty$ for a sequence (to distinguish it from the set $\{x_n\}_{n=1}^\infty$ in which the order of the elements is irrelevant); we will often abbreviate this to (x_n), including the index if this is required to prevent ambiguity, e.g. for a subsequence $(x_{n_k})_k$. We will also frequently abbreviate 'a sequence $(x_n)_{n=1}^\infty$ such that $x_n \in A$ for every $n \in \mathbb{N}$' to 'a sequence $(x_n) \in A$'.

Definition 2.7 A sequence $(x_n)_{n=1}^{\infty}$ in a metric space (X, d) converges in (X, d) to $x \in X$ if $d(x_n, x) \to 0$ as $n \to \infty$. We write $x_n \to x$ in (X, d) (or often simply 'in X').

For sequences in $\mathbb{K}$ we often use the fact that any convergent sequence is bounded, and the same is true in a metric space, given the following definition.

Definition 2.8 A subset Y of a metric space (X, d) is *bounded* if there exists[1] $a \in X$ and $r > 0$ such that $Y \subseteq B(a, r)$, i.e. $d(y, a) < r$ for every $y \in Y$.

Any convergent sequence is bounded, since if $x_n \to x$, then there exists $N \in \mathbb{N}$ such that $d(x_n, x) < 1$ for all $n \geq N$ and so

$$d(x_n, x) \leq \max \left(1, \max_{j=1,\dots,N-1} d(x_j, x) \right) \qquad \text{for every } n \in \mathbb{N}.$$

We now describe convergence in terms of open sets.

Lemma 2.9 *A sequence $(x_n) \in (X, d)$ converges to x if and only if for any open set U that contains x there exists an N such that $x_n \in U$ for every $n \geq N$.*

Proof Given any open set U that contains x there exists $\varepsilon > 0$ such that $B(x, \varepsilon) \subseteq U$, and so there exists N such that $x_n \in B(x, \varepsilon) \subseteq U$ for all $n \geq N$. For the other implication, just use the fact that for any $\varepsilon > 0$ the set $B(x, \varepsilon)$ is open and contains x. □

We can now characterise closed sets in terms of the limits of sequences.

Lemma 2.10 *A subset A of (X, d) is closed if and only if whenever $(x_n) \in A$ with $x_n \to x$ it follows that $x \in A$.*

Proof Suppose that A is closed and that $(x_n) \in A$ with $x_n \to x$, but $x \notin A$. Then $X \setminus A$ is open and contains x, and so there exists N such that $x_n \in X \setminus A$ for all $n \geq N$, a contradiction.

Now suppose that whenever $(x_n) \in A$ with $x_n \to x$ we have $x \in A$, but A is not closed. Then $X \setminus A$ is not open: there exists $y \in X \setminus A$ and a sequence $r_n \to 0$ such that $B(y, r_n) \cap A \neq \varnothing$. So there exist points $y_n \in B(y, r_n) \cap A$, i.e. a sequence $(y_n) \in A$ such that $y_n \to y$. But then, by assumption, $y \in A$, a contradiction once more. □

---

[1] We could require that $a \in Y$ in this definition, since if $Y \subseteq B(a', r)$ with $a' \in X$, then for any choice of $a \in Y$ we have $d(y, a) \leq d(y, a') + d(a', a) < r + d(a', a)$.

2.3 Continuity and Sequential Continuity

We now define what it means for a map $f : X \to Y$ to be continuous when (X, d_X) and (Y, d_Y) are two metric spaces. We begin with the ε–δ definition.

Definition 2.11 A function $f : (X, d_X) \to (Y, d_Y)$ is *continuous at* $x \in X$ if for every $\varepsilon > 0$ there exists $\delta > 0$ such that

$$d_X(x', x) < \delta \quad \Rightarrow \quad d_Y(f(x'), f(x)) < \varepsilon.$$

We say that f is *continuous (on X)* if f is continuous at every $x \in X$.

Note that strictly there is a distinction to be made between f as a map from a set X into a set Y, and the continuity of f, which depends on the metrics d_X and d_Y on X and Y; this distinction is often blurred in practice.

As with simple real-valued functions, continuity and sequential continuity are equivalent in metric spaces (for a proof see Exercise 2.8).

Lemma 2.12 *A function* $f : (X, d_X) \to (Y, d_Y)$ *is continuous at* x *if and only if* $f(x_n) \to f(x)$ *in* Y *whenever* $(x_n) \in X$ *with* $x_n \to x$ *in* X.

Continuity can also be characterised in terms of open sets, by requiring the preimage of an open set to be open. This allows the notion of continuity to be generalised to topological spaces (see Appendix C).

Lemma 2.13 *A function* $f : (X, d_X) \to (Y, d_Y)$ *is continuous on* X *if and only if whenever* U *is an open set in* (Y, d_Y), $f^{-1}(U)$ *is an open set in* (X, d_X), *where*

$$f^{-1}(U) := \{x \in X : f(x) \in U\}$$

is the preimage *of* U *under* f. *The same is true if we replace open sets by closed sets.*

Proof Suppose that f is continuous (in the sense of Definition 2.11). Take an open subset U of Y, and $z \in f^{-1}(U)$. Since $f(z) \in U$ and U is open in Y, there exists an $\varepsilon > 0$ such that $B_Y(f(z), \varepsilon) \subseteq U$. Since f is continuous, there exists a $\delta > 0$ such that $x \in B_X(z, \delta)$ implies that $f(x) \in B_Y(f(z), \varepsilon) \subseteq U$. So $B_X(z, \delta) \subseteq f^{-1}(U)$, i.e. $f^{-1}(U)$ is open.

For the opposite implication, take $x \in X$ and set $U := B_Y(f(x), \varepsilon)$, which is an open set in Y. It follows that $f^{-1}(U)$ is open in X, so in particular, $B_X(x, \delta) \subseteq f^{-1}(U)$ for some $\delta > 0$. So

$$f(B_X(x, \delta)) \subseteq B_Y(f(x), \varepsilon),$$

which implies that f is continuous.

The result for closed sets follows from the identity

$$f^{-1}(Y \setminus A) = X \setminus f^{-1}(A) \qquad \text{for all} \qquad A \subseteq Y. \qquad \qquad \square$$

One has to be a little careful with preimages. If $f \colon X \to Y$, then for $U \subseteq X$ and $V \subseteq Y$ we have

$$f^{-1}(f(U)) \supseteq U \qquad \text{and} \qquad f(f^{-1}(V)) \subseteq V.$$

However, both these inclusions can be strict, as the simple example $f \colon \mathbb{R} \to \mathbb{R}$ with $f(x) = 0$ for every $x \in \mathbb{R}$ shows: here we have

$$f^{-1}(f([-1, 1])) = f^{-1}(0) = \mathbb{R} \qquad \text{and} \qquad f(f^{-1}([-1, 1])) = f(\mathbb{R}) = 0.$$

Also be aware that in general the *image* of an open set under a continuous map need not be open, e.g. the image of $(-4, 4)$ under the map $x \mapsto \sin x$ is $[-1, 1]$.

2.4 Interior, Closure, Density, and Separability

We recall the definition of the interior A° and closure $\overline{A}$ of a subset of a metric space. The closure operation allows us to define what it means for a subset to be dense $(\overline{A} = X)$ and this in turn gives rise to the notion of separability (existence of a countable dense subset).

Definition 2.14 If $A \subseteq (X, d)$, then the *interior* of A, written A°, is the union of all open subsets of A.

Note that A° is open (since it is the union of open sets; see Lemma 2.6) and that $A^\circ = A$ if and only if A is open.

Lemma 2.15 *A point $x \in X$ is contained in A° if and only if*

$$B(x, \varepsilon) \subseteq A \qquad \text{for some } \varepsilon > 0.$$

Proof If $x \in A^\circ$, then it is an element of some open set $U \subseteq A$, and then $B(x, \varepsilon) \subseteq U \subseteq A$ for some $\varepsilon > 0$. Conversely, if $B(x, \varepsilon) \subseteq A$, then we have $B(x, \varepsilon) \subseteq A^\circ$, and so $x \in A^\circ$. $\qquad \square$

We will make significantly more use of the closure in what follows.

Definition 2.16 If $A \subseteq (X, d)$, then the *closure* of A in X, written $\overline{A}$, is the intersection of all closed subsets of X that contain A.

Note that $\overline{A}$ is closed (since it is the intersection of closed sets; see Lemma 2.6 again). Furthermore, A is closed if and only if $A = \overline{A}$ and hence $\overline{\overline{A}} = \overline{A}$.

Lemma 2.17 *A point $x \in X$ is contained in $\overline{A}$ if and only if*

$$B(x, \varepsilon) \cap A \neq \varnothing \qquad \text{for every } \varepsilon > 0. \tag{2.6}$$

It follows that $x \in \overline{A}$ if and only if there exists a sequence $(x_n) \in A$ such that $x_n \to x$.

Proof We prove the reverse, that $x \notin \overline{A}$ if and only if $B(x, \varepsilon) \cap A = \varnothing$ for every $\varepsilon > 0$.

If $x \notin \overline{A}$, then there is some closed set K that contains A such that $x \notin K$. Since K is closed, $X \setminus K$ is open, and so $B(x, \varepsilon) \cap K = \varnothing$ for some $\varepsilon > 0$, which shows that $B(x, \varepsilon) \cap A = \varnothing$ (since $K \supseteq A$).

Conversely, if there exists $\varepsilon > 0$ such that $B(x, \varepsilon) \cap A = \varnothing$, then x is not contained in the closed set $X \setminus B(x, \varepsilon)$, which contains A; so $x \notin \overline{A}$. This proves the 'if and only if' statement in the lemma.

To prove the final part, if $x \in \overline{A}$, then (2.6) implies that for any $n \in \mathbb{N}$ we have $B(x, 1/n) \cap A \neq \varnothing$, so we can find $x_n \in A$ such that $d(x_n, x) < 1/n$ and thus $x_n \to x$. Conversely, if $(x_n) \in A$ with $x_n \to x$, then $d(x_n, x) < \varepsilon$ for n sufficiently large, which gives (2.6). $\qquad \square$

Note that in a general metric space

$$\overline{B_X(a, r)} \neq \{x \in X : d(x, a) \leq r\}. \tag{2.7}$$

If we use the discrete metric from Example 2.3, then $B_X(a, 1) = \{a\}$ for any $a \in X$, and since $\{a\}$ is closed we have $\overline{B_X(a, 1)} = \{a\}$. However, $\{y \in X : d(x, a) \leq 1\} = X$.

Given the definition of the closure of the set, we can now define what it means for a set $A \subset X$ to be dense in (X, d).

Definition 2.18 A subset A of a metric space (X, d) is *dense* in X if $\overline{A} = X$.

Using Lemma 2.17 an equivalent definition is that A is dense in X if for every $x \in X$ and every $\varepsilon > 0$

$$B(x, \varepsilon) \cap A \neq \varnothing,$$

i.e. there exists $a \in A$ such that $d(a, x) < \varepsilon$. Another similar reformulation is that $A \cap U \neq \varnothing$ for every open subset U of X.

Definition 2.19 A metric space (X, d) is *separable* if it contains a countable dense subset.

Separability means that elements of X can be approximated arbitrarily closely by some countable collection $\{x_n\}_{n=1}^{\infty}$: given any $x \in X$ and $\varepsilon > 0$, there exists $j \in \mathbb{N}$ such that $d(x_j, x) < \varepsilon$.

For some familiar examples, $\mathbb{R}$ is separable, since $\mathbb{Q}$ is a countable dense subset; $\mathbb{C}$ is separable since the set '$\mathbb{Q} + i\mathbb{Q}$' of all complex numbers of the form $q_1 + iq_2$ with $q_1, q_2 \in \mathbb{Q}$ is countable and dense. Since separability of (X, d_X) and (Y, d_Y) implies separability of $X \times Y$ (with an appropriate metric; see Exercise 2.9), it follows that $\mathbb{R}^n$ and $\mathbb{C}^n$ are separable.

Separability is inherited by subsets (using the same metric, as in (2.3)). This is not trivial, since the original countable dense set could be entirely disjoint from the chosen subset (e.g. $\mathbb{Q}^2$ is dense in $\mathbb{R}^2$, but disjoint from the subset $\{\pi\} \times \mathbb{R}$).

Lemma 2.20 *If (X, d) is separable and $Y \subseteq X$, then (Y, d) is also separable.*

Proof We construct A, a countable dense subset of Y, as follows.

Suppose that $\{x_n\}_{n=1}^{\infty}$ is dense in X; then for each $n, k \in \mathbb{N}$, if

$$B(x_n, 1/k) \cap Y \neq \varnothing$$

then we choose one point from $B(x_n, 1/k) \cap Y$ and add it to A. Constructed in this way A is (at most) a countable set since we can have added at most $\mathbb{N} \times \mathbb{N}$ points.

To show that A is dense, take $z \in Y$ and $\varepsilon > 0$. Now choose k such that $1/k < \varepsilon/2$ and $x_n \in X$ with $d(x_n, z) < 1/k$. Since $z \in B(x_n, 1/k) \cap Y$, we have $B(x_n, 1/k) \cap Y \neq \varnothing$; because of this there must exist $y \in A$ such that $d(x_n, y) < 1/k$ and hence

$$d(y, z) \leq d(y, x_n) + d(x_n, z) < 2/k < \varepsilon. \qquad \square$$

2.5 Compactness

Compactness is an extremely useful property that is the key to many of the proofs that follow. The most familiar 'compactness' result is the Bolzano–Weierstrass Theorem: any bounded set of real numbers has a convergent subsequence.

The fundamental definition of compactness in terms of open sets makes the definition applicable in any topological space (see Appendix C). To state this definition we require the following terminology: a *cover* of a set K is any collection of sets whose union contains K; given a cover, a *subcover* is a subcollection of sets from the original cover whose union still contains K.

Definition 2.21 A subset K of a metric space (X, d) is *compact* if any cover of K by open sets has a finite subcover, i.e. if $\{O_\alpha\}_{\alpha \in \mathbb{A}}$ is a collection of open subsets of X such that

$$K \subseteq \bigcup_{\alpha \in \mathbb{A}} O_\alpha,$$

then there is a finite set $\{\alpha_j\}_{j=1}^n \subset \mathbb{A}$ such that

$$K \subseteq \bigcup_{j=1}^n O_{\alpha_j}.$$

In a metric space compactness in this sense is equivalent to 'sequential compactness', and it is in this form that we will most often make use of compactness in what follows. The equivalence of these two definitions in a metric space is not trivial; a proof is given in Appendix C (see Theorem C.14).

Definition 2.22 If K is a subset of (X, d), then K is *sequentially compact* if any sequence in K has a subsequence that converges and whose limit lies in K.

(Recall that a subsequence of $(x_n)_{n=1}^\infty$ is a sequence of the form $(x_{n_k})_{k=1}^\infty$ where $n_k \in \mathbb{N}$ with $n_{k+1} > n_k$.)

Using the Bolzano–Weierstrass Theorem we can easily prove the following basic compactness result.

Theorem 2.23 *Any closed bounded subset of $\mathbb{K}$ is compact.*

Proof First we prove the result for $\mathbb{K} = \mathbb{R}$. Take any closed bounded subset A of $\mathbb{R}$, and let (x_n) be a sequence in A. Since $(x_n) \in A$, we know that (x_n) is bounded, and so it has a convergent subsequence $x_{n_j} \to x$ for some $x \in \mathbb{R}$. Since $x_{n_j} \in A$ and A is closed, it follows that $x \in A$ and so A is compact.

Now let A be a closed bounded subset of $\mathbb{C}$ and (z_n) a sequence in A. If we write $z_n = x_n + iy_n$, then, since $|z_n|^2 = |x_n|^2 + |y_n|^2$, (x_n) and (y_n) are both bounded sequences in $\mathbb{R}$. First take a subsequence $(z_{n_j})_j$ such that x_{n_j} converges to some $x \in \mathbb{R}$. Then take a subsequence of $(z_{n_j})_j$, $(z_{n_j'})_j$ such that $y_{n'_j}$ converges to some $y \in \mathbb{R}$; we still have $x_{n'_j} \to x$. It follows that $z_{n'_j} \to x + iy$, and since $z_{n'_j} \in A$ and A is closed it follows that $x + iy \in A$, which shows that A is compact. □

Compact subsets of metric spaces are closed and bounded.

Lemma 2.24 *If K is a compact subset of a metric space (X, d), then K is closed and bounded.*

Proof If $(x_n) \in K$ and $x_n \to x$, then any subsequence of (x_n) also converges to x. Since K is compact, it has a subsequence $x_{n_j} \to x'$ with $x' \in K$. By uniqueness of limits it follows that $x' = x$ and so $x \in K$, which shows that K is closed (see Lemma 2.10).

If K is compact, then the cover of K by the open balls $\{B(k, 1) : k \in K\}$ has a finite subcover by balls centred at $\{k_1, \ldots, k_n\}$. Then for any $k \in K$ we have $k \in B(k_j, 1)$ for some $j = 1, \ldots, n$ and so

$$d(k, k_1) \le d(k, k_j) + d(k_j, k_1) < 1 + \max_{j=2,\ldots,n} d(x_j, x_1). \qquad \square$$

Lemma 2.25 *If (X, d) is a compact metric space, then a subset K of X is compact if and only if it is closed.*

Proof If K is a compact subset of (X, d), then it is closed by Lemma 2.24. If K is a closed subset of a compact metric space, then any sequence in K has a convergent subsequence; its limit must lie in K since K is closed, and thus K is compact. □

We will soon prove in Theorem 2.27 that being closed and bounded characterises compact subsets of $\mathbb{R}^n$, based on the following observation.

Theorem 2.26 *If K_1 is a compact subset of (X_1, d_1) and K_2 is a compact subset of (X_2, d_2), then $K_1 \times K_2$ is a compact subset of the product space $(X_1 \times X_2, \varrho_p)$, where ϱ_p is any of the product metrics from Exercise 2.2.*

Proof Suppose that $(x_n, y_n) \in K_1 \times K_2$. Then, since K_1 is compact, there is a subsequence (x_{n_j}, y_{n_j}) such that $x_{n_j} \to x$ for some $x \in K_1$. Now, using the fact that K_2 is compact, take a further subsequence, $(x_{n'_j}, y_{n'_j})$ such that we also have $y_{n'_j} \to y$ for some $y \in K_2$; because $x_{n'_j}$ is a subsequence of x_{n_j} we still have $x_{n'_j} \to x$. Since

$$\varrho_p((x_{n'_j}, y_{n'_j}), (x, y)) = \left[d_1(x_{n'_j}, x)^p + d_2(y_{n'_j}, y)^p \right]^{1/p}$$

it follows that $(x_{n'_j}, y_{n'_j}) \to (x, y) \in K_1 \times K_2$. □

By induction this shows that the product of any finite number of compact sets is compact. (Tychonoff's Theorem, proved in Appendix C, shows that the product of *any* collection of compact sets is compact when considered with an appropriate topology.)

Theorem 2.27 *A subset of $\mathbb{K}^n$ (with the usual metric) is compact if and only if it is closed and bounded.*

Note that $\mathbb{K}^n$ with the usual metric is given by the product $\mathbb{K} \times \cdots \times \mathbb{K}$, using ϱ_2 to construct the metric on the product.

Proof That any compact subset of $\mathbb{K}^n$ is closed and bounded follows immediately from Lemma 2.24.

For the converse, note that it follows from Theorem 2.26 that

$$Q^n_M := \{ x \in \mathbb{K}^n : |x_j| \leq M, \ j = 1, \ldots, n \}$$

is a compact subset of $\mathbb{K}^n$ for any $M > 0$. If K is a bounded subset of $\mathbb{K}^n$, then it is a subset of Q^n_M for some $M > 0$. If it is also closed, then it is a closed subset of a compact set, and hence compact (Lemma 2.25). $\qquad\square$

We now give three fundamental results about continuous functions on compact sets.

Theorem 2.28 *Suppose that K is a compact subset of (X, d_X) and that $f \colon (X, d_X) \to (Y, d_Y)$ is continuous. Then $f(K)$ is a compact subset of (Y, d_Y).*

Proof Let $(y_n) \in f(K)$. Then $y_n = f(x_n)$ for some $x_n \in K$. Since $(x_n) \in K$ and K is compact, there is a subsequence of x_n that converges to some $x^* \in K$, i.e. $x_{n_j} \to x^* \in K$. Since f is continuous it follows (using Lemma 2.12) that as $j \to \infty$

$$y_{n_j} = f(x_{n_j}) \to f(x^*) =: y^* \in f(K),$$

i.e. the subsequence y_{n_j} converges to some $y^* \in f(K)$. It follows that $f(K)$ is compact. $\qquad\square$

The following is an (almost) immediate corollary.

Proposition 2.29 *Let K be a compact subset of (X, d). Then any continuous function $f \colon K \to \mathbb{R}$ is bounded and attains its bounds, i.e. there exists an $M > 0$ such that $|f(x)| \leq M$ for all $x \in K$, and there exist $\underline{x}, \overline{x} \in K$ such that*

$$f(\underline{x}) = \inf_{x \in K} f(x) \qquad and \qquad f(\overline{x}) = \sup_{x \in K} f(x). \qquad (2.8)$$

Proof Since f is continuous and K is compact, $f(K)$ is a compact subset of $\mathbb{R}$, so $f(K)$ is closed and bounded (by Theorem 2.27); in particular, there exists $M > 0$ such that $|f(x)| \le M$ for every $x \in K$. Since $f(K)$ is closed, it follows that

$$\sup \{y : y \in f(K)\}$$

(see Exercise 2.12) and so there exists $\overline{x}$ as in (2.8). The argument for $\underline{x}$ is almost identical. $\qquad \square$

Finally, any continuous function on a compact set is also uniformly continuous.

Lemma 2.30 *If $f : (X, d_X) \to (Y, d_Y)$ is continuous and X is compact, then f is uniformly continuous on X: given $\varepsilon > 0$ there exists $\delta > 0$ such that*

$$d_X(x, y) < \delta \qquad \Rightarrow \qquad d_Y(f(x), f(y)) < \varepsilon \qquad x, y \in X.$$

Proof If f is not uniformly continuous, then there exists $\varepsilon > 0$ such that for every $\delta > 0$ we can find $x, y \in X$ with $d_X(x, y) < \varepsilon$ and $d_Y(f(x), f(y)) \ge \varepsilon$. Choosing x_n, y_n for each $\delta = 1/n$ we obtain x_n, y_n such that

$$d_X(x_n, y_n) < 1/n \qquad and \qquad d_Y(f(x_n), f(y_n)) \ge \varepsilon. \qquad (2.9)$$

Since X is compact, we can find a subsequence x_{n_j} such that $x_{n_j} \to x$ with $x \in X$. Since

$$d_X(y_{n_j}, x) \le d_X(y_{n_j}, x_{n_j}) + d_X(x_{n_j}, x),$$

it follows that $y_{n_j} \to x$ also. Since f is continuous at x, we can find $\delta > 0$ such that $d_X(z, x) < \delta$ ensures that $d_Y(f(z), f(x)) < \varepsilon/2$. But then for j sufficiently large we have $d_X(x_{n,j}, x) < \delta$ and $d_X(y_{n_j}, x) < \delta$, which implies that

$$d_Y(f(x_{n_j}), f(y_{n_j})) \le d_Y(f(x_{n_j}), f(x)) + d_Y(f(x), f(y_{n_j})) < \varepsilon,$$

contradicting (2.9). $\qquad \square$

We often apply this when $f : (K, d_X) \to (Y, d_Y)$ and K is a compact subset of a larger metric space (X, d_X).

Exercises

2.1   Using the fact that $\sum_{j=1}^{n}(\xi_j - \lambda\eta_j)^2 \geq 0$ for any $\boldsymbol{\xi}, \boldsymbol{\eta} \in \mathbb{R}^n$ and any $\lambda \in \mathbb{R}$, prove the Cauchy–Schwarz inequality

$$\left(\sum_{j=1}^{n}\xi_j\eta_j\right)^2 \leq \left(\sum_{j=1}^{n}\xi_j^2\right)\left(\sum_{j=1}^{n}\eta_j^2\right).$$

(For the usual dot product in $\mathbb{R}^n$ this shows that $|\boldsymbol{x} \cdot \boldsymbol{y}| \leq \|\boldsymbol{x}\|\|\boldsymbol{y}\|$.)

2.2   Suppose that (X_j, d_j), $j = 1,\ldots,n$ are metric spaces. Show that $X_1 \times \cdots \times X_n$ is a metric space when equipped with any of the product metrics ϱ_p defined by setting

$$\varrho_p((x_1,\ldots,x_n),(y_1,\ldots,y_n))$$

$$:= \begin{cases} \left\{\sum_{j=1}^{n} d_j(x_j,y_j)^p\right\}^{1/p}, & 1 \leq p < \infty \\ \max_{j=1,\ldots,n} d_j(x_j,y_j), & p = \infty. \end{cases}$$

(You should use the inequality

$$\left\{(\alpha_1+\beta_1)^p + (\alpha_2+\beta_2)^p\right\}^{1/p} \leq (\alpha_1^p+\alpha_2^p)^{1/p}+(\beta_1^p+\beta_2^p)^{1/p}, \quad (2.10)$$

which holds for all $\alpha_1, \alpha_2, \beta_1, \beta_2 \in \mathbb{R}$ and $1 \leq p < \infty$; we will prove this in Lemma 3.6 in the next chapter.)

2.3   Show that if d is a metric on X, then so is

$$\hat{d}(x,y) := \frac{d(x,y)}{1+d(x,y)}.$$

(This new metric makes $(X, \hat{d})$ into a bounded metric space.) [Hint: the map $t \mapsto t/(1+t)$ is monotonically increasing in t.]

2.4   Let $\mathfrak{s}(\mathbb{K})$ be the space of all sequences $\boldsymbol{x} = (x_j)_{j=1}^{\infty}$ with $x_j \in \mathbb{K}$ (bounded or unbounded). Show that

$$d(\boldsymbol{x}, \boldsymbol{y}) := \sum_{j=1}^{\infty} 2^{-j}\frac{|x_j - y_j|}{1+|x_j - y_j|}$$

defines a metric on $\mathfrak{s}$. If $(\boldsymbol{x}^{(n)})_n$ is any sequence in $\mathfrak{s}$ show that $\boldsymbol{x}^{(n)} \to \boldsymbol{y}$ in this metric if and only if $x_j^{(n)} \to y_j$ for each $j \in \mathbb{N}$. (Kreyszig, 1978)

2.5   Show that any finite union of closed sets is closed and any intersection of closed sets is closed.

2.6   Show that if $x \in X$ and $r > 0$, then $B(x,r)$ is open.

2.7 Show that any open subset of a metric space (X, d) can be written as the union of open balls. [Hint: if U is open, then for each $x \in U$ there exists $r(x) > 0$ such that $x \in B(x, r(x)) \subseteq U$.]

2.8 Show that a function $f : (X, d_X) \to (Y, d_Y)$ is continuous if and only if $f(x_n) \to f(x)$ whenever $(x_n) \in X$ with $x_n \to x$ in (X, d).

2.9 Show that if (X, d_X) and (Y, d_Y) are separable, then $(X \times Y, \varrho_p)$ is separable, where ϱ_p is any one of the metrics from Exercise 2.2.

2.10 Suppose that $\{F_\alpha\}_{\alpha \in \mathbb{A}}$ are a family of closed subsets of a compact metric space (X, d) with the property that the intersection of any finite number of the sets has non-empty intersection. Show that $\cap_{\alpha \in \mathbb{A}} F_\alpha$ is non-empty.

2.11 Suppose that (F_j) is a decreasing sequence $[F_{j+1} \subseteq F_j]$ of non-empty closed subsets of a compact metric space (X, d). Use the result of the previous exercise to show that $\cap_{j=1}^\infty F_j \neq \varnothing$.

2.12 Show that if S is a closed subset of $\mathbb{R}$, then $\sup(S) \in S$.

2.13 Show that if $f : (X, d_X) \to (Y, d_Y)$ is a continuous bijection and X is compact, then f^{-1} is also continuous (i.e. f is a homeomorphism).

2.14 Any compact metric space (X, d) is separable. Prove the stronger result that in any compact metric space there exists a countable subset $(x_j)_{j=1}^\infty$ with the following property: for any $\varepsilon > 0$ there is an $M(\varepsilon)$ such that for every $x \in X$ we have

$$d(x_j, x) < \varepsilon \qquad \text{for some } 1 \leq j \leq M(\varepsilon).$$

PART II

Normed Linear Spaces

3

Norms and Normed Spaces

In the first two chapters we considered spaces with a linear structure (vector spaces) and more general spaces in which we could define a notion of convergence (metric spaces). We now turn to the natural setting in which to combine these, i.e. in which to consider convergence in vector spaces.

3.1 Norms

The majority of the spaces that we will consider in the rest of the book will be normed spaces, i.e. vector spaces equipped with an appropriate norm. A norm $\| \cdot \|$ provides a generalised notion of 'length'.

Definition 3.1 A *norm* on a vector space X is a map $\| \cdot \| \colon X \to [0, \infty)$ such that

(i) $\|x\| = 0$ if and only if $x = 0$;
(ii) $\|\lambda x\| = |\lambda| \|x\|$ for every $\lambda \in \mathbb{K}$, $x \in X$; and
(iii) $\|x + y\| \le \|x\| + \|y\|$ for every $x, y \in X$ (the triangle inequality).

A *normed space* is a pair $(X, \| \cdot \|)$ where X is a vector space and $\| \cdot \|$ is a norm on X.

Any norm on X gives rise to a metric on X if we set $d(x, y) := \|x - y\|$. In this case we have $d(x, y) = 0$ if and only if $x = y$, $d(x, y) = d(y, x)$, and the triangle inequality holds since

$$d(x, z) = \|x - z\| \le \|x - y\| + \|y - z\| = d(x, y) + d(y, z).$$

This means that any normed space $(X, \| \cdot \|)$ can also be viewed as a metric space (X, d), and so all the concepts discussed in Chapter 2 are immediately

35

applicable to normed spaces. (It is easy to find examples of metrics that do not come from norms, such as the discrete metric from Example 2.3 or the metric on the space of all sequences from Exercise 2.4; see Exercise 3.1.)

In a normed space the open ball in X centred at y of radius r is

$$B_X(y, r) = \{x \in X : \|x - y\| < r\},$$

and the closed ball is[1]

$$\overline{B_X(y, r)} = \{x \in X : \|x - y\| \leq r\}.$$

When the space is obvious from the context we will drop the X subscript.

In a linear space the origin plays a special role, and the unit balls centred there are of particular interest. Since we will use these frequently, we will write B_X for $B_X(0, 1)$ ('the open unit ball in X') and $\mathbb{B}_X$ for $\overline{B_X(0, 1)}$ ('the closed unit ball in X'). Both of these unit balls are convex.

Definition 3.2 Let V be a vector space. A subset K of V is *convex* if whenever $x, y \in K$ the line segment joining x and y lies in K, i.e. for every $\lambda \in [0, 1]$ we have $\lambda x + (1 - \lambda)y \in K$; see Figure 3.1.

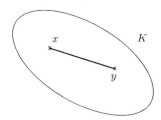

Figure 3.1 A set K is convex if the line joining any $x, y \in K$ is entirely contained in K.

Lemma 3.3 *In any normed space* $\mathbb{B}_X$ *and* B_X *are convex.*

Proof If $x, y \in \mathbb{B}_X$, then $\|x\| \leq 1$ and $\|y\| \leq 1$. So for $\lambda \in (0, 1)$

$$\|\lambda x + (1 - \lambda)y\| \leq |\lambda|\|x\| + |1 - \lambda|\|y\| \leq \lambda + (1 - \lambda) = 1,$$

so $\lambda x + (1 - \lambda y) \in \mathbb{B}_X$. The convexity of B_X follows similarly. $\square$

We now give a relatively simple way to check that a particular function defines a norm, based on the convexity of the resulting 'closed unit ball'.

[1] Note that in a normed space the closed ball is the closure of the open ball with the same radius, unlike in a general metric space; see (2.7).

Lemma 3.4 *Suppose that* $N : X \to \mathbb{R}$ *satisfies*

(i) $N(x) = 0$ *if and only if* $x = 0$;

(ii) $N(\lambda x) = |\lambda| N(x)$ *for every* $\lambda \in \mathbb{K}$, $x \in X$

(i.e. (i) *and* (ii) *from the definition of a norm) and, in addition, that the set* $\mathbb{B} := \{x : N(x) \le 1\}$ *is convex. Then* N *satisfies the triangle inequality*

$$N(x + y) \le N(x) + N(y) \qquad (3.1)$$

and so N *defines a norm on* X.

Proof We only need to prove (3.1). If $N(x) = 0$, then $x = 0$ and

$$N(x + y) = N(y) = N(x) + N(y),$$

so we can assume that $N(x) > 0$ and $N(y) > 0$.

In this case $x/N(x) \in \mathbb{B}$ and $y/N(y) \in \mathbb{B}$, so using the convexity of $\mathbb{B}$ we have

$$\underbrace{\frac{N(x)}{N(x) + N(y)}}_{\lambda} \left(\frac{x}{N(x)} \right) + \underbrace{\frac{N(y)}{N(x) + N(y)}}_{1 - \lambda} \left(\frac{y}{N(y)} \right) \in \mathbb{B}.$$

Therefore

$$\frac{x + y}{N(x) + N(y)} \in \mathbb{B},$$

which means, using property (ii) from Definition 3.1 that

$$1 \ge N \left(\frac{x + y}{N(x) + N(y)} \right) = \frac{N(x + y)}{N(x) + N(y)} \quad \Rightarrow \quad N(x + y) \le N(x) + N(y),$$

as required. $\qquad \qquad \square$

In fact any bounded convex symmetric subset in $\mathbb{R}^n$ can be the unit ball for some norm on $\mathbb{R}^n$; see Exercise 5.1.

To use Lemma 3.4 in examples we often use the convexity of some real-valued function. Recall that a function $f : [a, b] \to \mathbb{R}$ is *convex* if whenever $x, y \in [a, b]$ we have

$$f(\lambda x + (1 - \lambda)y) \le \lambda f(x) + (1 - \lambda)f(y) \qquad \text{for all } \lambda \in (0, 1).$$

If $f \in C^2(a, b) \cap C^1([a, b])$, then a sufficient condition for the convexity of f is that $f''(x) \ge 0$ for all $x \in (a, b)$; in particular, we will use the fact that $s \mapsto |s|^p$ is convex for all $1 \le p < \infty$ and that $s \mapsto e^s$ is convex (see Exercise 3.4).

3.2 Examples of Normed Spaces

Strictly, a normed space should be written $(X, \|\cdot\|)$ where $\|\cdot\|$ is the particular norm on X. However, many normed spaces have standard norms, and so often the norm is not specified. For example, unless otherwise stated, $\mathbb{K}^n$ is equipped with the norm

$$\|x\| = \|x\|_{\ell^2} := \left(\sum_{j=1}^{n} |x_j|^2 \right)^{1/2}, \qquad x \in \mathbb{K}^n.$$

Example 3.5 One can equip $\mathbb{K}^n$ with many other norms. For example, for $1 \le p < \infty$ the ℓ^p norms are given by

$$\|x\|_{\ell^p} := \left(\sum_{j=1}^{n} |x_j|^p \right)^{1/p} \qquad 1 \le p < \infty$$

(the standard norm corresponds to the choice $p = 2$) and the ℓ^∞ norm is given by

$$\|x\|_{\ell^\infty} := \max_{j=1,\ldots,n} |x_j|. \tag{3.2}$$

See Figure 3.2 for the unit balls in $(\mathbb{R}^2, \|\cdot\|_{\ell^p})$, and Exercise 3.5 for a justification of the notation ℓ^∞ for the norm in (3.2). (The corresponding metrics are those in Example 2.2.)

In order to show that these are indeed norms we need to check the triangle inequality.

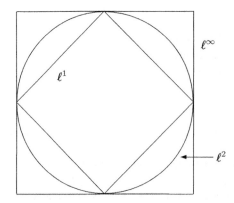

Figure 3.2 Closed unit balls in $\mathbb{R}^2$ with the ℓ^1, ℓ^2 (standard), and ℓ^∞ norms.

Lemma 3.6 (Minkowski's inequality in $\mathbb{K}^n$) *For all $1 \leq p \leq \infty$, if $x, y \in \mathbb{K}^n$, then*

$$\|x + y\|_{\ell^p} \leq \|x\|_{\ell^p} + \|y\|_{\ell^p}. \tag{3.3}$$

Proof If $p = \infty$, this is straightforward. For $p \in [1, \infty)$ we will use Lemma 3.4 and show that the set

$$\mathbb{B} := \{x \in \mathbb{K}^n : \|x\|_{\ell^p} \leq 1\} = \{x \in \mathbb{K}^n : \|x\|_{\ell^p}^p \leq 1\}$$

is convex. We use the fact that the function $t \mapsto |t|^p$ is convex (see Exercise 3.4) for all $1 \leq p < \infty$: if $x, y \in \mathbb{B}$, then

$$\begin{aligned}
\|\lambda x + (1 - \lambda)y\|_{\ell^p}^p &= \sum_{j=1}^{n} |\lambda x_j + (1 - \lambda)y_j|^p \\
&\leq \sum_{j=1}^{n} \Big|\lambda|x_j| + (1 - \lambda)|y_j|\Big|^p \\
&\leq \sum_{j=1}^{n} \lambda|x_j|^p + (1 - \lambda)|y_j|^p \leq 1,
\end{aligned}$$

and so $\lambda x + (1 - \lambda)y \in \mathbb{B}$ and $\mathbb{B}$ is convex; (3.3) now follows from Lemma 3.4. $\qquad\square$

We can always define a norm on any finite-dimensional space, which makes it look like $\mathbb{K}^n$.

Lemma 3.7 *Let V be any finite-dimensional vector space and suppose that $E := \{e_j\}_{j=1}^{n}$ is a basis for V. Then*

$$\|x\|_E := \left(\sum_{j=1}^{n} |\alpha_j|^2\right)^{1/2}, \qquad for \quad x = \sum_{j=1}^{n} \alpha_j e_j$$

is a norm on V.

Proof Clearly $\|x\|_E \geq 0$, and if $\|x\|_E = 0$, then $\alpha_j = 0$ for all j, and so $x = 0$. Since expansions in terms of a basis are unique, if $x = \sum_{j=1}^{n} \alpha_j e_j$, then $\lambda x = \sum_{j=1}^{n} (\lambda \alpha_j)e_j$, so property (ii) holds. Finally, the triangle inequality follows from the triangle inequality for the standard norm in $\mathbb{K}^n$, since if we have $x = \sum_{j=1}^{n} \alpha_j e_j$ and $y = \sum_{j=1}^{n} \beta_j e_j$, then

$$\|x + y\|_E = \|\alpha + \beta\|_{\ell^2} \leq \|\alpha\|_{\ell^2} + \|\beta\|_{\ell^2} = \|x\|_E + \|y\|_E,$$

where $\alpha = (\alpha_1, \ldots, \alpha_n)$ and $\beta = (\beta_1, \ldots, \beta_n)$. $\qquad\square$

We now consider some infinite-dimensional examples.

Example 3.8 The sequence space $\ell^p(\mathbb{K})$, $1 \le p < \infty$, consists of all $\mathbb{K}$-valued sequences $x = (x_j)_{j=1}^\infty$ such that the ℓ^p norm is finite, where

$$\|x\|_{\ell^p} := \left(\sum_{j=1}^\infty |x_j|^p \right)^{1/p} < \infty; \tag{3.4}$$

and $\ell^\infty(\mathbb{K})$ is the space of bounded sequences equipped with the norm

$$\|x\|_\infty = \sup_{j \in \mathbb{N}} |x_j|. \tag{3.5}$$

Note that we use the same notation for the norms on these sequences spaces as we do for the similar norms $\mathbb{K}^n$ (Example 3.5).

(These spaces were introduced in Example 1.5.) The space $c_0(\mathbb{K})$ is the subspace of $\ell^\infty(\mathbb{K})$ consisting of null sequences, i.e. all those $x \in \ell^\infty$ such that $x_j \to 0$ as $j \to \infty$; we use the ℓ^∞ norm from (3.5) as the norm on $c_0(\mathbb{K})$ unless otherwise stated.

We can deduce that (3.4) and (3.5) do indeed define norms on ℓ^p using Lemma 3.6.

Lemma 3.9 (Minkowski's inequality in $\ell^p(\mathbb{K})$) *For all $1 \le p \le \infty$, if we have $x, y \in \ell^p(\mathbb{K})$, then $x + y \in \ell^p(\mathbb{K})$ and*

$$\|x + y\|_{\ell^p} \le \|x\|_{\ell^p} + \|y\|_{\ell^p}. \tag{3.6}$$

Proof The proof when $p = \infty$ is straightforward, so we only give the proof for $1 \le p < \infty$. Given $x, y \in \ell^p(\mathbb{K})$, we can use (3.3) to guarantee that

$$\left(\sum_{j=1}^n |x_j + y_j|^p \right)^{1/p} \le \left(\sum_{j=1}^n |x_j|^p \right)^{1/p} + \left(\sum_{j=1}^n |y_j|^p \right)^{1/p}$$
$$\le \|x\|_{\ell^p} + \|y\|_{\ell^p};$$

now we can take the limit as $n \to \infty$ to deduce (3.6). $\qquad\square$

It is worth observing that these ℓ^p spaces are nested; the largest is $\ell^\infty(\mathbb{K})$ and the smallest $\ell^1(\mathbb{K})$.

Lemma 3.10 *If $1 \le p \le q \le \infty$, then $\ell^p(\mathbb{K}) \subset \ell^q(\mathbb{K})$ and*

$$\|x\|_{\ell^q} \le \|x\|_{\ell^p} \qquad \text{whenever} \quad x \in \ell^p(\mathbb{K}). \tag{3.7}$$

Proof Suppose that $x \in \ell^p$ with $\|x\|_{\ell^p} = 1$. Then in particular, $|x_j| \leq 1$ for every j, in which case, since $q \geq p$, it follows that $|x_j|^q \leq |x_j|^p$. Therefore

$$\|x\|_{\ell^q}^q = \sum_{j=1}^{\infty} |x_j|^q \leq \sum_{j=1}^{\infty} |x_j|^p = 1,$$

i.e. $\|x\|_{\ell^q} \leq 1$. Given a general non-zero $x \in \ell^p$, consider $y := x/\|x\|_{\ell^p}$; then $\|y\|_{\ell^p} = 1$ and so

$$\|y\|_{\ell^q} = \left\| \frac{x}{\|x\|_{\ell^p}} \right\|_{\ell^q} = \frac{\|x\|_{\ell^q}}{\|x\|_{\ell^p}} \leq 1,$$

from which (3.7) follows. □

There are no inclusions in the reverse direction: given $1 \leq p < q \leq \infty$, if we set $x_j = j^{-1/p}$, then $x \in \ell^q$ but $x \notin \ell^p$ (see also Exercise 3.7).

We now briefly consider various norms on spaces of continuous functions.

Example 3.11 On the space $C([a, b])$,

$$\|f\|_{\infty} := \max_{x \in [a,b]} |f(x)|$$

defines a norm (the 'maximum norm' or 'supremum norm' (see below)).

Proof If $\|f\|_{\infty} = 0$, then $|f(x)| = 0$ for every $x \in [a, b]$ and so $f = 0$. Clearly $\|\lambda f\|_{\infty} = |\lambda| \|f\|_{\infty}$ for any $\lambda \in \mathbb{K}$, and if $f, g \in C([a, b])$, then

$$\|f + g\|_{\infty} = \max_{x \in [a,b]} |(f + g)(x)| \leq \max_{x \in [a,b]} (|f(x)| + |g(x)|)$$

$$\leq \max_{x \in [a,b]} |f(x)| + \max_{x \in [a,b]} |g(x)| = \|f\|_{\infty} + \|g\|_{\infty}. \quad \square$$

We can generalise this example to spaces of continuous functions on metric spaces.

Example 3.12 If X is a metric space consider the space $C_b(X; \mathbb{K})$ of all bounded continuous functions from X into $\mathbb{K}$. Then

$$\|f\|_{\infty} := \sup_{x \in X} |f(x)|$$

defines a norm $C_b(X; \mathbb{K})$, the 'sup(remum) norm'. If X is compact, then $C(X; \mathbb{K}) = C_b(X; \mathbb{K})$, and this norm is the same as the 'maximum norm'

$$\|f\|_{\infty} = \max_{x \in X} |f(x)|,$$

since when X is compact, f attains its bounds on X (Proposition 2.29).

Example 3.13 On the space $C([0, 1])$ for any $1 \le p < \infty$ the expression

$$\|f\|_{L^p} := \left(\int_0^1 |f(x)|^p \, dx \right)^{1/p}$$

defines a norm ('the L^p norm').

Proof Clearly $\|f\|_{L^p} \ge 0$. To show that $\|f\|_{L^p} = 0$ implies that $f = 0$, suppose for a contradiction that $f \ne 0$. Then there exists $x_0 \in [0, 1]$ such that $|f(x_0)| > 0$. Since f is continuous, there exists $\delta > 0$ such that

$$x \in [0, 1] \text{ with } |x - x_0| < \delta \quad \Rightarrow \quad |f(x)| > \frac{|f(x_0)|}{2}.$$

Therefore, reducing δ so that at least one of the intervals $[x_0 - \delta, x_0]$ and $[x_0, x_0 + \delta]$ lies within $[0, 1]$, we have

$$\int_0^1 |f(x)|^p \, dx \ge \delta \left(\frac{|f(x_0)|}{2} \right)^p > 0.$$

It follows that we must have $f = 0$.

Property (ii) is clear, and the triangle inequality follows as for the ℓ^p norm, using Lemma 3.4: if $f, g \in C([0, 1])$ with $\|f\|_{L^p} \le 1$ and $\|g\|_{L^p} \le 1$, then

$$\begin{aligned}
\|\lambda f + (1 - \lambda g)\|_{L^p}^p &= \int_0^1 |\lambda f(x) + (1 - \lambda)g(x)|^p \, dx \\
&\le \int_0^1 \lambda |f(x)|^p + (1 - \lambda)|g(x)|^p \, dx \\
&\le 1.
\end{aligned}$$
$\qquad \square$

3.3 Convergence in Normed Spaces

As we already observed, any norm $\| \cdot \|$ on X gives rise to a metric on X by setting $d(x, y) := \|x - y\|$. The definitions of convergence, continuity, etc., from Chapter 2 therefore apply immediately to any normed space, but there are some particular properties of convergence that hold in normed spaces that we mention here.

First we show that the map $x \mapsto \|x\|$ is always continuous.

Lemma 3.14 *If $x, y \in X$, then*

$$\Big| \|x\| - \|y\| \Big| \le \|x - y\|; \qquad\qquad (3.8)$$

in particular, the map $x \mapsto \|x\|$ is continuous from $(X, \| \cdot \|)$ into $\mathbb{R}$.

Proof The triangle inequality gives

$$\|x\| \le \|y\| + \|y - x\| \qquad \text{and} \qquad \|y\| \le \|x\| + \|x - y\|$$

which implies (3.8). □

We use this to prove part (i) of the following lemma.

Lemma 3.15 *If* (x_n), $(y_n) \in X$ *with* $x_n \to x$ *and* $y_n \to y$, *then*

(i) $\|x_n\| \to \|x\|$;
(ii) $x_n + y_n \to x + y$; *and*
(iii) *if* $(\alpha_n) \in \mathbb{K}$ *with* $\alpha_n \to \alpha$, *then* $\alpha_n x_n \to \alpha x$.

Proof (i) follows from (3.8) and (ii) follows from the triangle inequality since

$$\|(x_n + y_n) - (x + y)\| \le \|x_n - x\| + \|y_n - y\|.$$

For (iii), since convergent sequences are bounded we have $|\alpha_n| \le M$ for some $M > 0$, and so

$$\begin{aligned}
\|\alpha_n x_n - \alpha x\| &\le \|\alpha_n(x_n - x)\| + \|(\alpha_n - \alpha)x\| \\
&\le M\|x_n - x\| + \|x\||\alpha_n - \alpha|.
\end{aligned}$$

□

3.4 Equivalent Norms

In general there are many possible norms on a vector space (as we saw with $\mathbb{K}^n$ in Example 3.5). However, there is a notion of equivalence of norms which, as we will soon see, ensures that they give rise to the same open sets and therefore make the same sequences convergent.

Definition 3.16 Two norms $\|\cdot\|_1$ and $\|\cdot\|_2$ on a vector space X are *equivalent* – we write $\|\cdot\|_1 \sim \|\cdot\|_2$ – if there exist constants $0 < c_1 \le c_2$ such that

$$c_1\|x\|_1 \le \|x\|_2 \le c_2\|x\|_1 \qquad \text{for all} \qquad x \in X.$$

It is clear that the above notion of 'equivalence' is reflexive ($\|\cdot\| \sim \|\cdot\|$) and symmetric ($\|\cdot\|_1 \sim \|\cdot\|_2$ if and only if $\|\cdot\|_2 \sim \|\cdot\|_1$). It is also transitive, so this is indeed an equivalence relation.

Lemma 3.17 *Suppose that* $\|\cdot\|_1$, $\|\cdot\|_2$, *and* $\|\cdot\|_3$ *are all norms on a vector space* V, *such that*

$$\|\cdot\|_1 \sim \|\cdot\|_2 \qquad and \qquad \|\cdot\|_2 \sim \|\cdot\|_3.$$

Then $\|\cdot\|_1 \sim \|\cdot\|_3$.

Proof There exist constants $0 < \alpha_1 \le \alpha_2$ and $0 < \beta_1 \le \beta_2$ such that

$$\alpha_1\|x\|_1 \le \|x\|_2 \le \alpha_2\|x\|_1 \qquad and \qquad \beta_1\|x\|_2 \le \|x\|_3 \le \beta_2\|x\|_2;$$

therefore

$$\alpha_1\beta_1\|x\|_1 \le \|x\|_3 \le \alpha_2\beta_2\|x\|_1$$

i.e. $\|\cdot\|_1 \sim \|\cdot\|_3$. $\square$

Exercise 3.10 shows that the norms $\|\cdot\|_{\ell^1}$, $\|\cdot\|_{\ell^2}$, and $\|\cdot\|_{\ell^\infty}$ on $\mathbb{K}^n$ are all equivalent. In fact all norms on any finite-dimensional vector space are equivalent, which we will prove in Theorem 5.1.

However, it is easy to find norms on infinite-dimensional spaces that are not equivalent. For example, Lemma 3.10 shows that we could use any choice of $\|\cdot\|_{\ell^p}$ as a norm on the vector space $\ell^1(\mathbb{K})$ if we wanted to (since if $x \in \ell^1$ all of its ℓ^p norms are finite). While that lemma shows that for $p < q$ we have

$$\|x\|_{\ell^q} \le \|x\|_{\ell^p} \qquad \text{for every} \qquad x \in \ell^1,$$

the same inequality does not hold in reverse: we do not have

$$\|x\|_{\ell^p} \le C\|x\|_{\ell^q} \qquad \text{for every} \qquad x \in \ell^1, \tag{3.9}$$

for any choice of $C > 0$.

To see this, consider the sequences[2] $x^{(n)} \in \ell^1$ given by

$$x_j^{(n)} = \begin{cases} j^{-1/p} & j = 1\ldots, n \\ 0 & j = n+1, \ldots. \end{cases}$$

Then $\|x^{(n)}\|_{\ell^q} < \sum_{j=1}^\infty j^{-q/p} < \infty$ for every n, but $\|x^{(n)}\|_{\ell^p} = \sum_{j=1}^n j^{-1}$ is unbounded, so (3.9) cannot hold for any $C > 0$.

The L^p norms on $C([0, 1])$ defined in Example 3.13 are also not equivalent: if $p > q$ there is no constant C such that

$$\|f\|_{L^p} \le C\|f\|_{L^q} \qquad \text{for every} \quad f \in C([0, 1]). \tag{3.10}$$

To see this, consider the functions $f_n(x) = x^n$ with norms

$$\|f_n\|_{L^p} = \left(\int_0^1 x^{np}\,dx\right)^{1/p} = \left(\frac{1}{np+1}\right)^{1/p}.$$

----

[2] Note that each $x^{(n)}$ is an element of ℓ^1 with $x^{(n)} = (x_1^{(n)}, x_2^{(n)}, x_3^{(n)}, \ldots)$, so $(x^{(n)})_{n=1}^\infty$ is a sequence of sequences.

If we had (3.10), then applying this inequality to each f_n would give

$$\left(\frac{1}{np+1}\right)^{1/p} \le C \left(\frac{1}{nq+1}\right)^{1/q},$$

i.e.

$$(nq+1)^{1/q}(np+1)^{-1/p} \le C$$

for every $n \in \mathbb{N}$. But since $p > q$ the left-hand side tends to infinity as $n \to \infty$, which shows that (3.10) cannot hold. Equivalent norms define the same open sets.

Lemma 3.18 *Suppose that* $\|\cdot\|_1$ *and* $\|\cdot\|_2$ *are two equivalent norms on a linear space* X. *Then a set is open/closed in* $(X, \|\cdot\|_1)$ *if and only if it is open/closed in* $(X, \|\cdot\|_2)$.

Proof We only need to prove the statement about open sets, since the statement for closed sets will follow by taking complements. We write $B_i(x, \varepsilon)$ for $\{y \in X : \|y - x\|_i < \varepsilon\}$.

Assume that there exist constants $0 < c_1 \le c_2$ such that

$$c_1\|x\|_1 \le \|x\|_2 \le c_2\|x\|_1 \qquad \text{for all} \qquad x \in X; \qquad (3.11)$$

then

$$B_2(x, c_1\varepsilon) \subseteq B_1(x, \varepsilon) \subseteq B_2(x, c_2\varepsilon).$$

It follows that if $U \subseteq X$ is open using $\|\cdot\|_1$, i.e. for every $x \in U$ we can find $\varepsilon > 0$ such that $B_1(x, \varepsilon) \subseteq U$, then it is open using $\|\cdot\|_2$, since we have $B_2(x, c_1\varepsilon) \subseteq B_1(x, \varepsilon) \subseteq U$. A similar argument applies to show that any set open using $\|\cdot\|_2$ is open using $\|\cdot\|_1$. □

We saw in Chapter 2 that it is possible to define convergence and continuity solely in terms of open sets, so equivalent norms produce the same definitions of convergence and continuity. For a less topological argument it is easy to prove this directly: assuming that $\|\cdot\|_1$ and $\|\cdot\|_2$ are equivalent as in (3.11), then

$$c_1\|x_n - x\|_1 \le \|x_n - x\|_2 \le c_2\|x_n - x\|_1,$$

which shows that $\|x_n - x\|_1 \to 0$ if and only if $\|x_n - x\|_2 \to 0$. Compactness is also preserved under an equivalent norm since this relies on covers by open sets (Definition 2.21) or notions of convergence (Definition 2.22).

3.5 Isomorphisms between Normed Spaces

If we want to show that two mathematical objects are 'the same' we require a bijection between them that also preserves the essential structures of the objects. To identify two normed spaces $(X, \| \cdot \|_X)$ and $(Y, \| \cdot \|_Y)$ we therefore require a map $T \colon X \to Y$ that is

  (i) a bijection (i.e. T is injective and surjective);

 (ii) linear, so that it preserves the linear (vector-space) structure; and

(iii) for some $0 < c_1 \le c_2$ we have

$$c_1 \|x\|_X \le \|Tx\|_Y \le c_2 \|x\|_X \qquad \text{for all} \qquad x \in X, \tag{3.12}$$

    so that the norms in X and Y are comparable.

We will refer to such a map as an *isomorphism* between two normed spaces X and Y, although perhaps one should say strictly that this is a 'normed-space isomorphism', since property (i) alone is enough to guarantee that T is an isomorphism in the set-theoretic sense and (i)+(ii) make X and Y isomorphic as vector spaces.

Definition 3.19 Two normed spaces $(X, \| \cdot \|_X)$ and $(Y, \| \cdot \|_Y)$ are said to be *isomorphic* if there exists a bijective linear map $T \colon X \to Y$ that satisfies (iii) and in this case we will write $X \simeq Y$.

In some cases we can strengthen (iii) and find a map $T \colon X \to Y$ that satisfies (i), (ii), and

(iii$'$) T is an *isometry*, i.e. we have

$$\|Tx\|_Y = \|x\|_X \qquad \text{for all} \qquad x \in X,$$

    so that T preserves the norm.

We will refer to such a map T as an *isometric isomorphism* between the spaces X and Y. (Pryce (1973) and some other authors use the term *congruence* in this case, but we will not adopt this here.)

Definition 3.20 We say that $(X, \| \cdot \|_X)$ and $(Y, \| \cdot \|_Y)$ are *isometrically isomorphic* if there exists a bijective linear isometry $T \colon X \to Y$ and in this case we will write $X \equiv Y$.

Note that the exact terms used to identify normed spaces, and the corresponding notation, can differ from one author to another, so it is important to check precisely which definition is being used when referring to other sources.

Observe that any linear map $T\colon X \to Y$ satisfying (3.12) (in particular, any linear isometry) is automatically injective. Indeed, if $Tx = Ty$, then

$$c_1\|x - y\|_X \le \|T(x - y)\|_Y = \|Tx - Ty\|_Y = 0$$

and so $x = y$. So to show that such a map T is bijective (and hence that $X \simeq Y$) we only need to show that T is surjective.

Example 3.21 The space $\mathbb{C}^n$ is isometrically isomorphic to $\mathbb{R}^{2n}$ via the map

$$(z_1, \ldots, z_n) \mapsto (x_1, y_1, \cdots, x_n, y_n),$$

where $z_i = x_i + iy_i$.

(This example shows very clearly that $\equiv$ is not the same as $=$.)

Proof The map is clearly a linear bijection, and

$$
\begin{aligned}
\|(z_1, \ldots, z_n)\|_{\mathbb{C}^n}^2 &= \sum_{j=1}^{n} |z_j|^2 \\
&= \sum_{j=1}^{n} |x_j|^2 + |y_j|^2 = \|(x_1, y_1, \cdots, x_n, y_n)\|_{\mathbb{R}^{2n}}^2. \quad \square
\end{aligned}
$$

We can also show that with the norm we constructed in Lemma 3.7 any finite-dimensional space is isometrically isomorphic to $\mathbb{K}^n$ with its standard norm.

Lemma 3.22 *Let V be a finite-dimensional vector space and $\|\cdot\|_E$ the norm on V defined in Lemma 3.7. Then $(\mathbb{K}^n, \|\cdot\|_{\ell^2}) \equiv (V, \|\cdot\|_E)$.*

Proof We take a basis $E = \{e_j\}_{j=1}^n$ for V. Writing $\alpha = (\alpha_1, \ldots, \alpha_n)$, the map $\Phi\colon \mathbb{K}^n \to V$ given by

$$\Phi(\alpha) = \sum_{j=1}^{n} \alpha_j e_j$$

is a linear bijection from $\mathbb{K}^n$ onto $(V, \|\cdot\|_E)$: injectivity and surjectivity follow from the fact that $\{e_j\}$ is a basis, and the definition of $\|\cdot\|_E$ ensures that $\|\Phi(\alpha)\|_E = \|\alpha\|_{\ell^2}$. $\square$

3.6 Separability of Normed Spaces

We briefly discussed separability in the context of metric spaces in Section 2.4. We now investigate this a little more given the additional structure available in a normed space.

If E is a subset of a normed space X, recall that the linear span of E, Span(E), is the set of all finite linear combinations of elements of E (Definition 1.9). The closure of Span(E) is called the *closed linear span of E*, and is the set of all elements of X that can be approximated arbitrarily closely by finite linear combinations of elements of E (see Lemma 2.17). We will denote it by clin(E), so

$$\text{clin}(E) := \overline{\text{Span}(E)}.$$

Lemma 3.23 *If X is a normed space, then the following three statements are equivalent:*

 (i) *X is separable (i.e. X contains a countable dense subset);*
 (ii) *the unit sphere in X, $S_X := \{x \in X : \|x\| = 1\}$, is separable; and*
 (iii) *X contains a countable set $\{x_j\}_{j=1}^{\infty}$ whose linear span is dense, i.e. whose closed linear span is all of X,*

$$\text{clin}(\{x_j\}) = X.$$

Proof Lemma 2.20 shows that (i) $\Rightarrow$ (ii). For (ii) $\Rightarrow$ (iii) choose a countable dense subset $\{x_1, x_2, x_3, \ldots\}$ of S_X: then for any non-zero $x \in X$ we have $x/\|x\| \in S_X$, and so for any $\varepsilon > 0$ there exists an x_k such that

$$\left\| x_k - \frac{x}{\|x\|} \right\| < \frac{\varepsilon}{\|x\|}.$$

It follows that

$$\left\| x - \|x\| x_k \right\| < \varepsilon,$$

and since $\|x\| x_k$ is contained in the linear span of the $\{x_j\}$ this gives (iii).

To show that (iii) implies (i) note that the collection of finite linear combinations of the $\{x_j\}$ with rational coefficients is countable (when $\mathbb{K} = \mathbb{C}$ by 'rational' we mean an element of $\mathbb{Q} + i\mathbb{Q}$). This countable collection is dense: given $x \in X$ and $\varepsilon > 0$, choose an element in the linear span of $\{x_1, x_2, \ldots\}$ such that

$$\left\| x - \sum_{j=1}^{n} \alpha_j x_j \right\| < \frac{\varepsilon}{2}, \qquad \alpha_j \in \mathbb{K},$$

and then for each $j = 1, \ldots, n$ choose rational q_j such that

$$|q_j - \alpha_j| < \frac{\varepsilon}{2n\|x_j\|}.$$

It then follows from the triangle inequality that

$$\left\| x - \sum_{j=1}^{n} q_j x_j \right\| \leq \left\| x - \sum_{j=1}^{n} \alpha_j x_j \right\| + \left\| \sum_{j=1}^{n} \alpha_j x_j - \sum_{j=1}^{n} q_j x_j \right\|$$

$$\leq \frac{\varepsilon}{2} + \sum_{j=1}^{n} |\alpha_j - q_j| \|x_j\| < \frac{\varepsilon}{2} + \frac{\varepsilon}{2} = \varepsilon. \qquad \square$$

We have already remarked that $\mathbb{R}^n$ and $\mathbb{C}^n$ are separable (with countable dense subsets $\mathbb{Q}^n$ and $\mathbb{Q}^n + i\mathbb{Q}^n$). Less trivial examples are the sequence spaces ℓ^p for $1 \leq p < \infty$; along with the fact that ℓ^∞ is not separable this result will help furnish a number of useful counterexamples.

Lemma 3.24 *For $1 \leq p < \infty$ the space $\ell^p(\mathbb{K})$ is separable, but $\ell^\infty(\mathbb{K})$ is not separable.*

Proof If $1 \leq p < \infty$, then the linear span of $\{e^{(j)}\}_{j=1}^{\infty}$, where $e_i^{(j)} = \delta_{ij}$ (see (1.3)), is dense in $\ell^p(\mathbb{K})$: given any $x \in \ell^p$ and any $\varepsilon > 0$ there exists N such that

$$\sum_{j=n+1}^{\infty} |x_j|^p < \varepsilon^p \qquad \text{for every } n \geq N.$$

It follows that

$$\left\| x - \sum_{j=1}^{n} x_j e^{(j)} \right\|_{\ell^p} = \|(0, \ldots, 0, x_{n+1}, x_{n+2}, \cdots)\|_{\ell^p}$$

$$= \left(\sum_{j=n+1}^{\infty} |x_j|^p \right)^{1/p} < \varepsilon. \qquad (3.13)$$

However, in the space $\ell^\infty(\mathbb{K})$ consider the set

$$\mathfrak{b} := \{x \in \ell^\infty : x_j = 0 \text{ or } 1 \text{ for each } j \in \mathbb{N}\}.$$

This set $\mathfrak{b}$ is uncountable (this can be shown using Cantor's diagonal argument; see Exercise 3.15) and any two distinct elements x and y in $\mathfrak{b}$ satisfy

$$\|x - y\|_{\ell^\infty} = 1$$

since they must differ by 1 in at least one term. Any dense set A must therefore contain an uncountable number of elements: since A is dense, for every $x \in \mathfrak{b}$ there must be some $x' \in A$ such that $\|x' - x\|_{\ell^\infty} < 1/3$. But if x, y are distinct elements of $\mathfrak{b}$, it then follows that x' and y' are also distinct, since

$$\|x' - y'\|_{\ell^\infty} \geq \|x - y\|_{\ell^\infty} - \|x' - x\|_{\ell^\infty} - \|y' - y\|_{\ell^\infty} > 1/3.$$

Since $\mathfrak{b}$ contains an uncountable number of elements, so must A.   □

Note that the sequence space $c_0(\mathbb{K})$, which is a subspace of $\ell^\infty(\mathbb{K})$, *is* separable (see Example 3.8 for the definition of $c_0(\mathbb{K})$ and Exercise 3.12 for its separability). Note also that the linear span of $\{e^{(j)}\}_{j=1}^\infty$ is precisely the space c_{00} of sequences with only a finite number of non-zero terms; the proof of the above lemma shows that this space is dense in ℓ^p for all $1 \leq p < \infty$.

Finally, we observe that linear subspaces of normed spaces may or may not be closed, although any finite-dimensional linear subspace is closed (see Exercise 5.3). As a simple example, for any $p \in [1, \infty)$ consider the subset c_{00} of ℓ^p consisting of sequences with only a finite number of non-zero terms (Example 1.8). While this is a subspace of ℓ^p, any $x \in \ell^p$ can be approximated by elements of c_{00} arbitrarily closely: we did precisely this in the proof of Lemma 3.24. It follows that c_{00} is not a closed subspace of ℓ^p.

Any open linear subspace of a normed space is the whole space (see Exercise 3.8) and the closure of any linear subspace yields a closed linear subspace (see Exercise 3.9).

Exercises

3.1   Show that if d is a metric on a vector space X derived from a norm $\|\cdot\|$, i.e. $d(x, y) = \|x - y\|$, then d is translation invariant and homogeneous, i.e.

$$d(x + z, y + z) = d(x, y) \quad \text{and} \quad d(\alpha x, \alpha y) = \alpha d(x, y).$$

Deduce that the metric on $\mathfrak{s}$ in Exercise 2.4 does not come from a norm. (Kreyszig, 1978)

3.2   If A and B are subsets of a vector space, then we can define

$$A + B := \{a + b : a \in A, \, b \in B\}.$$

Show that if A and B are both convex, then so is $A + B$.

3.3   If C is a closed subset of a vector space X show that C is convex if and only if $a, b \in C$ implies that $(a + b)/2 \in C$.

3.4   Show that if $f \colon [a, b] \to \mathbb{R}$ is C^2 on (a, b) and C^1 on $[a, b]$, then f is convex on $[a, b]$ if $f''(x) \geq 0$. Deduce that $f(x) = e^x$ and $f(x) = |x|^p$, $1 \leq p < \infty$, are convex functions on $\mathbb{R}$.

3.5   Show that if $m = \max_{j=1,\dots,n} |x_j|$, then for any $p \in [1, \infty)$,

$$m^p \leq \sum_{j=1}^{n} |x_j|^p \leq nm^p$$

and deduce that for any $x \in \mathbb{K}^n$, $\|x\|_{\ell^p} \to \|x\|_{\ell^\infty}$ as $p \to \infty$.

3.6   Show that if $x \in \ell^1$, then

$$\|x\|_{\ell^\infty} = \lim_{p \to \infty} \|x\|_{\ell^p}.$$

(Show that for every $\varepsilon > 0$ there exists N such that

$$\|x\|_{\ell^p} - \varepsilon \leq \|(x_1, \dots, x_n)\|_{\ell^p} \leq \|x\|_{\ell^p} \qquad \text{for all } n \geq N;$$

treat $p \in [1, \infty)$ together using the result of Lemma 3.10, and then $p = \infty$ separately. To finish the proof use the result of Exercise 3.5.)

3.7   Given any $1 < p < \infty$, find a sequence x such that $x \in \ell^p$ but $x \notin \ell^q$ for all $1 \leq q < p$.

3.8   Show that if U is an open subspace of a normed space X then $U = X$. (Naylor and Sell, 1982)

3.9   Show that if U is a linear subspace of a normed space X then $\overline{U}$ is a closed linear subspace of X.

3.10 Show that the norms $\|\cdot\|_{\ell^1}$, $\|\cdot\|_{\ell^2}$, and $\|\cdot\|_{\ell^\infty}$ on $\mathbb{R}^n$ are all equivalent.

3.11 Show that if (f_n), $f \in C([0, 1])$ and $\|f_n - f\|_\infty \to 0$ (uniform convergence) then $\|f_n - f\|_{L^p} \to 0$ (convergence in L^p) and $f_n(x) \to f(x)$ for every $x \in [0, 1]$ ('pointwise convergence'). (For related results see Exercises 7.1–7.3.)

3.12 Show that $c_0(\mathbb{K})$ is separable.

3.13 If $(X, \|\cdot\|_X) \simeq (Y, \|\cdot\|_Y)$ show that X is separable if and only if Y is separable.

3.14 Let $(X, \|\cdot\|)$ be a normed space. If $A \subseteq B \subseteq X$ with A dense in B show that $\operatorname{clin}(A) = \operatorname{clin}(B)$. Deduce that if A is dense in B and B is dense in X then A is dense in X (equivalently that $\operatorname{clin}(A) = X$).

3.15 Show that the set

$$\mathfrak{b} := \{x \in \ell^\infty : x_j = 0 \text{ or } 1 \text{ for each } j \in \mathbb{N}\}$$

is uncountable.

3.16 Show that a normed space $(X, \| \cdot \|)$ is separable if and only if

$$X = \bigcup_{j=1}^{\infty} X_j,$$

where the $\{X_j\}$ are finite-dimensional subspaces of X. (Zeidler, 1995)

3.17 Show that a normed space $(X, \| \cdot \|)$ is separable if and only if there exists a compact set $K \subset X$ such that $X = \operatorname{clin}(K)$. (Megginson, 1998)

4

Complete Normed Spaces

If a sequence (x_n) in a metric space converges, then it must be *Cauchy*, i.e. for every $\varepsilon > 0$ there exists an N such that

$$d(x_n, x_m) < \varepsilon \qquad \text{for all } n, m \geq N.$$

Indeed, if (x_n) converges to some $x \in X$, then there exists N such that $d(x_n, x) < \varepsilon/2$ for every $n \geq N$, and then for $n, m \geq N$ the triangle inequality yields

$$d(x_n, x_m) \leq d(x_n, x) + d(x, x_m) < \frac{\varepsilon}{2} + \frac{\varepsilon}{2} = \varepsilon.$$

A metric space in which any Cauchy sequence converges is called *complete*.

Normed spaces with this completeness property are called 'Banach spaces' and are central to functional analysis; almost all of the spaces we consider throughout the rest of this book will be Banach spaces.[1] Banach himself called them 'spaces of type (B)'.

In the first section of this chapter we prove some abstract results about complete normed spaces, which will help us to show that various important normed spaces are complete in Section 4.2. We then discuss convergent series in Banach spaces: in this case the completeness will allow us to obtain results that parallel those available for sequences in $\mathbb{R}$. The chapter ends with a proof of the Contraction Mapping Theorem.

4.1 Banach Spaces

Banach spaces, which are complete normed spaces, are one of the central topics of what follows.

---

[1] This is the reason for treating completeness in the less general context of normed spaces, rather than introducing the definition for metric spaces and then delaying all our examples for two chapters.

Definition 4.1 A normed space $(X, \|\cdot\|)$ is *complete* if every Cauchy sequence in X converges in X (to a limit that lies in X). A *Banach space* is a complete normed space.

It is a fundamental property of $\mathbb{R}$ (and $\mathbb{C}$) that it is complete. This carries over to $\mathbb{R}^n$ and $\mathbb{C}^n$ with their standard norms, as we now prove. A much neater (but more abstract) proof of the completeness of $\mathbb{K}^n$ is given in Exercise 4.3, but the more methodical proof given here serves as a useful prototype for completeness proofs in more general situations.

Such 'completeness arguments' usually follow similar lines:

   (i) use the definition of what it means for a sequence to be Cauchy to identify a possible limit;
   (ii) show that the original sequence converges to this 'possible limit' in the appropriate norm;
  (iii) check that the 'limit' lies in the correct space.

We label these steps in the proof of the following theorem, but will not be so explicit in the examples in Section 4.2.

Theorem 4.2 *The space $\mathbb{K}^d$ is complete (with its standard norm).*

Proof (i) *Identify a possible limit.* Let $(\boldsymbol{x}^{(k)})_{k=1}^{\infty}$ be a Cauchy sequence in $\mathbb{K}^d$. Then for every $\varepsilon > 0$ there exists $N(\varepsilon)$ such that

$$\|\boldsymbol{x}^{(n)} - \boldsymbol{x}^{(m)}\|_{\ell^2} = \left(\sum_{i=1}^{d} |x_i^{(n)} - x_i^{(m)}|^2 \right)^{1/2} < \varepsilon \qquad \text{for } m, n \geq N(\varepsilon). \quad (4.1)$$

In particular, for each $i = 1, \ldots, d$ we have

$$|x_i^{(n)} - x_i^{(m)}| < \varepsilon \qquad \text{for } m, n \geq N(\varepsilon),$$

so $(x_i^{(n)})_{n=1}^{\infty}$ is a Cauchy sequence in $\mathbb{K}$. Since $\mathbb{K}$ is complete, Cauchy sequences in $\mathbb{K}$ converge, so $x_i^{(n)} \to a_i$ for some $a_i \in \mathbb{K}$. Set $\boldsymbol{a} = (a_1, \ldots, a_d)$, which is our 'possible limit'.

(ii) *Show that our sequence converges to this 'limit'.* We can take the limit as $m \to \infty$ in (4.1) since there are only a finite number of terms, and obtain

$$\|\boldsymbol{x}^{(n)} - \boldsymbol{a}\|_{\ell^2} = \left(\sum_{i=1}^{d} |x_i^{(n)} - a_i|^2 \right)^{1/2} < \varepsilon \qquad \text{for } n \geq N(\varepsilon),$$

which shows that $\lim_{n\to\infty} \|\boldsymbol{x}^{(n)} - \boldsymbol{a}\|_{\ell^2} = 0$, so $\boldsymbol{x}^{(n)} \to \boldsymbol{a}$ as $n \to \infty$.

(iii) *Check that the limit lies in the correct space.* This is trivial here, since any n-component vector is an element of $\mathbb{K}^n$. □

We have already observed that linear subspaces of normed spaces need not be closed. However, if a subspace is closed, then it inherits the completeness of the parent space. We will use this result a number of times in the next section when we look at some particular examples and want to prove that they are complete.

Lemma 4.3 *If $(X, \| \cdot \|)$ is a Banach space and Y is a linear subspace of X, then $(Y, \| \cdot \|)$ is a Banach space if and only if Y is closed.*

Proof If $(Y, \| \cdot \|)$ is complete and $(x_n) \in Y$ is Cauchy in Y it is also Cauchy in X. Since X is complete, we know that $a_n \to a \in X$. If Y is closed, then $a \in Y$ and so $(Y, \| \cdot \|)$ is complete. For the opposite implication, if $(Y, \| \cdot \|)$ is complete and $a_n \to a$ for some $a \in X$, then (a_n) is Cauchy, and therefore $a = \lim_{n \to \infty} a_n \in Y$, which shows that Y is closed. □

Completeness is also preserved under isomorphisms.

Lemma 4.4 *If $(X, \| \cdot \|_X) \simeq (Y, \| \cdot \|_Y)$, then $(X, \| \cdot \|_X)$ is complete if and only if $(Y, \| \cdot \|_Y)$ is complete.*

Proof Let $T : X \to Y$ be an isomorphism between X and Y with

$$c_1 \|x\|_X \le \|Tx\|_Y \le c_2 \|x\|_X \qquad \text{for all} \qquad x \in X,$$

for some $0 < c_1 \le c_2$. We show that completeness of $(Y, \| \cdot \|_Y)$ implies completeness of $(X, \| \cdot \|_X)$.

If (x_n) is a Cauchy sequence in X, then Tx_n is a Cauchy sequence in Y, since

$$\|Tx_n - Tx_m\|_Y \le c_2 \|x_n - x_m\|_X.$$

Since Y is complete, $Tx_n \to y$ for some $y \in Y$. Setting $x = T^{-1}y$ it follows that $x_n \to x$, since

$$\|x_n - x\|_X \le \frac{1}{c_1} \|Tx_n - Tx\|_Y = \frac{1}{c_1} \|Tx_n - y\|_Y \to 0. \qquad □$$

The following corollary is unsurprising.

Corollary 4.5 *If X is a normed space and $\| \cdot \|_1$ and $\| \cdot \|_2$ are two equivalent norms on X, then $(X, \| \cdot \|_1)$ is complete if and only if $(X, \| \cdot \|_2)$ is complete.*

Proof Take T to be the identity map $I_X \colon (X, \| \cdot \|_1) \to (X, \| \cdot \|_2)$ in the previous lemma. $\qquad\square$

It is also the case that products of Banach spaces are Banach spaces (given an appropriate norm on the product).

Lemma 4.6 *If $(X, \| \cdot \|_X)$ and $(Y, \| \cdot \|_Y)$ are complete, then $X \times Y$ is complete for the norm*

$$\|(x, y)\|_1 := \|x\|_X + \|y\|_Y$$

(the standard norm on $X \times Y$) and also for the (equivalent) norm $\| \cdot \|_2$ defined by

$$\|(x, y)\|_2 := \left(\|x\|_X^2 + \|y\|_Y^2 \right)^{1/2}. \tag{4.2}$$

Proof If $((x_n, y_n))_n$ is a Cauchy sequence in $X \times Y$ using the norm $\| \cdot \|_p$, then for every $\varepsilon > 0$ there exists N such that

$$\|(x_n, y_n) - (x_m, y_m)\|_p^p = \|x_n - x\|_X^p + \|y_n - y\|_Y^p < \varepsilon.$$

It follows that (x_n) is a Cauchy sequence in X and (y_n) is a Cauchy sequence in Y, so since X and Y are complete there exist $x \in X$ and $y \in Y$ such that $x_n \to x$ in $(X, \| \cdot \|_X)$ and $y_n \to y$ in $(Y, \| \cdot \|_Y)$. Since

$$\|(x_n, y_n) - (x, y)\|_p^p = \|x_n - x\|_X^p + \|y_n - y\|_Y^p,$$

it follows that $(x_n, y_n) \to (x, y)$ in $(X \times Y, \| \cdot \|_p)$, and so this space is complete. $\qquad\square$

We have just given a number of results that allow us to construct new complete spaces from known complete spaces. We will see in Chapter 7 that when a normed space $(X, \| \cdot \|)$ is not complete there is a well-defined way to construct a larger space $(\mathscr{X}, \| \cdot \|_{\mathscr{X}})$ that is complete and has X as a dense subspace.

4.2 Examples of Banach Spaces

In this section we give some particular examples of Banach spaces, and prove their completeness. We look first at spaces of sequences, and then at spaces of functions.

4.2.1 Sequence Spaces

First we generalise the argument we used to show that $\mathbb{K}^n$ is complete with the standard norm to show that the ℓ^p sequences spaces are complete (and therefore Banach spaces). We can then use Lemma 4.3 (completeness of closed subspaces) to prove the completeness of $c_0(\mathbb{K})$.

Theorem 4.7 (Completeness of ℓ^p) *For each* $1 \le p \le \infty$ *the sequence space* $\ell^p(\mathbb{K})$ *is complete when equipped with its standard norm.*

Proof We give the proof for $1 \le p < \infty$, leaving the case $p = \infty$ for Exercise 4.2.

Suppose that $\boldsymbol{x}^{(k)} = (x_1^{(k)}, x_2^{(k)}, \cdots)$ is a Cauchy sequence in $\ell^p(\mathbb{K})$. Then for every $\varepsilon > 0$ there exists an N_ε such that

$$\|\boldsymbol{x}^{(n)} - \boldsymbol{x}^{(m)}\|_{\ell^p}^p = \sum_{j=1}^\infty |x_j^{(n)} - x_j^{(m)}|^p < \varepsilon^p \qquad \text{for all} \qquad n, m \ge N_\varepsilon. \quad (4.3)$$

In particular, $(x_j^{(n)})_{n=1}^\infty$ is a Cauchy sequence in $\mathbb{K}$ for every fixed j. Since $\mathbb{K}$ is complete, it follows that for each $j \in \mathbb{N}$

$$x_j^{(n)} \to a_j \qquad \text{as} \qquad n \to \infty$$

for some $a_j \in \mathbb{K}$. Set $\boldsymbol{a} = (a_1, a_2, \cdots)$.

We need to show that $\boldsymbol{a} \in \ell^p$ and that $\|\boldsymbol{x}^{(n)} - \boldsymbol{a}\|_{\ell^p} \to 0$ as $n \to \infty$. From (4.3) it follows that for any $k \in \mathbb{N}$ we have

$$\sum_{j=1}^k |x_j^{(n)} - x_j^{(m)}|^p \le \sum_{j=1}^\infty |x_j^{(n)} - x_j^{(m)}|^p < \varepsilon^p \qquad \text{for all} \qquad n, m \ge N_\varepsilon.$$

Letting $m \to \infty$ (which we can do since the left-hand side contains only a finite number of terms) we obtain

$$\sum_{j=1}^k |x_j^{(n)} - a_j|^p \le \varepsilon^p \qquad \text{for all} \qquad n \ge N_\varepsilon,$$

since $x_j^{(m)} \to a_j$ as $m \to \infty$. Since this holds for every $k \in \mathbb{N}$, it follows that

$$\sum_{j=1}^\infty |x_j^{(n)} - a_j|^p \le \varepsilon^p \qquad \text{for all} \qquad n \ge N_\varepsilon, \quad (4.4)$$

and so $\boldsymbol{x}^{(n)} - \boldsymbol{a} \in \ell^p$ provided that $n \ge N_\varepsilon$. But since ℓ^p is a vector space and $\boldsymbol{x}^{(n)} \in \ell^p$ for every $n \in \mathbb{N}$, this implies that $\boldsymbol{a} \in \ell^p$ and (4.4) shows that

$$\|x^{(n)} - a\|_{\ell^p} \leq \varepsilon \qquad \text{for all} \qquad n \geq N_\varepsilon,$$

and so $x^{(n)} \to a$ in $\ell^p(\mathbb{K})$, as required. $\qquad\qquad\qquad\qquad\square$

The space $c_0(\mathbb{K})$ is a closed subspace of $\ell^\infty(\mathbb{K})$: for a proof see Exercise 4.4; it is therefore complete.

Corollary 4.8 *The space $c_0(\mathbb{K})$ of null sequences is complete when equipped with ℓ^∞ norm.*

4.2.2 Spaces of Functions

In this section we prove the completeness of various spaces of bounded functions in the supremum norm. We begin with a large space (the collection of all bounded functions on a metric space X), and then use the 'closed subspace' method to prove the completeness of the space of all bounded continuous functions on X. With a little more work we can use the completeness of $C([a, b])$ to prove the completeness of the space $C^1([a, b])$ of continuously differentiable functions with an appropriate norm.

Theorem 4.9 *Let X be a metric space and let $\mathcal{F}_b(X; \mathbb{K})$ be the collection of all functions $f : X \to \mathbb{K}$ that are bounded, i.e. $\sup_{x \in X} |f(x)| < \infty$. Then $\mathcal{F}_b(X; \mathbb{K})$ is complete with the supremum norm*

$$\|f\|_\infty := \sup_{x \in X} |f(x)|.$$

Convergence of functions in the supremum norm, i.e. $\|f_n - f\|_\infty \to 0$, is known as uniform convergence, since it implies that given any $\varepsilon > 0$, there exists N such that

$$\sup_{x \in X} |f_n(x) - f(x)| < \varepsilon \qquad \text{for all } n \geq N,$$

i.e. $|f_n(x) - f(x)| < \varepsilon$ for every $x \in X$.

Proof If $(f_k)_{k=1}^\infty$ is a Cauchy sequence in $\mathcal{F}_b(X; \mathbb{K})$, then given any $\varepsilon > 0$ there exists an N such that

$$\|f_n - f_m\|_\infty = \sup_{x \in X} |f_n(x) - f_m(x)| < \varepsilon \qquad \text{for all} \qquad n, m \geq N. \quad (4.5)$$

In particular, for each fixed $x \in X$ the sequence $(f_k(x))$ is Cauchy in $\mathbb{K}$, and so converges (because $\mathbb{K}$ is complete); therefore we can set

$$f(x) = \lim_{k \to \infty} f_k(x).$$

Now we need to show that $f \in \mathcal{F}_b(X; \mathbb{K})$ and that $f_k \to f$ uniformly on X. Using (4.5) for every $x \in X$ we have

$$|f_n(x) - f_m(x)| < \varepsilon \qquad \text{for all} \qquad n, m \geq N,$$

where N does not depend on x. Letting $m \to \infty$ in this expression we obtain

$$|f_n(x) - f(x)| < \varepsilon \qquad \text{for all} \qquad n \geq N,$$

where again N does not depend on x. It follows that

$$\sup_{x \in X} |f_n(x) - f(x)| < \varepsilon \qquad \text{for all} \qquad n \geq N, \qquad (4.6)$$

i.e. f_n converges uniformly to f on X.

It now only remains to show that this limiting function f is bounded; this follows from (4.6), since f_N is bounded and

$$\|f_N - f\|_\infty < \varepsilon. \qquad \square$$

We can now use the fact that closed subspaces of Banach spaces are complete to prove the completeness of the space of all bounded continuous functions from X into $\mathbb{K}$.

Corollary 4.10 *If (X, d) is any metric space, then $C_b(X; \mathbb{K})$, the space of all bounded continuous functions from X into $\mathbb{K}$, is complete when equipped with the supremum norm*

$$\|f\|_\infty := \sup_{x \in X} |f(x)|.$$

Proof The space $C_b(X; \mathbb{K})$ is a subspace of $\mathcal{F}_b(X; \mathbb{K})$, which is complete, so by Lemma 4.3 we need only show that $C_b(X; \mathbb{K})$ is a closed subspace of $\mathcal{F}_b(X; \mathbb{K})$, i.e. if $(f_k) \in C_b(X; \mathbb{K})$ and $\|f_k - f\|_\infty \to 0$, then f is continuous.

Given any $\varepsilon > 0$ there exists N such that $\|f_n - f\|_\infty < \varepsilon/3$ for every $n \geq N$. In particular, $\|f_N - f\|_\infty < \varepsilon/3$. Now fix $x \in X$. Since $f_N \in C_b(X; \mathbb{K})$, there exists $\delta > 0$ such that $d(y, x) < \delta$ implies that $|f_N(y) - f_N(x)| < \varepsilon/3$. It follows that if $d(y, x) < \delta$, then

$$|f(y) - f(x)| \leq |f(y) - f_N(y)| + |f_N(y) - f_N(x)| + |f_N(x) - f(x)|$$
$$< \frac{\varepsilon}{3} + \frac{\varepsilon}{3} + \frac{\varepsilon}{3} = \varepsilon,$$

and so f is continuous at x. Since this holds for any $x \in X$, $f \in C_b(X; \mathbb{K})$ as required. $\qquad \square$

Corollary 4.11 *If (X, d) is any compact metric space, then $C(X; \mathbb{K})$ is a Banach space when equipped with the maximum norm*

$$\|f\|_\infty := \max_{x \in X} |f(x)|.$$

Proof Proposition 2.29 guarantees that a continuous function on a compact metric space is bounded, so $C(X; \mathbb{K}) = C_b(X; \mathbb{K})$. That the supremum norm is in fact the same as the maximum norm follows from the same proposition, since every continuous function on a compact metric space attains its bounds. $\qquad\qquad\square$

Since the supremum norm makes $C(X; \mathbb{K})$ complete, it is the 'standard norm' on this space. However, there are other (useful) norms on $C(X; \mathbb{K})$ that are not equivalent to the supremum norm (see the next chapter), and $C(X; \mathbb{K})$ is not complete in these norms.

We end this section by proving the completeness of the space of all continuously differentiable functions on $[a, b]$ with an appropriate norm.

Theorem 4.12 *The space $C^1([a, b])$ of all continuously differentiable functions on $[a, b]$ is complete with the C^1 norm*

$$\|f\|_{C^1} := \|f\|_\infty + \|f'\|_\infty.$$

Proof Let $(f_n) \in C^1([a, b])$ be a Cauchy sequence in the C^1 norm. Then for any $\varepsilon > 0$ there exists N such that

$$\|f_n - f_m\|_\infty + \|f_n' - f_m'\|_\infty < \varepsilon \qquad n, m \geq N. \tag{4.7}$$

Therefore (f_n) and (f_n') are both Cauchy sequences in $C([a, b])$ with the supremum norm, and since this space is complete there exist $f, g \in C([a, b])$ such that $f_n \to f$ and $f_n' \to g$. We need to show that $g = f'$. To do this we use the Fundamental Theorem of Calculus: for every n we have

$$f_n(x) = f_n(a) + \int_a^x f_n'(t)\, dt, \qquad x \in [a, b], \tag{4.8}$$

and now we take limits as $n \to \infty$. Since

$$\left| \int_a^x f_n'(t)\, dt - \int_a^x g(t)\, dt \right| \leq \int_a^x |f_n'(t) - g(t)|\, dt \leq (b - a)\|f_n' - g\|_\infty$$

and $\|f_n' - g\|_\infty \to 0$ as $n \to \infty$, it follows that

$$\int_a^x f_n'(t)\, dt \to \int_a^x g(t)\, dt$$

uniformly on $[a, b]$. Taking limits in (4.8) we therefore obtain

$$f(x) = f(a) + \int_a^x g(t) \, dt,$$

which shows that $f' = g$; in particular, $f \in C^1([a, b])$. We can now take $m \to \infty$ in (4.7) to deduce that

$$\|f_n - f\|_\infty + \|f_n' - f'\|_\infty < \varepsilon \qquad n \geq N,$$

which shows that $f_n \to f$ in $C^1([a, b])$ as required. $\qquad\qquad\square$

In a similar way one can prove the completeness of the space $C^k([a, b])$ of all k times continuously differentiable functions on $[a, b]$ with the C^k norm

$$\|f\|_{C^k} = \sum_{j=0}^{k} \|f^{(j)}\|_\infty,$$

where $f^{(j)} = \mathrm{d}^j f / \mathrm{d}x^j$.

4.3 Sequences in Banach Spaces

In any normed space we have both a notion of convergence and a linear structure, so it is possible to consider the convergence of series. As in the case of real numbers, if $(x_j) \in X$ we say that

$$\sum_{j=1}^{\infty} x_j = x$$

if the partial sums $\sum_{j=1}^{k} x_j$ converge to x (in X) as $k \to \infty$, i.e.

$$\left\| x - \sum_{j=1}^{k} x_j \right\| \to 0 \qquad \text{as} \qquad k \to \infty.$$

While we can consider convergence of series in any normed space (whether or not it is complete), the theory is much more satisfactory in Banach spaces. For example, for real sequences we know that any absolutely convergent sequence converges, and there is a corresponding result in Banach spaces.

Lemma 4.13 *Let X be a Banach space and (x_j) a sequence in X. Then*

$$\sum_{j=1}^{\infty} \|x_j\| < \infty \qquad \Rightarrow \qquad \sum_{j=1}^{\infty} x_j \text{ converges in } X.$$

Proof Since $\sum_{j=1}^{\infty} \|x_j\|$ converges, its partial sums form a Cauchy sequence. So for any $\varepsilon > 0$ there exists N such that for $m > n \geq N$

$$\left| \sum_{j=1}^{m} \|x_j\| - \sum_{j=1}^{n} \|x_j\| \right| = \sum_{j=n+1}^{m} \|x_j\| < \varepsilon.$$

Then

$$\left\| \sum_{j=1}^{m} x_j - \sum_{j=1}^{n} x_j \right\| = \left\| \sum_{j=n+1}^{m} x_j \right\| \leq \sum_{j=n+1}^{m} \|x_j\| < \varepsilon,$$

and so the partial sums $\sum_{j=1}^{k} x_j$ form a Cauchy sequence in X. Since X is complete, these partial sums converge, i.e. the series converges. $\square$

The statement of this lemma can in some sense be reversed, providing a useful test for completeness.

Lemma 4.14 *If $(X, \|\cdot\|)$ is a normed space with the property that whenever $\sum_{j=1}^{\infty} \|x_j\| < \infty$ the sum $\sum_{j=1}^{\infty} x_j$ converges in X, then X is complete.*

Proof Suppose that (y_j) is a Cauchy sequence in X. Using Exercise 4.1 (iii) it is enough to show that some subsequence of (y_j) converges to some $y \in X$.

Find n_0 such that $\|y_i - y_j\| < 1$ for $i, j \geq n_0$, and then choose $n_k \in \mathbb{N}$ inductively such that $n_{k+1} > n_k$ and

$$\|y_i - y_j\| < 2^{-k} \qquad i, j \geq n_k;$$

we can do this because (y_j) is Cauchy.

Let $x_1 = y_{n_1}$ and set $x_j = y_{n_j} - y_{n_{j-1}}$ for $j \geq 2$. Then for $j \geq 2$ we have

$$\|x_j\| = \|y_{n_j} - y_{n_{j-1}}\| < 2^{-(j-1)}$$

and so $\sum_{j=1}^{\infty} \|x_j\| < \infty$, which implies that $\sum_{j=1}^{\infty} x_j$ converges to some element $y \in X$. However,

$$\sum_{j=1}^{n} x_j = y_1 + \sum_{j=2}^{n} \left(y_{n_j} - y_{n_{j-1}} \right) = y_{n_j},$$

and so $y_{n_j} \to y$ as $j \to \infty$. It follows from Exercise 4.1 (iii) that $y_k \to y$ as $k \to \infty$, and so X is complete. $\square$

4.4 The Contraction Mapping Theorem

In a complete normed[2] space $(X, \|\cdot\|)$ the Contraction Mapping Theorem (also known as Banach's Fixed Point Theorem) enables us to find a fixed point of any map that is a contraction. We will use this result to prove the existence of solutions for ordinary differential equations in Exercise 4.8, that the collection of all invertible operators is open among all bounded operators (Lemma 11.16), and the Lax–Milgram Lemma (Exercise 12.4).

Theorem 4.15 (Contraction Mapping Theorem) *Let K be a non-empty closed subset of a complete normed space $(X, \|\cdot\|)$ and $f \colon K \to K$ a contraction, i.e. a map such that*

$$\|f(x) - f(y)\| \le \kappa \|x - y\|, \qquad x, y \in K, \tag{4.9}$$

for some $\kappa < 1$. Then f has a unique fixed point in K, i.e. there exists a unique $x \in K$ such that $f(x) = x$.

Proof Choose any $x_0 \in K$ and set $x_{n+1} = f(x_n)$. Then

$$\|x_{j+1} - x_j\| \le \kappa \|x_j - x_{j-1}\| \le \kappa^2 \|x_{j-1} - x_{j-2}\| \le \cdots \le \kappa^j \|x_1 - x_0\|;$$

so if $k > j$, using the triangle inequality repeatedly, we have

$$\|x_k - x_j\| \le \sum_{i=j}^{k-1} \|x_{i+1} - x_i\| \le \sum_{i=j}^{k-1} \kappa^i \|x_1 - x_0\| \le \frac{\kappa^j}{1 - \kappa}\|x_1 - x_0\|.$$

It follows that (x_n) is a Cauchy sequence in X. Since X is complete, $x_n \to x$ for some $x \in X$, and since $(x_n) \in K$ and K is closed we have $x \in K$.

Since (4.9) implies that f is continuous, we have $f(x_n) \to f(x)$ (using Lemma 2.12) and so if we take limits on both sides of

$$x_{n+1} = f(x_n)$$

we obtain $x = f(x)$.

Any such x must be unique, since if $x, y \in K$ with $f(x) = x$ and $f(y) = y$ it follows that

$$\|x - y\| = \|f(x) - f(y)\| \le \kappa \|x - y\| \qquad \Rightarrow \qquad (1 - \kappa)\|x - y\| = 0,$$

so $x = y$. $\qquad\qquad\qquad\qquad\qquad\qquad\qquad\qquad\qquad\qquad\qquad\qquad\qquad\square$

The conclusion of the theorem is no longer valid if we only have

$$\|f(x) - f(y)\| < \|x - y\|, \qquad x, y \in K$$

unless K is compact (see Exercise 4.6).

---

[2] The theorem also holds in any complete metric space (X, d) with the obvious changes.

Exercises

4.1   Show that in a normed space $(X, \| \cdot \|)$
      (i) any Cauchy sequence is bounded; and
      (ii) if (x_n) is Cauchy and $x_{n_k} \to x$ for some subsequence (x_{n_k}), then
      $x_n \to x$.

4.2   Show that $\ell^\infty(\mathbb{K})$ is complete.

4.3   Show, using Lemma 4.6, that $\mathbb{K}^n$ is complete.

4.4   Show that $c_0(\mathbb{K})$ is a closed subspace of $\ell^\infty(\mathbb{K})$, and deduce using
      Lemma 4.3 that $c_0(\mathbb{K})$ is complete.

4.5   Show that if X is a normed space and U is a closed linear subspace of
      U, then

$$\|[x]\|_{X/U} = \inf_{u \in U} \|x + u\|_X$$

      defines a norm on the quotient space X/U introduced in Exercise 1.6.
      Show that if X is complete, then X/U is complete with this norm. (Use
      Lemma 4.14 for the completeness.)

4.6   Show that the conclusion of the Contraction Mapping Theorem (Theo-
      rem 4.15) need not hold if

$$\|f(x) - f(y)\|_X < \|x - y\|_X, \qquad x, y \in K,$$

      but that the result does still hold under this weakened condition if K is
      compact.

4.7   Show that if $f \in C(\mathbb{R}; \mathbb{R})$ and $x \in C([0, T]; \mathbb{R})$, then

$$\dot{x} = f(x), \qquad \text{with} \qquad x(0) = x_0, \qquad \text{for all } t \in [0, T] \quad (4.10)$$

      if and only if

$$x(t) = x_0 + \int_0^t f(x(s)) \, ds \qquad \text{for all } t \in [0, T]. \quad (4.11)$$

4.8   Suppose that $f : \mathbb{R} \to \mathbb{R}$ is Lipschitz continuous, i.e. satisfies

$$|f(x) - f(y)| \le L|x - y|, \qquad x, y \in \mathbb{R},$$

      for some $L > 0$. Use the Contraction Mapping Theorem in the space
      $C([0, T]; \mathbb{R})$ on the mapping

$$(\mathcal{J}x)(t) = x_0 + \int_0^t f(x(s)) \, ds$$

      to show that (4.10) has a unique solution on any time interval $[0, T]$ with
      $LT < 1$. By considering the solution to (4.10) starting at $x(T)$ deduce
      that in fact (4.10) has a unique solution that exists for all $t \ge 0$.

4.9 Show that the space $C(\mathbb{R}; \mathbb{R})$ is complete when equipped with the metric

$$d(f, g) = \sum_{n=1}^{\infty} \frac{1}{2^n} \frac{\|f - g\|_{[-n,n]}}{1 + \|f - g\|_{[-n,n]}},$$

where $\|f\|_{[-n,n]} := \max_{x \in [-n,n]} |f(x)|$. Show that this metric does not correspond to a norm. (Goffman and Pedrick, 1983)

5

Finite-Dimensional Normed Spaces

In this chapter we will (briefly) investigate finite-dimensional normed spaces. We show that in a finite-dimensional space all norms are equivalent, and that being compact is the same as being closed and bounded. We also show that a normed space is finite-dimensional if and only if its closed unit ball is compact; this proof relies on Riesz's Lemma, which will prove extremely useful later (in Chapter 24) when we discuss the spectral properties of compact operators.

5.1 Equivalence of Norms on Finite-Dimensional Spaces

First we show that all norms are equivalent on finite-dimensional spaces. We make use of the norm defined in Lemma 3.7.

Theorem 5.1 *If V is a finite-dimensional space, then all norms on V are equivalent.*

(To prove that all norms on $\mathbb{K}^n$ are equivalent is a little easier, since we can just use the standard norm on $\mathbb{K}^n$ rather than having to construct the norm $\| \cdot \|_E$ for V.)

Proof Let $E := (e_j)_{j=1}^n$ be a basis for V, and let

$$\|x\|_E := \left(\sum_{j=1}^n |\alpha_j|^2 \right)^{1/2} \qquad \text{when} \qquad x = \sum_{j=1}^n \alpha_j e_j$$

be the norm on V defined in Lemma 3.7. Recall for use later that we showed in Lemma 3.22 that $(\mathbb{K}^n, \| \cdot \|_{\ell^2})$ is isometrically isomorphic to $(V, \| \cdot \|_E)$ via the mapping $\Phi \colon \mathbb{K}^n \to V$ given by

$$\Phi(\alpha_1, \ldots, \alpha_n) = \sum_{j=1}^{n} \alpha_j e_j. \tag{5.1}$$

We show that $\| \cdot \|$ and $\| \cdot \|_E$ are equivalent, which is sufficient to prove the result since equivalence of norms is an equivalence relation (Lemma 3.17).

First, using the triangle inequality and the Cauchy–Schwarz inequality from Exercise 2.1, when $x = \sum_{j=1}^{n} \alpha_j e_j$ we have

$$\|x\| = \left\| \sum_{j=1}^{n} \alpha_j e_j \right\| \leq \sum_{j=1}^{n} |\alpha_j| \|e_j\|$$

$$\leq \left(\sum_{j=1}^{n} \|e_j\|^2 \right)^{1/2} \left(\sum_{j=1}^{n} |\alpha_j|^2 \right)^{1/2} = C \|x\|_E, \tag{5.2}$$

where $C := \left(\sum_{j=1}^{n} \|e_j\|^2 \right)^{1/2}$ is a constant that does not depend on x.

This estimate implies that

$$\big| \|x\| - \|y\| \big| \leq \|x - y\| \leq C \|x - y\|_E,$$

and so the map $x \mapsto \|x\|$ is continuous from $(V, \| \cdot \|_E)$ into $\mathbb{R}$.

Now, the set

$$\{ u \in \mathbb{K}^n \text{ with } \|u\|_{\ell^2} = 1 \}$$

is a closed and bounded subset of $\mathbb{K}^n$, so compact (Theorem 2.27); therefore

$$S := \{ x \in V : \|x\|_E = 1 \} = \{ \Phi(u) : u \in \mathbb{K}^n \text{ with } \|u\|_{\ell^2} = 1 \}$$

is compact, since it is the continuous image of a compact set (see Theorem 2.28).

Proposition 2.29 therefore guarantees that the map $x \mapsto \|x\|$ from S into $\mathbb{R}$ is bounded below and attains its lower bound on S: there exists $\alpha \in \mathbb{R}$ such that $\|x\| \geq \alpha$ for every $x \in S$, and $\|x\| = \alpha$ for some $x \in S$. This second fact means that $\alpha > 0$, since otherwise $x = 0$ (since $\| \cdot \|$ is a norm), and this is impossible since $\|x\|_E = 1$. It follows that

$$\|x\| = \|x\|_E \left\| \frac{x}{\|x\|_E} \right\| \geq \alpha \|x\|_E.$$

Combining this with the inequality $\|x\| \leq C \|x\|_E$ we proved above in (5.2) shows that $\| \cdot \|$ and $\| \cdot \|_E$ are equivalent. $\qquad\square$

Using this result we can immediately deduce two important consequences.

Theorem 5.2 *Any finite-dimensional normed space* $(V, \| \cdot \|)$ *is complete.*

Proof The complete space $(\mathbb{K}^n, \|\cdot\|_{\ell^2})$ is isometrically isomorphic to the space $(V, \|\cdot\|_E)$ so $(V, \|\cdot\|_E)$ is complete by Lemma 4.4. Since $\|\cdot\|_E$ is equivalent to $\|\cdot\|$, it follows from Corollary 4.5 that $(V, \|\cdot\|)$ is complete.   □

Theorem 5.3 *A subset of a finite-dimensional normed space is compact if and only if it is closed and bounded.*

Proof We showed in Lemma 2.24 that any compact subset of a metric space must be closed and bounded, and any normed space can be considered as a metric space with metric $d(x, y) = \|x - y\|$.

Suppose that K is a closed, bounded, subset of $(V, \|\cdot\|)$. Then it is also a closed bounded subset of $(V, \|\cdot\|_E)$, since closure and boundedness are preserved under equivalent norms. If we can show that K is a compact subset of $(V, \|\cdot\|_E)$, then it is also compact subset of $(V, \|\cdot\|)$, since (5.2) shows that the identity map from $(V, \|\cdot\|_E)$ into $(V, \|\cdot\|)$ is continuous.

To show that K is a compact subset of $(V, \|\cdot\|_E)$ we once again use the map $\Phi\colon \mathbb{K}^n \to V$ from (5.1) that is an isometric isomorphism between $(\mathbb{K}^n, \|\cdot\|_{\ell^2})$ and $(V, \|\cdot\|_E)$, and observe that $\Phi^{-1}(K)$ is closed (since it is the preimage of a closed set under the continuous map Φ) and bounded (since Φ is an isometry). It follows that $\Phi^{-1}(K)$ is compact, since it is a closed bounded subset of $\mathbb{K}^n$ (see Theorem 2.27). Since Φ is continuous, it follows that $K = \Phi(\Phi^{-1}(K))$ is the continuous image of a compact set, and so compact.   □

5.2 Compactness of the Closed Unit Ball

We end this chapter by showing that a normed space is finite-dimensional if and only if its closed unit ball is compact. This means that Theorem 5.3 cannot hold in an infinite-dimensional space.

Lemma 5.4 (Riesz's Lemma) *Let $(X, \|\cdot\|)$ be a normed space and Y a proper[1] closed subspace of X. Then there exists $x \in X$ with $\|x\| = 1$ such that $\|x - y\| \geq 1/2$ for every $y \in Y$.*

Proof Choose $x_0 \in X \setminus Y$ and set

$$d = \mathrm{dist}(x_0, Y) := \inf_{y \in Y} \|x_0 - y\|.$$

---

[1] A proper subspace of X is a subspace that is not equal to X.

Since Y is closed, $d > 0$. Indeed, if dist$(x_0, Y) = 0$, then there exists a sequence $(y_n) \in Y$ such that $\|y_n - x_0\| \to 0$, i.e. such that $y_n \to x_0$, and then, since Y is closed, we would have $x_0 \in Y$.

Now choose $y_0 \in Y$ such that

$$d \leq \|x_0 - y_0\| \leq 2d$$

and set

$$x = \frac{x_0 - y_0}{\|x_0 - y_0\|}.$$

Clearly $\|x\| = 1$, and for any $y \in Y$ we have

$$\|x - y\| = \left\| \frac{x_0 - y_0}{\|x_0 - y_0\|} - y \right\|$$

$$= \frac{1}{\|x_0 - y_0\|} \left\| (x_0 - [y_0 + y\|x_0 - y_0\|]) \right\|.$$

Since $y_0 + y\|x_0 - y_0\| \in Y$, it follows that

$$\|x - y\| \geq \frac{d}{\|x_0 - y_0\|} \geq \frac{d}{2d} = \frac{1}{2}. \qquad \square$$

Using this we can prove our promised result characterising finite-dimensional spaces.

Theorem 5.5 *A normed space X is finite-dimensional if and only if its closed unit ball is compact.*

Proof If X is finite-dimensional, then its closed unit ball is compact, by Theorem 5.3. So we suppose that X is infinite-dimensional and show that its closed unit ball is not compact.

Take any $x_1 \in X$ with $\|x_1\| = 1$. Then the linear span of $\{x_1\}$ is a proper closed linear subspace of X, so by Lemma 5.4 there exists $x_2 \in X$ with $\|x_2\| = 1$ and $\|x_2 - x_1\| \geq 1/2$. Now Span$\{x_1, x_2\}$ is a proper closed linear subspace of X, so there exists $x_3 \in X$ with $\|x_3\| = 1$ and $\|x_3 - x_j\| \geq 1/2$. One can continue inductively to obtain a sequence (x_n) with $\|x_n\| = 1$ and $\|x_i - x_j\| \geq 1/2$ whenever $i \neq j$. No subsequence of the (x_n) can be Cauchy, so no subsequence can converge, from which it follows that the closed unit ball in X is not compact. $\qquad \square$

We can see this non-compactness easily in the ℓ^p sequence spaces, which we have already shown are infinite-dimensional (Example 1.16). Whatever value of p we choose the elements $(e^{(j)})_{j=1}^{\infty}$ from (1.3) all have $\|e^{(j)}\|_{\ell^p} = 1$ so form a sequence in the closed unit ball of ℓ^p. However, if $i \neq j$, then

$$\|e^{(i)} - e^{(j)}\|_{\ell^p} = 2^{1/p};$$

no subsequence of $(e^{(j)})$ can be Cauchy, so the closed unit ball is not compact in any of these spaces.

Exercises

5.1   Show that if K is a non-empty open subset of $\mathbb{R}^n$ that is (i) convex, (ii) symmetric ($x \in K$ implies that $-x \in K$), and (iii) bounded, then

$$N(x) := \inf\{M > 0 : M^{-1}x \in K\}$$

defines a norm on $\mathbb{R}^n$ ('K is the unit ball for N').

5.2   Show that if $(V, \|\cdot\|_V)$ is a normed space, W a vector space, and $T : W \to V$ a linear bijection, then

$$\|x\|_W := \|Tx\|_V$$

defines a norm on W, and $T : (W, \|\cdot\|_W) \to (V, \|\cdot\|_V)$ is an isometry. (So $(W, \|\cdot\|_W)$ and $(V, \|\cdot\|_V)$ are isometrically isomorphic.)

5.3   Show that any finite-dimensional subspace of a Banach space is closed.

5.4   Show that if Y is a finite-dimensional subspace of a normed space X and $x \in X \setminus Y$, then there exists $y \in Y$ with $\|x - y\| = \mathrm{dist}(x, Y)$.

5.5   Show that if Y is a subspace of a normed space X and $x \in X \setminus Y$, then

$$\mathrm{dist}(\alpha x, Y) = |\alpha|\, \mathrm{dist}(x, Y) \quad \text{for any} \quad \alpha \in \mathbb{K},$$

and

$$\mathrm{dist}(x + w, Y) = \mathrm{dist}(x, Y) \quad \text{for any} \quad w \in Y.$$

5.6   Suppose that Y is a proper finite-dimensional subspace of a normed space X. Show that for any $y \in Y$ and $r > 0$ there exists $x \in X$ such that $\|x - y\| = \mathrm{dist}(x, Y) = r$. (Megginson, 1998)

5.7   Use the result of the previous exercise to show that no infinite-dimensional Banach space can have a countable Hamel basis. (Megginson, 1998) [Hint: given a Hamel basis $\{e_j\}_{j=1}^{\infty}$ for X, let $X_n = \mathrm{Span}(e_1, \ldots, e_n)$. Now find a sequence (y_n) with $y_n \in X_n$ such that

$$\|y_n - y_{n-1}\| = \mathrm{dist}(y_n, X_{n-1}) = 3^{-n}$$

and show that (y_n) is Cauchy in X but its limit cannot lie in any of the X_n.] For another, simpler, proof using the Baire Category Theorem, see Exercise 22.1.

6

Spaces of Continuous Functions

We showed in Corollary 4.11 that the space $C(X; \mathbb{K})$ of continuous functions from a compact metric space X into $\mathbb{K}$ is complete with the supremum norm

$$\|f\|_\infty = \sup_{x \in X} |f(x)|.$$

In this chapter we prove some key results about such spaces of continuous functions.

First we show that continuous functions on an interval can be uniformly approximated by polynomials (the 'Weierstrass Approximation Theorem'), which has interesting applications to Fourier series. Then we prove the Stone–Weierstrass Theorem, which generalises this to continuous functions on compact metric spaces and other collections of approximating functions. We end with a proof of the Arzelà–Ascoli Theorem, which characterises compact subsets of $C(X; \mathbb{K})$.

6.1 The Weierstrass Approximation Theorem

In this section we show that any continuous real-valued function on $[0, 1]$ can be uniformly approximated by polynomials. As a consequence we prove the same result for continuous functions on $[a, b]$ and deduce that $C([a, b]; \mathbb{R})$ is separable. We also consider the approximation of continuous functions by Fourier sine and cosine series.

We will prove a more general version of this result, the Stone–Weierstrass Theorem, later in this chapter.

Theorem 6.1 (Weierstrass Approximation Theorem) *If $f \in C([0, 1])$, then the sequence of polynomials*[1]

$$P_n(x) = \sum_{k=0}^{n} \binom{n}{k} f\left(\frac{k}{n}\right) x^k (1 - x)^{n-k} \tag{6.1}$$

converges uniformly to $f(x)$ on $[0, 1]$. (The sequence of polynomials in (6.1) are known as the Bernstein polynomials.)

For the proof we will need the following result.

Lemma 6.2 *If we define*

$$r_k(x) := \binom{n}{k} x^k (1 - x)^{n-k},$$

then

$$\sum_{k=0}^{n} r_k(x) = 1 \qquad and \qquad \sum_{k=0}^{n} (k - nx)^2 r_k(x) = nx(1 - x).$$

Proof We start with the binomial identity

$$(x + y)^n = \sum_{k=0}^{n} \binom{n}{k} x^k y^{n-k}. \tag{6.2}$$

Differentiate with respect to x and multiply by x to give

$$nx(x + y)^{n-1} = \sum_{k=0}^{n} k \binom{n}{k} x^k y^{n-k};$$

differentiate twice with respect to x and multiply by x^2 to give

$$n(n - 1)x^2(x + y)^{n-2} = \sum_{k=0}^{n} k(k - 1)\binom{n}{k} x^k y^{n-k}.$$

It follows, since $r_k(x)$ is the right-hand side of (6.2) when we set $y = 1 - x$, that

$$\sum_{k=0}^{n} r_k(x) = 1, \quad \sum_{k=0}^{n} k r_k(x) = nx, \quad \text{and} \quad \sum_{k=0}^{n} k(k - 1)r_k(x) = n(n - 1)x^2.$$

---

[1] As is standard, we define $n! = 1 \cdot 2 \cdots n$ and $\binom{n}{k} = \frac{n!}{(n-k)!k!}$.

Therefore

$$\sum_{k=0}^{n}(k-nx)^2 r_k(x) = n^2 x^2 \sum_{k=0}^{n} r_k(x) - 2nx \sum_{k=0}^{n} k r_k(x) + \sum_{k=0}^{n} k^2 r_k(x)$$

$$= n^2 x^2 - 2nx \cdot nx + (nx + n(n-1)x^2)$$

$$= nx(1-x). \qquad \qquad \square$$

Using this we can now prove the Weierstrass Approximation Theorem.

Proof of Theorem 6.1 Since f is continuous on the compact set $[0, 1]$, it is bounded with $|f(x)| \le M$ for some $M > 0$ (Proposition 2.29). It also follows (see Lemma 2.30) that f is uniformly continuous on $[0, 1]$, so for any $\varepsilon > 0$ there exists a $\delta > 0$ such that

$$|x - y| < \delta \qquad \Rightarrow \qquad |f(x) - f(y)| < \varepsilon.$$

Noting that we can write

$$P_n(x) = \sum_{k=0}^{n} f\left(\frac{k}{n}\right) r_k(x)$$

and using the fact that $\sum_{k=0}^{n} r_k(x) = 1$ we have

$$\left| f(x) - \sum_{k=0}^{n} f\left(\frac{k}{n}\right) r_k(x) \right| = \left| \sum_{k=0}^{n} \left\{ f(x) - f\left(\frac{k}{n}\right) \right\} r_k(x) \right|.$$

This expression is bounded by

$$\left| \sum_{\substack{k=0,\dots,n \\ |(k/n)-x| \le \delta}} \left\{ f(x) - f\left(\frac{k}{n}\right) \right\} r_k(x) \right| + \left| \sum_{\substack{k=0,\dots,n \\ |(k/n)-x| > \delta}} \left\{ f(x) - f\left(\frac{k}{n}\right) \right\} r_k(x) \right|;$$

writing this as $R_1 + R_2$ we have

$$|R_1| \le \sum_{\substack{k=0,\dots,n \\ |(k/n)-x| \le \delta}} \left| f(x) - f\left(\frac{k}{n}\right) \right| r_k(x) \le \varepsilon \sum_{k=0}^{n} r_k(x) = \varepsilon,$$

and for R_2 we use Lemma 6.2 to obtain

$$|R_2| \le \sum_{\substack{k=0,\ldots,n \\ |(k/n)-x|>\delta}} \left| f(x) - f\left(\frac{k}{n}\right) \right| r_k(x) \le 2M \sum_{\substack{k=0,\ldots,n \\ |(k/n)-x|>\delta}} r_k(x)$$

$$\le \frac{2M}{n^2\delta^2} \sum_{k=0}^{n} (k - nx)^2 r_k(x)$$

$$= \frac{2Mx(1-x)}{n\delta^2} \le \frac{2M}{n\delta^2},$$

since $x \in [0, 1]$, and this tends to zero as $n \to \infty$. $\qquad\square$

Note that for a fixed k the coefficient of x^k in $P_n(x)$ depends on n. For example, the first approximations of $f(x) = 1/2 - |x - 1/2|$ are

$$P_1(x) = 0, \quad P_2(x) = P_3(x) = x - x^2, \quad P_4(x) = P_5(x) = x - 2x^3 + x^4,$$

and $P_6(x) = x - 5x^4 + 6x^5 - 2x^6$; see Figure 6.1.

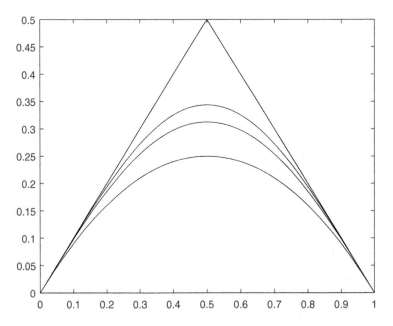

Figure 6.1 Approximation of $f(x) = \frac{1}{2} - |x - \frac{1}{2}|$ by Bernstein polynomials: shown are P_2, P_4, P_6, and the function f.

Corollary 6.3 *Any* $f \in C([a, b]; \mathbb{R})$ *can be uniformly approximated by polynomials.*

Proof The function $g(x) := f((x - a)/(b - a))$ is a continuous function on $[0, 1]$. If it is approximated within ε on $[0, 1]$ by a degree n polynomial $q(x)$, then $p(x) := q((x - a)(b - a))$ is a degree n polynomial that approximates f on $[a, b]$ to within ε. □

An immediate consequence of this result is that $C([a, b]; \mathbb{R})$ is separable.

Corollary 6.4 *The space $C([a, b]; \mathbb{R})$ (equipped with the supremum norm) is separable.*

Proof The previous corollary guarantees that the linear span of the countable set $\{x^k\}_{k=0}^{\infty}$ is dense in $C([a, b]; \mathbb{R})$; the result now follows from Lemma 3.23. □

We now deduce from Theorem 6.1 that we can approximate any continuous function by a Fourier cosine series. The proofs of the following two corollaries follow ideas in Renardy and Rogers (1993).

Corollary 6.5 *Every continuous function on $[0, 1]$ can be uniformly approximated arbitrarily closely by an expression of the form*

$$\sum_{k=0}^{n} c_k \cos(k\pi x)$$

for some $n \in \mathbb{N}$ and $\{c_k\}_{k=0}^{n} \in \mathbb{R}$.

Proof Given $f \in C([0, 1])$, consider the function $g : [-1, 1] \to \mathbb{R}$ defined by setting $g(\cos \pi x) = f(x)$. Since the resulting function g is continuous, given any $\varepsilon > 0$ we can use the Weierstrass Approximation Theorem to find n and $\{a_k\}_{k=0}^{n}$ such that

$$\sup_{u \in [-1, 1]} \left| g(u) - \sum_{k=0}^{n} a_k u^n \right| < \varepsilon.$$

It follows that

$$\sup_{x \in [0, 1]} \left| f(x) - \sum_{k=0}^{n} a_k (\cos \pi x)^k \right| < \varepsilon,$$

and since elementary trigonometric identities can be used to show that $\cos(\pi x)^n$ is given by an expression of the form $\sum_{j=0}^{n} b_j \cos(j\pi x)$ the result follows as stated. □

(We will give a more elegant (but less 'elementary') proof of this result in the next section as a consequence of the Stone–Weierstrass Theorem.)

With the restriction that $f(0) = f(1) = 0$ we can also approximate continuous functions on $[0, 1]$ using a sine series.

Corollary 6.6 *Every $f \in C([0, 1])$ with $f(0) = f(1) = 0$ can be uniformly approximated arbitrarily closely by an expression of the form*

$$\sum_{k=1}^{n} b_k \sin(k\pi x) \qquad (6.3)$$

for some $n \in \mathbb{N}$ and some $\{b_k\}_{k=1}^{n} \in \mathbb{R}$.

Note that if we are to approximate f by a series like (6.3) we must have $f(0) = f(1) = 0$, since any expression of this form is zero at $x = 0$ and $x = 1$.

Proof First note that given such a function f and $\varepsilon > 0$ we can find a function $g \in C([0, 1])$ such that $\|f - g\|_\infty < \varepsilon/2$ and $g = 0$ in a neighbourhood of $x = 0$ and $x = 1$: use the continuity of f at $x = 0$ and $x = 1$ to find $\delta > 0$ such that $|f(x)| < \varepsilon$ for all $0 \le x < \delta$ and all $1 - \delta < x \le 1$ and set

$$g(x) = \begin{cases} 0 & 0 \le x < \delta \\ \frac{x-\delta}{\delta} f(2\delta) & \delta \le x < 2\delta \\ f(x) & 2\delta \le x \le 1 - 2\delta \\ \frac{1-\delta-x}{\delta} f(1 - 2\delta) & 1 - 2\delta < x \le 1 - \delta \\ 0 & 1 - \delta < x \le 1. \end{cases}$$

It suffices to approximate g to within $\varepsilon/2$ by a series of the form in (6.3).

Now since $g(x) = 0$ near the endpoints, we can define a function $h \in C([0, 1])$ by setting $h(x) := g(x)/\sin \pi x$. We use Corollary 6.5 to approximate h uniformly to within $\varepsilon/2$ by a cosine series:

$$\sup_{x \in [0,1]} \left| h(x) - \sum_{k=0}^{n} c_k \cos(k\pi x) \right| = \sup_{x \in [0,1]} \left| \frac{g(x)}{\sin \pi x} - \sum_{k=0}^{n} c_k \cos(k\pi x) \right| < \varepsilon/2.$$

It follows, since $|\sin \pi x| \le 1$, that

$$\sup_{x \in [0,1]} \left| g(x) - \sum_{k=0}^{n} c_k \cos(k\pi x) \sin \pi x \right| < \varepsilon/2.$$

Since

$$\cos(k\pi x) \sin \pi x = \frac{1}{2} [\sin((k + 1)\pi x) - \sin((k - 1)\pi x)],$$

the result follows. $\qquad\qquad\qquad\qquad\qquad\qquad\qquad\qquad\qquad\qquad\qquad\square$

6.2 The Stone–Weierstrass Theorem

We now prove a generalisation of the Weierstrass Approximation Theorem that can be applied to $C(X; \mathbb{K})$ when X is a compact metric space.

Given $f, g \in C(X; \mathbb{K})$ we can define

$$fg \in C(X; \mathbb{K}) \qquad \text{by} \qquad (fg)(x) = f(x)g(x).$$

A linear subspace A of $C(X; \mathbb{K})$ is called an *algebra* (or a subalgebra of $C(X; \mathbb{K})$) if

- $1 \in A$ and
- $f, g \in A$ implies that $fg \in A$.

Note that

(i) if $f \in A$, then any polynomial in f is an element of A, since $f^n \in A$ for any $n \in \mathbb{N}$ (by the second algebra property) and linear combinations of elements of A is in A since it is a linear subspace of $C(X; \mathbb{K})$; and

(ii) if A is an algebra, then so is $\overline{A}$, the closure of A in the supremum norm. Exercise 3.9 shows that $\overline{A}$ is still a subspace of $C(X; \mathbb{K})$. If $f, g \in \overline{A}$, then there exist $(f_n) \in A$ and $(g_n) \in A$ such that $f_n \to f$ and $g_n \to g$. Then $f_n g_n \in A$ and $f_n g_n \to fg$, since

$$\begin{aligned}
\|f_n g_n - fg\|_\infty &= \|f_n g_n - fg_n + fg_n - fg\|_\infty \\
&\leq \|(f - f_n)g_n\|_\infty + \|f(g_n - g)\|_\infty \\
&\leq \|f - f_n\|_\infty \|g_n\|_\infty + \|f\|_\infty \|g_n - g\|_\infty,
\end{aligned}$$

which tends to zero since $g_n \to g$ implies that $\|g_n\|_\infty$ is bounded. This shows that $fg \in \overline{A}$ and so $\overline{A}$ is also an algebra.

The following simple lemma will allow us to show that these conditions are enough to ensure that when A is a subalgebra of the space of *real-valued* continuous functions $C(X; \mathbb{R})$, if $f, g \in A$, then $\max(f, g)$ and $\min(f, g)$ are elements of $\overline{A}$. This is the basis of the proof of the Stone–Weierstrass Theorem.

Lemma 6.7 *Suppose that X is compact and that A is a subalgebra of $C(X; \mathbb{R})$. If $f, g \in A$, then $\max(f, g) \in \overline{A}$ and $\min(f, g) \in \overline{A}$.*

Since $\overline{A}$ is an algebra whenever A is, it follows that if $f, g \in \overline{A}$, then we still have $\max(f, g) \in \overline{A}$ and $\min(f, g) \in \overline{A}$.

Proof First we show that if $f \in A$ with $f \geq 0$, then $\sqrt{f} \in \overline{A}$.

Suppose that $0 \leq f \leq 1$. Set $g = 1 - f$; then $g \in A$, $0 \leq g \leq 1$, and $f = 1 - g$.

The Taylor expansion of $\sqrt{1-x}$ about $x = 0$ gives

$$\sqrt{1-x} = 1 - \sum_{k=0}^{\infty} \frac{2}{k+1} \binom{2k}{k} \left(\frac{x}{4}\right)^{k+1} \tag{6.4}$$

and this converges for all $|x| < 1$ (see Exercise 6.2). In fact the right-hand side converges uniformly on $[0, 1]$, since for every $x \in [0, 1]$ we have

$$\sum_{k=n+1}^{\infty} \frac{2}{k+1} \binom{2k}{k} \left(\frac{x}{4}\right)^{k+1} \leq \sum_{k=n+1}^{\infty} \frac{2}{k+1} \binom{2k}{k} \left(\frac{1}{4}\right)^{k+1}$$

$$\leq c \sum_{k=n+1}^{\infty} \frac{2}{k+1} \frac{2^{2k}}{\sqrt{k}} \left(\frac{1}{4}\right)^{k+1}$$

$$= \frac{c}{2} \sum_{k=n+1}^{\infty} \frac{1}{\sqrt{k}(k+1)},$$

which tends to zero as $n \to \infty$. In going from the first to the second line we have used the fact that

$$\binom{2k}{k} \leq c \frac{2^{2k}}{\sqrt{k}},$$

where c is a constant that does not depend on k (see Exercise 6.1). It follows that both the left- and right-hand sides of (6.4) are continuous on $[0, 1]$, and since they agree on $[0, 1)$ they also agree at $x = 1$.

Therefore, rewriting (6.4) for simplicity as $\sqrt{1-x} = 1 - \sum_{k=1}^{\infty} a_k x^k$, it follows that

$$\sqrt{f(x)} = \sqrt{1 - g(x)} = 1 - \sum_{n=1}^{\infty} a_n (g(x)^n);$$

since the sum in (6.4) converges uniformly on $[0, 1]$ and $0 \leq g(x) \leq 1$ for every $x \in X$, this expression converges uniformly for $x \in X$. Since $g \in A$, the partial sums are all polynomials in g and so elements of $A \subseteq \overline{A}$; since $\overline{A}$ is closed it follows that $\sqrt{f} \in \overline{A}$.

For a general $f \in A$ with $f \geq 0$ consider instead λf with $\lambda \geq 0$ chosen so that $0 \leq \lambda f \leq 1$. Then $\sqrt{\lambda f} = \sqrt{\lambda}\sqrt{f} \in \overline{A}$, which again implies that $\sqrt{f} \in \overline{A}$ as required.

As a consequence, if $f \in A$, then $|f| \in \overline{A}$, since $|f| = \sqrt{f^2}$. We can now prove the result as stated, since if $f, g \in A$ we have

$$\min(f, g) = \frac{1}{2}(f + g - |f - g|) \in \overline{A}$$

and

$$\max(f, g) = \frac{1}{2}(f + g + |f - g|) \in \overline{A}. \qquad \square$$

For a simpler version of this proof that uses the Weierstrass Approximation Theorem see Exercise 6.3, and for a third proof using an iterative method see Exercise 6.5.

We now use this to prove the real version of the Stone–Weierstrass Theorem.

Theorem 6.8 (Real Stone–Weierstrass Theorem) *If X is a compact metric space and A is a subalgebra of $C(X; \mathbb{R})$ that separates points of X, i.e. for any $x, y \in X$ with $x \neq y$ there exists $f \in A$ such that $f(x) \neq f(y)$, then $\overline{A} = C(X; \mathbb{R})$. (We say that A is 'uniformly dense' in $C(X; \mathbb{R})$.)*

Proof We first show that given any two distinct points $x, y \in X$ and $\lambda, \mu \in \mathbb{R}$, there exists a map $g \in A$ such that $g(x) = \lambda$ and $g(y) = \mu$. Indeed, since A separates points there exists a function $h \in A$ such that $h(x) \neq h(y)$, and then we can set

$$g(z) = \mu + (\lambda - \mu)\frac{h(z) - h(y)}{h(x) - h(y)} \qquad \text{for each } z \in X.$$

Now, fix $f \in C(X)$ and $\varepsilon > 0$. For each pair $x, y \in X$ there exists $f_{x,y} \in A$ such that

$$f_{x,y}(x) = f(x) \qquad \text{and} \qquad f_{x,y}(y) = f(y).$$

If $x \neq y$ we use the construction in the first paragraph of the proof, while if $x = y$ we can just take $f_{x,y}(z) = f(x)$ for every $z \in X$.

For each fixed $x \in X$ the set

$$U_{x,y} = \{\xi \in X : f_{x,y}(\xi) < f(\xi) + \varepsilon\} \qquad (6.5)$$

is open and contains y, since $y \in U_{x,y}$, both f and $f_{x,y}$ are continuous, and $U_{x,y} = (f_{xy} - f)^{-1}(-\infty, \varepsilon)$. Therefore for each fixed x these sets form an open cover of X:

$$X = \bigcup_{y \in X} U_{x,y}.$$

Since X is compact, this cover has a finite subcover, so there exist $y_1, \ldots, y_n$ such that

$$X = \bigcup_{j=1}^{n} U_{x,y_j}.$$

Now set

$$h_x := \min(f_{x,y_j}),$$

i.e. $h_x(\xi) = \min_{j=1,\ldots,n} f_{x,y_j}(\xi)$ for each $\xi \in X$. Then

- $h_x \in \overline{A}$ using Lemma 6.7;
- $h_x(x) = f(x)$ for every $x \in X$ (since $f_{x,y_j}(x) = x$ for every j); and
- $h_x < f + \varepsilon$, since for every $z \in X$ we have $z \in U_{x,y_j}$ for some j and then

$$h_x(z) \le f_{x,y_j}(z) < f(z) + \varepsilon,$$

using (6.5).

Now for each $x \in X$ let

$$V_x := \{\xi \in X : f(\xi) - \varepsilon < h_x(\xi)\}; \tag{6.6}$$

then V_x is open and contains x, since $x \in V_x$, h_x and f are continuous, and $V_x = (f - h_x)^{-1}(-\infty, \varepsilon)$. The collection of all these sets provides an open cover of X; since X is compact it follows that there exists a finite collection $x_1, \ldots, x_m$ such that

$$X = \bigcup_{j=1}^{m} V_{x_j}.$$

Finally, we set

$$F := \max_j h_{x_j}.$$

Then

- $F \in \overline{A}$ using Lemma 6.7;
- $F < f + \varepsilon$ since $h_{x_j} < f + \varepsilon$ for each j; and
- $f - \varepsilon < F$, since for every $z \in X$ we have $z \in V_{x_j}$ for some j, and then

$$f(z) - \varepsilon < h_{x_j}(z) \le F(z)$$

from (6.6).

In other words, we have found $F \in \overline{A}$ such that

$$f - \varepsilon < F < f + \varepsilon,$$

i.e. $\|F - f\|_\infty < \varepsilon$. This shows that $f \in \overline{\overline{A}} = \overline{A}$. □

We now show that this result is indeed a generalisation of the Weierstrass Approximation Theorem (Theorem 6.1). Take $X = [0, 1]$ and let A be the set of all polynomials; this is an algebra, since it is a linear subspace of $C([0, 1])$, 1 is a polynomial, and the product of two polynomials is another polynomial. Moreover, the polynomial x separates points, so $\overline{A} = C([0, 1]; \mathbb{R})$, i.e. any element of $C([0, 1]; \mathbb{R})$ can be approximated arbitrary closely by polynomials. (Despite the fact that the proof of Lemma 6.7

using the Weierstrass Approximation Theorem is easier, this argument would be circular if we had proved Lemma 6.7 this way.)

For another proof of the Fourier cosine series result in Corollary 6.5, we can take $X = [0, 1]$ and let

$$A = \left\{ \sum_{k=0}^{n} a_k \cos(\pi k x) : a_k \in \mathbb{R}, \, n \geq 0 \right\}.$$

This is a subalgebra of $C([0, 1])$, since $1 \in A$ and

$$\cos(\pi k x) \cos(\pi j x) = \frac{1}{2} \left[\cos(\pi (k + j)x) + \cos(\pi (k - j)x) \right].$$

Furthermore, it separates points, since if x, y are distinct points in $[0, 1]$, then $\cos(\pi x) \neq \cos(\pi y)$.

We now prove a complex version of the Stone–Weierstrass Theorem. Note that for this we will require the additional assumption that A is closed under complex conjugation: if $f \in A$, then $\overline{f} \in A$. To see that something additional must be required, note that a general continuous complex-valued function cannot be approximated by complex polynomials (expressions of the form $\sum_{j=0}^{n} a_j z^j$ with $a_j \in \mathbb{C}$). For example $\overline{z}$ cannot be approximated by polynomials, since

$$\int_{|z|=1} \overline{z} \, dz = 2\pi i$$

while

$$\int_{|z|=1} p(z) \, dz = 0$$

for any complex polynomial. (The set of all complex polynomials is not closed under conjugation.)

Theorem 6.9 (Complex Stone–Weierstrass Theorem) *Suppose that X is compact and A is a subalgebra of $C(X; \mathbb{C})$ that separates points in X and is closed under conjugation, i.e. $f \in A$ implies that $\overline{f} \in A$. Then $\overline{A} = C(X; \mathbb{C})$.*

Proof We want to show that $A_{\mathbb{R}}$, the elements of A that are real-valued, satisfy the requirements of the real Stone–Weierstrass Theorem. The algebra property is inherited from A itself; we need to show that $A_{\mathbb{R}}$ still separates points. So suppose that x, $y \in X$ and $f \in A$ separates points. Then we have either $\mathrm{Re} f(x) \neq \mathrm{Re} f(y)$ or $\mathrm{Im} f(x) \neq \mathrm{Im} f(y)$. Since f is closed under conjugation, we have

$$\mathrm{Re}(f) = \frac{1}{2}(f + \overline{f}) \in A \qquad \text{and} \qquad \mathrm{Im}(f) = \frac{1}{2i}(f - \overline{f}) \in A;$$

these are both elements of $C(X; \mathbb{R})$, and hence of $A_{\mathbb{R}}$. So $A_{\mathbb{R}}$ separates points.

Now since any element $f \in C(X; \mathbb{C})$ can be written as $f_1 + i f_2$, for appropriate $f_1, f_2 \in C(X; \mathbb{R})$, it follows that we can approximate any element of $C(X; \mathbb{C})$ by elements of the form $\phi + i\psi$ with $\phi, \psi \in A_{\mathbb{R}}$; and thus by $\phi + i\psi \in A$.   □

As one application, if we take $X = S^1 := \{z \in \mathbb{C} : |z| = 1\} \subset \mathbb{C}$ and let

$$A = \left\{ \sum_{j=-n}^{n} a_j z^j : a_j \in \mathbb{C} \right\},$$

then $1 \in A$, A is closed under conjugation (since $\bar{z} = 1/z$ if $|z| = 1$), and $z \in A$ so A separates points. It follows that $\overline{A} = C(S^1; \mathbb{C})$.

Corollary 6.10 *Any* $f \in C([-\pi, \pi])$ *for which* $f(-\pi) = f(\pi)$ *can be approximated uniformly on* $[-\pi, \pi]$ *by an expression of the form*

$$\sum_{k=-n}^{n} a_k e^{ikx}.$$

Proof Given any $f \in C([-\pi, \pi])$ with $f(-\pi) = f(\pi)$ we can define a continuous function $g \colon S^1 \to \mathbb{R}$ by setting $g(e^{ix}) = f(x)$ (see Exercise 6.9). This continuous g can be approximated uniformly by expressions of the form

$$\sum_{j=-n}^{n} a_j z^j;$$

it follows that f can be approximated uniformly by expressions

$$\sum_{j=-n}^{n} a_j e^{ijx} \tag{6.7}$$

as claimed.   □

Since the expression in (6.7) is equal to

$$a_0 + \sum_{k=1}^{n} a_k \cos(kx) + \sum_{k=1}^{n} b_k \sin(kx) \tag{6.8}$$

for some choice of coefficients, this corollary shows that

$$A = \left\{ a_0 + \sum_{k=1}^{n} a_k \cos(kx) + \sum_{k=1}^{n} b_k \sin(kx) : a_k, b_k \in \mathbb{R}, n \geq 1 \right\}$$

is uniformly dense in

$$\{ f \in C([-\pi, \pi]) : f(-\pi) = f(\pi) \}.$$

(Note that although a priori the coefficients a_k, b_k in (6.8) can be complex, if f is real, then we can assume that these coefficients are real too, since for any complex function $p(x)$

$$
\begin{aligned}
|f(x) - p(x)|^2 &= |\mathrm{Re}(f(x) - p(x))|^2 + |\mathrm{Im}(f(x) - p(x))|^2 \\
&= |f(x) - \mathrm{Re}\, p(x)|^2 + |\mathrm{Im}\, p(x)|^2 \\
&\geq |f(x) - \mathrm{Re}\, p(x)|^2;
\end{aligned}
$$

by taking the real part of (6.8) we do not increase the distance to f.)

We return to the topic of Fourier series in Lemma 9.16. For much more on this subject see the book by Körner (1989), for example.

6.3 The Arzelà–Ascoli Theorem

The Arzelà–Ascoli Theorem characterises precompact (and therefore compact) subsets of $C(X; \mathbb{K})$, where we say that a set is *precompact* if its closure is compact. We will use the following general result about precompact sets in the proof.

Lemma 6.11 *A subset A of a complete normed space $(X, \|\cdot\|)$ is precompact if and only if any sequence in A has a Cauchy subsequence.*

Proof Suppose that A is precompact, and that $(x_n) \in A$. Then since $(x_n) \in \overline{A}$ and $\overline{A}$ is compact, (x_n) has a convergent subsequence, and any convergent sequence is Cauchy.

For the other implication, suppose that any sequence in A has a Cauchy subsequence. Take a sequence $(y_n) \in \overline{A}$; then, using Lemma 2.17, there exist $(x_n) \in A$ such that $\|x_n - y_n\| < 1/n$. By assumption, (x_n) has a Cauchy subsequence (x_{n_k}); it follows that (y_{n_k}) is Cauchy too, and so converges to a limit y, which is contained in $\overline{A}$ since this set is closed. This shows that $\overline{A}$ is compact. $\qquad\square$

Theorem 6.12 (Arzelà–Ascoli Theorem: precompact version) *If X is a compact metric space then $A \subset C(X; \mathbb{R})$ is precompact if and only if it is bounded (there exists $R > 0$ such that $\|f\|_\infty \leq R$ for all $f \in A$) and equicontinuous, that is for each $\varepsilon > 0$ there exists a $\delta > 0$ such that*

$$
d_X(x, y) < \delta \quad \Rightarrow \quad |f(x) - f(y)| < \varepsilon \quad \text{for every} \quad f \in A, \; x, y \in X.
$$

f_n	x	x	x	x	x	x	x	x	x	x	x	x	x	x	x
$f_{n_1,j}$	⊗	x	x		x	x	x			x	x	x	x	x
$f_{n_2,j}$		x	⊗		x	x	x			x	x	x	x
$f_{n_3,j}$			x		x	⊗	x			x	x		x
$f_{n_4,j}$			x		x		x			⊗		x
$f_{n_5,j}$			x		x		x			x		⊗

Figure 6.2 The argument used to show that (f_j) has a subsequence $(f_{m,m})$ such that $f_{m,m}(x_i)$ converges for every i. The figure gives an idea of the indices used for each subsequence; the circled elements are the 'diagonal sequence' $f_{k,k}$.

Proof Using the result of Exercise 2.14 we can find a countable set (x_k) with the following property: given any $\delta > 0$, we can guarantee that there is an $M(\delta)$ such that for every $x \in X$, $d(x, x_k) < \delta$ for some $1 \le k \le M(\delta)$.

Since (f_j) is bounded, we can use a 'diagonal argument' to find a subsequence (which we relabel) such that $f_j(x_k)$ converges for every k. The idea is to repeatedly extract subsequences to ensure that $f_j(x_i)$ converges for more and more of the (x_k).

Since (f_j) is bounded, we know that $f_j(x_1)$ is a bounded sequence in $\mathbb{K}$, so we can find a subsequence $f_{1,j}$ such that $f_{1,j}(x_1)$ converges. Now we consider $f_{1,j}(x_2)$, which is again a bounded sequence in $\mathbb{K}$, so there is a subsequence of $f_{1,j}$, which we will label $f_{2,j}$, such that $f_{2,j}(x_2)$ converges. Since $f_{2,j}(x_1)$ is a subsequence of $f_{1,j}(x_1)$, this still converges.

We continue in this way, extracting subsequences of subsequences, so that

$$f_{n,j}(x_i) \qquad \text{converges for all } i = 1, \ldots, n.$$

This almost gives what we need, but not quite: although we can find a subsequence that converges at as many of the (x_i) as we wish, in order to obtain a subsequence that converges at all of the (x_i) simultaneously we use a 'diagonal trick' and consider the sequence $(f_{m,m})_{m=1}^{\infty}$. This is a subsequence of the original sequence (f_j), and a subsequence of $(f_{n,j})$ once $m \ge n$; see Figure 6.2.

It follows that

$$f_{m,m}(x_i) \qquad \text{converges for all } i \in \mathbb{N}.$$

Set $f_m^* = f_{m,m}$. We now show that (f_m^*) must be Cauchy in the supremum norm.

Given $\varepsilon > 0$, since (f_j) is equicontinuous there exists a $\delta > 0$ such that

$$d(x, y) < \delta \qquad \Rightarrow \qquad |f_j(x) - f_j(y)| < \varepsilon/3$$

for every j.

By our construction of the (x_i) there exists an M (depending on δ) such that for every $x \in X$ there exists an x_i with $1 \le i \le M$ such that $|x - x_i| < \delta$.

Since $f_n^*(x_i)$ converges for every i, there exists N such that for $n, m \ge N$ we have

$$|f_n^*(x_i) - f_m^*(x_i)| < \varepsilon/3 \qquad 1 \le i \le M,$$

and we can use the triangle inequality to obtain

$$
\begin{aligned}
|f_n^*(x) - f_m^*(x)| &= |f_n^*(x) - f_n^*(x_i) + f_n^*(x_i) - f_m^*(x_i) + f_m^*(x_i) - f_m^*(x)| \\
&\le |f_n^*(x) - f_n^*(x_i)| + |f_n^*(x_i) - f_m^*(x_i)| + |f_m^*(x_i) - f_m^*(x)| \\
&< \varepsilon,
\end{aligned}
$$

which shows that (f_n^*) is Cauchy in the supremum norm, and so A is precompact.

To show the converse, first note that if A is precompact, then it is bounded since its closure is compact. It remains to prove that A is equicontinuous.

Since A is compact, for any $\varepsilon > 0$ there exist $\{f_1, \ldots, f_n\}$ such that for every $f \in A$ we have

$$\|f - f_i\|_\infty < \varepsilon/3 \qquad \text{for some } i \in \{1, \ldots, n\}.$$

Since the f_i are all continuous functions on the compact set X they are each uniformly continuous (Lemma 2.30), so there exists $\delta > 0$ such that for every $i = 1, \ldots, n$

$$d(x, y) < \delta \qquad \Rightarrow \qquad |f_i(x) - f_i(y)| < \varepsilon/3 \qquad \text{for all } x \in X.$$

Then for any $f \in A$ choose j such that $\|f_j - f\|_\infty < \varepsilon/3$; then whenever $d(x, y) < \delta$ we have

$$
\begin{aligned}
|f(x) - f(y)| &\le |f(x) - f_j(x)| + |f_j(x) - f_j(y)| + |f_j(y) - f(y)| \\
&\le \|f - f_j\|_\infty + |f_j(x) - f_j(y)| + \|f_j - f\|_\infty < \varepsilon,
\end{aligned}
$$

so A is equicontinuous. $\qquad\qquad\qquad\qquad\qquad\qquad\qquad\qquad\qquad\quad\square$

Since a compact set is the closure of a precompact set, the following corollary is immediate.

Corollary 6.13 (Arzelà–Ascoli Theorem: compact version) *If X is a compact metric space then $A \subset C(X; \mathbb{K})$ is compact if and only if it is closed, bounded, and equicontinuous.*

As an example, for any $L > 0$ the collection of all L-Lipschitz functions on any compact metric space X,

$$\mathrm{Lip}_L(X; \mathbb{K}) := \{f \in C(X; \mathbb{K}) : |f(x) - f(y)| \le L|x - y|\},$$

is a compact subset of $C(X; \mathbb{K})$. More generally if $\omega: [0, \infty) \to [0, \infty)$ is any 'modulus of continuity', i.e. a non-decreasing function that satisfies

$$\lim_{r \to 0} \omega(r) = \omega(0) = 0,$$

then

$$\{f \in C(X; \mathbb{K}) : |f(x) - f(y)| \le \omega(|x - y|)\}$$

is a compact subset of $C(X; \mathbb{K})$.

Exercises

6.1   Using the upper and lower bounds on the factorial

$$\sqrt{2\pi} n^{n+1/2} e^{-n} \le n! \le e n^{n+1/2} e^{-n}$$

(see e.g. Holland, 2016) show that

$$\binom{2k}{k} \le c \frac{2^{2k}}{\sqrt{k}},$$

where c is a constant that does not depend on k.

6.2   Show that the Taylor expansion of $\sqrt{1 - x}$ about $x = 0$ is

$$\sqrt{1 - x} = 1 - \sum_{k=0}^{\infty} \frac{2}{k+1} \binom{2k}{k} \left(\frac{x}{4}\right)^{k+1}$$

and that this converges for all $x \in [0, 1)$. (Use Taylor's Theorem in the form

$$f(x) = \sum_{k=0}^{n} \frac{f^{(k)}(0)}{k!} x^k + \frac{f^{(n+1)}(c)}{n!} x(x - c)^n, \qquad x > 0,$$

for some $c \in (0, x)$, where $f^{(k)}(x)$ is the kth derivative of f at x.)

6.3   Use the Weierstrass Approximation Theorem to prove Lemma 6.7. [Hint: approximate $|x|$ uniformly on $[-1, 1]$ using polynomials.]

6.4   Prove Dini's Theorem: suppose that $(f_n) \in C([a, b]; \mathbb{R})$ is an increasing sequence $(f_{n+1}(x) \ge f_n(x))$ that converges pointwise to some $f \in C([a, b]; \mathbb{R})$. Show that f_n converges uniformly to f. (For each $\varepsilon > 0$ consider the sets

$$E_n := \{x \in [a, b] : f(x) - f_n(x) < \varepsilon\}$$

and use the compactness of $[a, b]$.)

6.5   Define a sequence of polynomials $p_n \colon [-1, 1] \to \mathbb{R}$ iteratively, by setting $p_0(x) = 0$ and

$$p_{n+1}(x) = \frac{1}{2}x^2 + p_n(x) - \frac{1}{2}p_n(x)^2.$$

Using the result of the previous exercise show that $p_n(x) \to |x|$ uniformly on $[-1, 1]$ as $n \to \infty$. (First consider the map $f \colon [-1, 1] \to \mathbb{R}$ give by $p \mapsto \frac{1}{2}x^2 + p - \frac{1}{2}p^2$. This gives another way to prove Lemma 6.7.) (Pryce, 1973)

6.6   Let X and Y be compact metric spaces. Use the Stone–Weierstrass Theorem to show that any function $f \in C(X \times Y)$ can be uniformly approximated by functions of the form

$$F(x, y) = \sum_{i=1}^{n} f_i(x) g_i(y),$$

where $f_i \in C(X)$ and $g_i \in C(Y), i = 1 \ldots, n$.

6.7   Use the Stone–Weierstrass Theorem to show that any continuous function $f \in C([a, b] \times [c, d])$ can be uniformly approximated by functions of the form

$$F(x, y) = \sum_{i,j=0}^{n} a_{ij} x^i y^j.$$

6.8   Suppose that $f \in C([a, b] \times [c, d]; \mathbb{R})$. Use the result of the previous exercise to show that

$$\int_a^b \int_c^d f(x, y)\, dx\, dy = \int_c^d \int_a^b f(x, y)\, dy\, dx$$

(Pryce, 1973).

6.9   Show that if $f \in C([-\pi, \pi])$ with $f(-\pi) = f(\pi)$, then the function $g \colon S^1 \to \mathbb{R}$ defined by setting $g(e^{ix}) = f(x)$ is continuous on S^1.

6.10   A subset A of a normed space is called *totally bounded* if for every $\varepsilon > 0$ it is possible to cover A with a finite collection of open balls of radius ε. Show that

   (i) if A is totally bounded, then $\overline{A}$ is totally bounded;

  (ii) any sequence in a totally bounded set has a Cauchy subsequence;

 (iii) if A is a subset of a complete normed space, then $\overline{A}$ is compact if and only if A is totally bounded.

(Part (iii) could be rephrased as 'a subset of a complete normed space is precompact if and only if it is totally bounded'.)

6.11 Use the Arzelà–Ascoli Theorem repeatedly to show that if a sequence $(f_n) \in C_b(\mathbb{R})$ is bounded and equicontinuous on $\mathbb{R}$, then it has a subsequence that converges uniformly on all compact subsets of $\mathbb{R}$.

6.12 Suppose that $f \in C_b(\mathbb{R})$. Show that for every $\delta > 0$ the function

$$f_\delta(x) := \frac{1}{2\delta} \int_{x-\delta}^{x+\delta} f(y)\,dy$$

is Lipschitz with $\|f_\delta\|_\infty \le \|f\|_\infty$. Show furthermore that f_δ converges uniformly to f on every bounded interval.

6.13 Suppose that $f \in C_b(\mathbb{R})$ with $\|f\|_\infty \le M$, and that $(f_n) \in C_b(\mathbb{R})$ is a sequence with $\|f_n\|_\infty \le M$ such that $f_n \to f$ uniformly on every bounded subinterval in $\mathbb{R}$. Use the Arzelà–Ascoli Theorem to show that if there exist $(x_n) \in C([0, T])$ such that

$$x_n(t) = x_0 + \int_0^t f_n(x_n(s))\,ds \qquad \text{for all } t \in [0, T] \qquad (6.9)$$

for each n then there exists $x \in C([0, T])$ such that

$$x(t) = x_0 + \int_0^t f(x(s))\,ds \qquad \text{for all } t \in [0, T]. \qquad (6.10)$$

6.14 Suppose that $f \in C_b(\mathbb{R})$ (this global boundedness condition can be relaxed). Combine the results of the previous two exercises with that of Exercise 4.8 to deduce that the ordinary differential equation

$$\dot{x} = f(x), \qquad \text{with} \qquad x(0) = x_0,$$

has at least one solution on $[0, T]$ for any $T > 0$.

7

Completions and the Lebesgue Spaces $L^p(\Omega)$

In Chapter 4 we saw that $C([0, 1])$ is complete when we use the supremum norm. But there are other natural norms on this space with which it is not complete. In this chapter we look at one particular example, the L^1 norm (and then at the whole family of L^p norms). We use this to motivate the abstract completion of a normed space and the Lebesgue integral, and then define the L^p spaces of Lebesgue integrable functions as completions of the space of continuous functions in the L^p norm.

7.1 Non-completeness of $C([0, 1])$ with the L^1 Norm

We have already seen in Example 3.13 that for any $1 \le p < \infty$ the integral expression

$$\|f\|_{L^p} := \left(\int_0^1 |f(x)|^p \, dx \right)^{1/p}$$

defines a norm ('the L^p norm') on $C([0, 1])$. In this first section we concentrate on the L^1 norm $\|f\|_{L^1} = \int_0^1 |f(x)| \, dx$, and show that $C^0([0, 1])$ is not complete with this norm.

To do this we find a sequence $(f_n) \in C([0, 1])$ that is Cauchy in the L^1 norm but that does not converge to a function in $C([0, 1])$. Indeed, consider the sequence for $n \ge 2$ given by

$$f_n(x) = \begin{cases} 0 & 0 \le x < 1/2 - 1/n \\ 1 - n(1/2 - x) & 1/2 - 1/n \le x \le 1/2 \\ 1 & 1/2 < x \le 1, \end{cases}$$

see Figure 7.1.

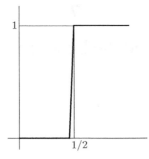

Figure 7.1 The function f_{10}.

Then for $n > m$, since the integrands agree everywhere except on the interval $(1/2 - 1/m, 1/2)$ we have

$$\| f_n - f_m \|_{L^1} = \int_0^1 |f_n(x) - f_m(x)|\, dx$$

$$\leq \int_{1/2-1/m}^{1/2} |f_n(x) - f_m(x)|\, dx \leq 2/m;$$

so this sequence is Cauchy. Suppose that it converges in the L^1 norm to some continuous function f. Then for all $n \geq m$ we have

$$\| f_n - f \|_{L^1} = \int_0^{1/2} |f(x) - f_n(x)|\, dx + \int_{1/2}^1 |f(x) - 1|\, dx$$

$$\geq \int_0^{1/2-1/m} |f(x)|\, dx + \int_{1/2}^1 |f(x) - 1|\, dx.$$

Letting $n \to \infty$ it follows that

$$\int_0^{1/2-1/m} |f(x)|\, dx + \int_{1/2}^1 |f(x) - 1|\, dx = 0.$$

Arguing as for Example 3.13 it follows that $f(x) = 0$ for $0 \leq x \leq 1/2 - 1/m$ and $f(x) = 1$ for $1/2 \leq x \leq 1$. Since this holds for all m, the limit function must satisfy

$$f(x) = \begin{cases} 0 & 0 \leq x < 1/2 \\ 1 & 1/2 \leq x \leq 1. \end{cases} \tag{7.1}$$

But this function is not continuous, so $C([0, 1])$ is not complete in the L^1 norm.

7.2 The Completion of a Normed Space

We have just seen that $C([0, 1])$ is not complete in the L^1 norm. In order to obtain a complete space it should be enough to add to $C([0, 1])$ the limits of all sequences that are Cauchy in the L^1 norm. In such a way we can hope to obtain a complete space in which $C([0, 1])$ is dense.

We have seen a simple example of this earlier: we showed in the proof of Lemma 3.24 that for $1 \le p < \infty$ the space c_{00} (of sequences with only a finite number of non-zero terms) is a dense subspace of ℓ^p; so ℓ^p is the 'completion' of $(c_{00}, \| \cdot \|_{\ell^p})$.

However, in general we cannot just 'add the limits of Cauchy sequences' to X, since, given an abstract normed space, there is 'nowhere else' for the limits of these Cauchy sequences to lie. To circumvent this problem we introduce a more abstract notion of completion.

Definition 7.1 If $(X, \| \cdot \|)$ is a normed space, a *completion* of X is a complete normed space $(\mathscr{X}, \| \cdot \|_{\mathscr{X}})$ along with a map $i \colon X \to \mathscr{X}$ that is an isometric isomorphism of X onto a dense subspace of $\mathscr{X}$. For simplicity we will often write $(i, \mathscr{X})$ in what follows, suppressing the norm.

In simple examples X is a subset of $\mathscr{X}$, and the map i is the identity; this is the case when we say that ℓ^p is the completion of c_{00} in the ℓ^p norm. For clarity, we spell this out in the following simple lemma.

Lemma 7.2 *If $(\mathscr{X}, \| \cdot \|_{\mathscr{X}})$ is complete and X is a dense subspace of $\mathscr{X}$, then* $(\mathrm{id}, \mathscr{X})$ *is a completion of X.*

Proof The identity map $\mathrm{id} \colon X \to \mathscr{X}$ is an isometric isomorphism of X onto itself, and X is a dense subspace of $(\mathscr{X}, \| \cdot \|_{\mathscr{X}})$. $\qquad\square$

We now show that every normed space has a completion.

Theorem 7.3 *Every normed space $(X, \| \cdot \|)$ has a completion* $(i, \mathscr{X})$.

The space $(\mathscr{X}, \| \cdot \|_{\mathscr{X}})$ that completes X is an abstract one: it consists of equivalence classes of Cauchy sequences in X, where $(x_n) \sim (y_n)$ if $\lim_{n\to\infty} \|x_n - y_n\|_X = 0$; the norm on $\mathscr{X}$ is defined by setting

$$\|[(x_n)]\|_{\mathscr{X}} = \lim_{n \to \infty} \|x_n\|_X.$$

In the proof we have to show that $(\mathscr{X}, \| \cdot \|_{\mathscr{X}})$ really is a Banach space, and that it contains an isometrically isomorphic copy of X as a dense subset.

Proof We consider Cauchy sequences in X, writing

$$\boldsymbol{x} = (x_1, x_2, \ldots), \qquad x_j \in X,$$

for a sequence in X. We say that two Cauchy sequences $\boldsymbol{x}$ and $\boldsymbol{y}$ are equivalent, $\boldsymbol{x} \sim \boldsymbol{y}$, if

$$\lim_{n \to \infty} \|x_n - y_n\|_X = 0. \tag{7.2}$$

We let $\mathscr{X}$ be the space of equivalence classes of Cauchy sequences in X; any element $\xi \in \mathscr{X}$ can be written as $[\boldsymbol{x}]$, where this notation denotes the equivalence class of a Cauchy sequence $\boldsymbol{x} = (x_n)$.

It is clear that $\mathscr{X}$ is a vector space, since the sum of two Cauchy sequences in X is again a Cauchy sequence in X. We define a candidate for our norm on $\mathscr{X}$: if $\xi \in \mathscr{X}$ then

$$\|\xi\|_{\mathscr{X}} := \lim_{n \to \infty} \|x_n\|_X, \tag{7.3}$$

for any $\boldsymbol{x} \in \xi$ (recall that ξ is an equivalence class of Cauchy sequences).

Note that (i) if $\boldsymbol{x} = (x_n)$ is a Cauchy sequence in X, then $(\|x_n\|)$ forms a Cauchy sequence in $\mathbb{R}$, so for any particular choice of $\boldsymbol{x} \in \xi$ the right-hand side of (7.3) exists, and (ii) if $\boldsymbol{x}, \boldsymbol{y} \in \xi$ then

$$\left| \lim_{n \to \infty} \|x_n\| - \lim_{n \to \infty} \|y_n\| \right| = \left| \lim_{n \to \infty} (\|x_n\| - \|y_n\|) \right|$$

$$= \lim_{n \to \infty} \left| \|x_n\| - \|y_n\| \right|$$

$$\leq \lim_{n \to \infty} \|x_n - y_n\| = 0$$

since $\boldsymbol{x} \sim \boldsymbol{y}$ (see (7.2)). So the expression in (7.3) is well defined, and it is easy to check that it satisfies the three requirements of a norm.

Now we define a map $\mathrm{i}\colon X \to \mathscr{X}$, by setting

$$\mathrm{i}(x) = [(x, x, x, x, x, x, \ldots)].$$

Clearly i is linear, and is a bijective isometry between X and its image. We want to show that $\mathrm{i}(X)$ is a dense subset of $\mathscr{X}$.

For any given $\xi \in \mathscr{X}$, choose some $\boldsymbol{x} \in \xi$. Since $\boldsymbol{x} = (x_n)$ is Cauchy, for any given $\varepsilon > 0$ there exists an N such that

$$\|x_n - x_m\|_X < \varepsilon \qquad \text{for all} \quad n, m \geq N.$$

In particular, $\|x_n - x_N\|_X < \varepsilon$ for all $n \geq N$, and so

$$\|\xi - \mathrm{i}(x_N)\|_{\mathscr{X}} = \lim_{n \to \infty} \|x_n - x_N\|_X \leq \varepsilon,$$

which shows that $\mathrm{i}(X)$ is dense in $\mathscr{X}$.

Finally, we have to show that $\mathscr{X}$ is complete, i.e. that any Cauchy sequence in $\mathscr{X}$ converges to another element of $\mathscr{X}$. (A Cauchy sequence in $\mathscr{X}$ is a Cauchy sequence of equivalence classes of Cauchy sequences in X!) Take such a Cauchy sequence, $(\xi_k)_{k=1}^{\infty}$. For each k, find $x_k \in X$ such that

$$\|\mathrm{i}(x_k) - \xi_k\|_{\mathscr{X}} < 1/k, \qquad (7.4)$$

using the density of $\mathrm{i}(X)$ in $\mathscr{X}$. Now let $x = (x_n)$; we will show (i) that x is a Cauchy sequence in X, and so $[x] \in \mathscr{X}$, and (ii) that ξ_k converges to $[x]$. This will show that $\mathscr{X}$ is complete.

(i) To show that x is Cauchy, observe that

$$\begin{aligned}
\|x_n - x_m\|_X &= \|\mathrm{i}(x_n) - \mathrm{i}(x_m)\|_{\mathscr{X}} \\
&= \|\mathrm{i}(x_n) - \xi_n + \xi_n - \xi_m + \xi_m - \mathrm{i}(x_m)\|_{\mathscr{X}} \\
&\leq \|\mathrm{i}(x_n) - \xi_n\|_{\mathscr{X}} + \|\xi_n - \xi_m\|_{\mathscr{X}} + \|\xi_m - \mathrm{i}(x_m)\|_{\mathscr{X}} \\
&\leq \frac{1}{n} + \|\xi_n - \xi_m\|_{\mathscr{X}} + \frac{1}{m}.
\end{aligned}$$

So now given $\varepsilon > 0$ choose N such that $\|\xi_n - \xi_m\|_{\mathscr{X}} < \varepsilon/3$ for $n, m \geq N$. If $N' = \max(N, 3/\varepsilon)$, it follows that

$$\|x_n - x_m\|_X < \varepsilon \quad \text{for all} \quad n, m \geq N',$$

i.e. x is Cauchy. So $[x] \in \mathscr{X}$.

(ii) To show that $\xi_k \to [x]$, first write

$$\|[x] - \xi_k\|_{\mathscr{X}} \leq \|[x] - \mathrm{i}(x_k)\|_{\mathscr{X}} + \|\mathrm{i}(x_k) - \xi_k\|_{\mathscr{X}}.$$

Given $\varepsilon > 0$ choose N large enough that $\|x_n - x_m\|_X < \varepsilon/2$ for all $n, m \geq N$, and then set $N' = \max(N, 2/\varepsilon)$. It follows that for $k \geq N'$,

$$\|[x] - \mathrm{i}(x_k)\|_{\mathscr{X}} = \lim_{n \to \infty} \|x_n - x_k\| < \varepsilon/2$$

and $\|\mathrm{i}(x_k) - \xi_k\|_{\mathscr{X}} < \varepsilon/2$ by (7.4); therefore

$$\|[x] - \xi_k\|_{\mathscr{X}} < \varepsilon,$$

and so $\xi_k \to [x]$ as claimed. $\qquad\qquad\square$

The completion we have just constructed is unique 'up to isometric isomorphism'. We state the result here, and leave the proof to Exercise 7.4.

Lemma 7.4 *The completion of X is unique up to isometric isomorphisms: if $(\mathrm{i}, \mathscr{X})$ and $(\mathrm{i}', \mathscr{X}')$ are two completions of X there exists an isometric isomorphism $\mathrm{j} \colon \mathscr{X} \to \mathscr{X}'$ such that $\mathrm{i}' = \mathrm{j} \circ \mathrm{i}$.*

7.3 Definition of the L^p Spaces as Completions

Fortunately in many situations there is a more concrete description of the completion. For example, we saw above that ℓ^p is the completion of c_{00} in the ℓ^p norm. We now return to the example of $C([0, 1])$ with the L^1 norm from the beginning of the chapter, which motivated our more abstract discussion of completions.

First we need to introduce some basic terminology from measure theory. A subset E of $\mathbb{R}$ has *measure zero* if for every $\varepsilon > 0$ there exists a countable collection $\{I_j\}_{j=1}^{\infty}$ of open intervals such that

$$E \subseteq \bigcup_{j=1}^{\infty} I_j \quad \text{and} \quad \sum_{j=1}^{\infty} |I_j| < \varepsilon,$$

where by $|I|$ we denote the length of the interval I. A property holds *almost everywhere* if it fails on a set of measure zero (which could be empty).

Now, suppose we view $C([0, 1])$ as a subset of $\mathcal{F}([0, 1]; \mathbb{R})$, the set of all real-valued functions on $[0, 1]$. Since $\mathcal{F}([0, 1]; \mathbb{R})$ is a much larger space than $C([0, 1])$, we can hope that sequences $(f_n) \in C([0, 1])$ that are Cauchy in the L^1 norm have limits that lie in $\mathcal{F}([0, 1]; \mathbb{R})$, even if they need not lie in $C([0, 1])$.

It is possible to show (see Corollary B.12) that if (f_n) is Cauchy in the L^1 norm, then it has a subsequence (f_{n_j}) that converges 'almost everywhere', i.e. at every point in $[0, 1]$ apart from some set E of measure zero: we can therefore define a limiting function by setting

$$f(x) := \lim_{j \to \infty} f_{n_j}(x)$$

at the points where $(f_{n_j}(x))$ converges, and however we like at the points of E. This gives a candidate $f \in \mathcal{F}([0, 1]; \mathbb{R})$ for the limit of the sequence (f_n), and one can show that

$$\| f_n - f \|_{L^1} \to 0$$

as $n \to \infty$. The set of all f constructed in this way we denote by $\mathcal{L}^1(0, 1)$.

While this gives us a way to construct a 'limit function' for any 'L^1-Cauchy' sequence (f_n), note that different subsequences may require a different measure zero set E, and that the limiting function is arbitrarily defined on E. If we want to ensure that this procedure defines a unique limit we therefore have to identify any two functions in $\mathcal{L}^1(0, 1)$ that differ only on a set of measure zero; this identification gives us the Lebesgue space $L^1(0, 1)$ (so, strictly speaking, elements of $L^1(0, 1)$ are equivalence classes of functions that agree on sets of measure zero).

Our construction also gives us a way of defining $\int_a^b f \, dx$ for any element $f \in L^1(0, 1)$: if f is the 'limit' of a Cauchy sequence $(f_n) \in C([a, b])$ we set

$$\int_a^b f(x) \, dx = \lim_{n \to \infty} \int_a^b f_n(x) \, dx;$$

this limit exists since

$$\left| \int_a^b f_n(x) \, dx - \int_a^b f_m(x) \, dx \right| \leq \int_a^b |f_n(x) - f_m(x)| \, dx = \|f_n - f_m\|_{L^1},$$

and we know that (f_n) is Cauchy in the L^1 norm.

The space $L^1(0, 1)$ can also be constructed in a more intrinsic way as the space of all 'Lebesgue integrable functions' on $(0, 1)$, i.e. measurable functions such that

$$\int_0^1 |f(x)| \, dx$$

is finite, where the integral is understood in the Lebesgue sense. (The Lebesgue integral is also defined by taking limits, but of 'simple' functions that take only a finite number of non-zero values rather than continuous functions; see Appendix B for details.)

If we want $\| \cdot \|_{L^1}$ to be a norm on $L^1(0, 1)$, then we must have $f = g$ in $L^1(0, 1)$ whenever

$$\int_0^1 |f(x) - g(x)| \, dx = 0;$$

so we deem two elements of L^1 to be equal if they are equal almost everywhere. (The requirement that we identify functions that agree on sets of measure zero therefore arises whether we define L^1 as a completion or using the theory of Lebesgue integration.)

In our discussion above there was nothing particularly special about our choice of the L^1 norm, and we can follow a very similar procedure for the L^p norm for any choice of $p \in [1, \infty)$. We can also do the same starting with $C(\overline{\Omega})$ for any open subset Ω of $\mathbb{R}^n$; we therefore give the following definition.

Definition 7.5 The space $L^p(\Omega)$ is the completion of $C(\overline{\Omega})$ in the L^p norm

$$\|f\|_{L^p(\Omega)} := \left(\int_\Omega |f(x)|^p \, dx \right)^{1/p},$$

where we identify limits of L^p-Cauchy sequences that agree almost everywhere.

The following two properties of $L^p(\Omega)$ are immediate consequences of Definition 7.5.

Lemma 7.6 *For any $p \in [1, \infty)$*

(i) *the space $L^p(\Omega)$ is complete; and*
(ii) *$C(\overline{\Omega})$ is dense in $L^p(\Omega)$.*

Defined intrinsically, the space $L^p(\Omega)$ consists of equivalence classes of measurable functions for which the L^p norm is finite (with the integral considered in the Lebesgue sense). In this case the proof of Lemma 7.6 is far from trivial; see Section B.3 in Appendix B.

Using part (ii) we can prove the separability of $L^p(a, b)$ as corollary of the Weierstrass Approximation Theorem (Theorem 6.1). For the separability of $L^p(\Omega)$ when $\Omega \subset \mathbb{R}^n$ see Lemma B.14.

Lemma 7.7 *The set of all polynomials $\mathcal{P}([a, b])$ is dense in $L^p(a, b)$ for any $1 \le p < \infty$. In particular, $L^p(a, b)$ is separable.*

Proof We know that $\mathcal{P}([a, b])$ is dense in $C([a, b])$ in the supremum norm; since

$$\|f - g\|_{L^p}^p = \int_a^b |f(x) - g(x)|^p \, \mathrm{d}x \le (b - a)\|f - g\|_\infty^p$$

it follows that $\mathcal{P}([a, b])$ is dense in $C([a, b])$ in the L^p norm, and the density of $\mathcal{P}([a, b])$ in $L^p(a, b)$ now follows from Exercise 3.14. The separability of $L^p(a, b)$ follows immediately, since $\mathcal{P}([a, b])$ is the linear span of the countable collection $\{1, x, x^2, \ldots\}$. $\qquad\qquad\square$

There is one final member of the family of the L^p spaces that we cannot define as a completion of the space of continuous functions, namely $L^\infty(\Omega)$. This is the equivalence classes of measurable functions that agree almost everywhere such that

$$\|f\|_{L^\infty} := \inf\{M : |f(x)| \le M \text{ almost everywhere}\} < \infty. \qquad (7.5)$$

Note that the L^∞ norm is the same as the supremum norm for continuous functions on Ω (see Exercise 7.6) and the space $C(\overline{\Omega})$ is complete. So $L^\infty(\Omega)$ cannot be the completion of $C(\overline{\Omega})$ in the L^∞ norm, since it contains functions that are not continuous, e.g. the function f in (7.1) is in $L^\infty(0, 1)$ but is not continuous.

The space $L^\infty(\Omega)$ is not separable: to see this for $L^\infty(0,1)$, consider the collection $\mathcal{U}$ of functions

$$\mathcal{U} := \{1_{[0,t]} : t \in [0,1]\},$$

where $1_{[0,t]}(x) = 1$ if $0 \le x \le t$ and is zero otherwise. This is an uncountable collection of functions, and any two elements $f, g \in \mathcal{U}$ with $f \neq g$ have $\|f - g\|_{L^\infty} = 1$. As in the earlier proof that ℓ^∞ is not separable (Lemma 3.24) it follows that L^∞ is not separable.

Exercises

7.1 Suppose that $(f_n) \in C([a,b])$ and that $f_n \to f$ in the supremum norm. Show that $\|f_n - f\|_{L^1} \to 0$ as $n \to \infty$.

7.2 Consider the functions $(f_n) \in C([0,1])$ defined by setting

$$f_n(x) = \begin{cases} 1 - nx & 0 \le x \le 1/n \\ 0 & 1/n < x \le 1, \end{cases}$$

see Figure 7.2. Show that $f_k \to 0$ in the L^1 norm, but that f_k does not converge to zero pointwise on [0,1].

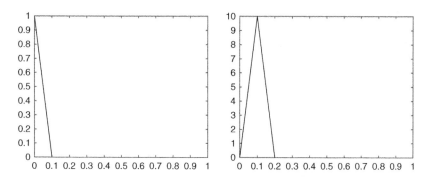

Figure 7.2 The functions f_{10} (left) and g_{10} (right).

7.3 Consider the functions $(g_n) \in C([0,1])$ defined by setting

$$g_n(x) = \begin{cases} n^2 x & 0 \le x \le 1/n, \\ n(2 - nx) & 1/n < x \le 2/n, \\ 0 & 2/n < x \le 1; \end{cases}$$

see Figure 7.2. Show that $g_n \to 0$ pointwise on $[0,1]$ but $\|g_n\|_{L^1} = 1$, so g_n does not converge to zero in L^1.

7.4 Prove Lemma 7.4. [Hint: if $x \in \mathscr{X}$ with $x = \lim_{n\to\infty} i(x_n)$, then set $j(x) = \lim_{n\to\infty} i'(x)$; show this map $j \colon \mathscr{X} \to \mathscr{X}'$ has the required properties.]

7.5 Suppose that (X, d) is a metric space, and denote by $\mathcal{F}_b(X; \mathbb{R})$ the normed space of all bounded maps from X into $\mathbb{R}$, equipped with the supremum norm. Show that for any choice of $x_0 \in X$ the map $i \colon X \to \mathcal{F}_b(X; \mathbb{R})$ given by

$$[i(x)](y) = d(y, x) - d(y, x_0)$$

is an isometry from X onto a subset of $\mathcal{F}_b(X; \mathbb{R})$. (It follows that if we let $\mathscr{X}$ be the closure of $i(X)$ in $\mathcal{F}_b(X; \mathbb{R})$, then as a closed subspace of a complete space, $(\mathscr{X}, \|\cdot\|_\infty)$ is complete; this construction therefore provides a completion $(i, \mathscr{X})$ of any metric space.)

7.6 Show that if $f \in C(\overline{\Omega})$, then $\|f\|_{L^\infty} = \|f\|_\infty$. (Note that the norm on the left-hand side is the L^∞ norm from (7.5); the norm on the right-hand side is the usual supremum norm.)

PART III

Hilbert Spaces

8

Hilbert Spaces

Hilbert spaces – which form the main topic of this part of the book – are a particular class of Banach spaces in which the norm is derived from an inner product. As such they share many properties with the familiar Euclidean spaces $\mathbb{R}^n$.

8.1 Inner Products

If $x = (x_1, \ldots, x_n)$ and $y = (y_1, \ldots, y_n)$ are two elements of $\mathbb{R}^n$, then their dot (scalar) product is

$$x \cdot y = x_1 y_1 + \cdots + x_n y_n. \tag{8.1}$$

This is one concrete example of an inner product on a vector space.

Definition 8.1 An inner product $(\cdot, \cdot)$ on a vector space V over $\mathbb{K}$ is a map $(\cdot, \cdot) : V \times V \to \mathbb{K}$ such that for all $x, y, z \in V$ and for all $\alpha \in \mathbb{K}$

(i) $(x, x) \geq 0$ with equality if and only if $x = 0$,

(ii) $(x + y, z) = (x, z) + (y, z)$,

(iii) $(\alpha x, y) = \alpha(x, y)$, and

(iv) $(x, y) = \overline{(y, x)}$.

Note that

- in a real vector space the complex conjugate in (iv) is unnecessary;
- in the complex case the restriction that $(y, x) = \overline{(x, y)}$ implies in particular that $(x, x) = \overline{(x, x)}$, i.e. that (x, x) is real, and so the requirement that $(x, x) \geq 0$ makes sense; and

101

- (ii) and (iii) imply that the inner product is linear in its first argument, but because of (iv) it is *conjugate-linear* in its second argument when $\mathbb{K} = \mathbb{C}$, i.e. $(x, \alpha y) = \overline{\alpha}(x, y)$.

An *inner-product space* is a vector space with an inner product.

Example 8.2 The canonical examples are $\mathbb{R}^n$ equipped with the standard dot product as in (8.1) and $\mathbb{C}^n$ equipped with the inner product

$$(x, y) = \sum_{j=1}^{n} x_j \overline{y_j}. \tag{8.2}$$

Example 8.3 In the space $\ell^2(\mathbb{K})$ of square summable sequences one can define the inner product of $x = (x_1, x_2, \ldots)$ and $y = (y_1, y_2, \ldots)$ by setting

$$(x, y) = \sum_{j=1}^{\infty} x_j \overline{y_j}. \tag{8.3}$$

This is the key example of an infinite-dimensional inner-product space: we will see why in Theorem 9.18.

Proof First observe that the expression on the right-hand side converges:

$$\sum_{j=1}^{\infty} |x_j \overline{y_j}| \le \frac{1}{2}\left(\sum_{j=1}^{\infty} |x_j|^2 + |y_j|^2\right) = \frac{1}{2}\left[\|x\|_{\ell^2}^2 + \|y\|_{\ell^2}^2\right] < \infty.$$

Now we check properties (i)–(iv) from Definition 8.1. For (i) we have

$$(x, x) = \sum_{j=1}^{\infty} |x_j|^2 \ge 0$$

and $(x, x) = 0$ implies that $x_j = 0$ for every j, i.e. $x = 0$. For (ii)

$$(x + y, z) = \sum_{j=1}^{\infty} (x_j + y_j)\overline{z_j} = \sum_{j=1}^{\infty} x_j \overline{z_j} + y_j \overline{z_j} = (x, z) + (y, z).$$

For (iii)

$$(\alpha x, y) = \sum_{j=1}^{\infty} \alpha x_j \overline{y_j} = \alpha \sum_{j=1}^{\infty} x_j \overline{y_j} = \alpha(x, y),$$

and for (iv)

$$(x, y) = \sum_{j=1}^{\infty} x_j \overline{y_j} = \overline{\sum_{j=1}^{\infty} \overline{x_j} y_j} = \overline{(y, x)}. \qquad \square$$

Example 8.4 The expression

$$(f, g) = \int_\Omega f(x)\overline{g(x)}\,dx \tag{8.4}$$

defines an inner product on the space $L^2(\Omega)$; see Exercise 8.1.

8.2 The Cauchy–Schwarz Inequality

Given an inner product on V we can define a map[1] $\| \cdot \|: V \to [0, \infty)$ by setting

$$\|v\|^2 := (v, v). \tag{8.5}$$

Note that for ℓ^2 and $L^2(\Omega)$ the inner products defined in (8.3) and (8.4) produce the usual norms in ℓ^2 and $L^2(\Omega)$ via (8.5):

$$(x, x)_{\ell^2} = \sum_{j=1}^\infty |x_j|^2 \qquad \text{and} \qquad (f, f)_{L^2} = \int_\Omega |f(x)|^2\,dx.$$

We will soon show that $\| \cdot \|$ always defines a norm; we say that it is the *norm induced by the inner product* $(\cdot, \cdot)$. As a first step we prove the Cauchy–Schwarz inequality for inner products, a generalisation of the familiar inequality $|x \cdot y| \le \|x\|_{\ell^2} \|y\|_{\ell^2}$ for $x, y \in \mathbb{R}^n$ (see Exercise 2.1).

Lemma 8.5 *Any inner product satisfies the Cauchy–Schwarz inequality*

$$|(x, y)| \le \|x\|\|y\| \qquad \text{for all} \qquad x, y \in V, \tag{8.6}$$

where $\| \cdot \|$ is defined in (8.5).

Proof If $x = 0$ or $y = 0$, then (8.6) is clear; so suppose that $x \ne 0$ and $y \ne 0$. For any $\lambda \in \mathbb{K}$ we have

$$0 \le (x - \lambda y, x - \lambda y) = (x, x) - \lambda(y, x) - \overline{\lambda}(x, y) + |\lambda|^2(y, y).$$

Setting $\lambda = (x, y)/\|y\|^2$ we obtain

$$0 \le \|x\|^2 - 2\frac{|(x, y)|^2}{\|y\|^2} + \frac{|(x, y)|^2}{\|y\|^2}$$

$$= \|x\|^2 - \frac{|(x, y)|^2}{\|y\|^2},$$

which implies (8.6). $\qquad\qquad\square$

---

[1] We will show that this defines a norm in Lemma 8.6, but beware of 'proof by notation': just because we have denoted the quantity $(v, v)^{1/2}$ as $\|v\|$ does not guarantee (without proof) that this really is a norm.

In the sequence space ℓ^2 the Cauchy–Schwarz inequality gives

$$|(\mathbf{x}, \mathbf{y})| \le \|\mathbf{x}\|_{\ell^2} \|\mathbf{y}\|_{\ell^2};$$

by considering instead $\mathbf{x}'$ and $\mathbf{y}'$ with $x'_j = |x_j|$ and $y'_j = |y_j|$ this yields

$$\sum_{j=1}^{\infty} |x_j y_j| \le \|\mathbf{x}\|_{\ell^2} \|\mathbf{y}\|_{\ell^2}.$$

We will obtain a more general inequality involving other ℓ^p spaces later (see Lemma 18.3).

The Cauchy–Schwarz inequality in $L^2(a, b)$ gives

$$\left| \int_a^b f(x)\overline{g(x)}\, dx \right| \le \left(\int_a^b |f(x)|^2\, dx \right)^{1/2} \left(\int_a^b |g(x)|^2\, dx \right)^{1/2},$$

for $f, g \in L^2(a, b)$. Again, if we take $|f|$ and $|g|$ rather than f and g this shows in particular that if $f, g \in L^2(a, b)$, then $fg \in L^1(a, b)$ with

$$\|fg\|_{L^1} \le \|f\|_{L^2} \|g\|_{L^2}.$$

We will see something more general in L^p spaces later (Theorem 18.4).

The Cauchy–Schwarz inequality now allows us to show easily that the map $x \mapsto \|x\|$ is a norm on V.

Lemma 8.6 *If V is an inner-product space with inner product $(\cdot, \cdot)$, then the map $\| \cdot \| : V \to [0, \infty)$ defined by setting*

$$\|v\| = (v, v)^{1/2}$$

defines a norm on V, which we call the norm induced by the inner product. In this way every inner-product space is also a normed space.

Proof We check that $\| \cdot \|$ satisfies the requirements of a norm in Definition 3.1. Property (i) is clear, since $\|x\| \ge 0$ and if $\|x\|^2 = (x, x) = 0$ then $x = 0$. Property (ii) is also clear, since

$$\|\alpha x\|^2 = (\alpha x, \alpha x) = \alpha\overline{\alpha}(x, x) = |\alpha|^2 \|x\|^2.$$

Property (iii), the triangle inequality, follows from the Cauchy–Schwarz inequality (8.6), since

$$\begin{aligned}
\|x + y\|^2 &= (x + y, x + y) \\
&= \|x\|^2 + (x, y) + (y, x) + \|y\|^2 \\
&= \|x\|^2 + 2\operatorname{Re}(x, y) + \|y\|^2
\end{aligned}$$

$$\leq \|x\|^2 + 2\|x\|\|y\| + \|y\|^2$$
$$= (\|x\| + \|y\|)^2,$$

i.e. $\|x + y\| \leq \|x\| + \|y\|$.   □

8.3 Properties of the Induced Norms

Norms derived from inner products must satisfy the 'parallelogram law'.

Lemma 8.7 (Parallelogram law/identity) *If V is an inner-product space with induced norm $\| \cdot \|$, then*

$$\|x + y\|^2 + \|x - y\|^2 = 2(\|x\|^2 + \|y\|^2) \qquad \text{for all} \qquad x, y \in V \quad (8.7)$$

(Figure 8.1).

Proof Simply expand the inner products:

$$\|x + y\|^2 + \|x - y\|^2 = (x + y, x + y) + (x - y, x - y)$$
$$= \|x\|^2 + (y, x) + (x, y) + \|y\|^2$$
$$+ \|x\|^2 - (y, x) - (x, y) + \|y\|^2$$
$$= 2(\|x\|^2 + \|y\|^2). \qquad □$$

It follows that if a norm on space does not satisfy the parallelogram law, it cannot have come from an inner product. For example, the supremum norm on $C([0, 1])$ does not come from an inner product, since if we consider $f(x) = x$ and $g(x) = 1 - x$ (for example), then $(f + g)(x) = 1$ and $(f - g)(x) = 2x - 1$, and so

$$\|f + g\|_\infty = \|f - g\|_\infty = \|f\|_\infty = \|g\|_\infty = 1$$

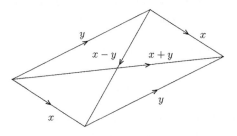

Figure 8.1 The parallelogram law: $\|x + y\|^2 + \|x - y\|^2 = 2(\|x\|^2 + \|y\|^2)$.

and (8.7) is not satisfied. Similarly, one can show that the ℓ^p and L^p norms do not come from an inner product when $p \neq 2$ (see Exercises 8.5 and 8.6).

If a norm is derived from an inner product, we can reconstruct the inner product as follows.

Lemma 8.8 (Polarisation identity) *Let V be an inner-product space with induced norm $\| \cdot \|$. Then if V is real,*

$$4(x, y) = \|x + y\|^2 - \|x - y\|^2, \tag{8.8}$$

while if V is complex,

$$4(x, y) = \|x + y\|^2 - \|x - y\|^2 + i\|x + iy\|^2 - i\|x - iy\|^2$$

$$= \sum_{n=0}^{3} i^n \|x + i^n y\|^2. \tag{8.9}$$

Proof Rewrite the right-hand sides as inner products, multiply out, and simplify: for example, in the real case

$$\|x + y\|^2 - \|x - y\|^2 = (x + y, x + y) - (x - y, x - y)$$

$$= [\|x\|^2 + (x, y) + (y, x) + \|y\|^2]$$

$$- [\|x\|^2 - (x, y) - (y, x) + \|y\|^2]$$

$$= 4(x, y). \qquad \square$$

If V is a real/complex inner-product space and $\| \cdot \|$ is a norm on V that satisfies the parallelogram law, then (8.8) or (8.9) defines an inner product on V. In other words, the parallelogram law characterises those norms that can be derived from inner products; this is the Jordan–von Neumann Theorem (for the proof in the case of a real Hilbert space see Exercise 8.8).

We end this chapter with a very useful result that deals with the interaction between inner products and convergent sequences.

Lemma 8.9 *If V is an inner-product space, then $x_n \to x$ and $y_n \to y$ implies that*

$$(x_n, y_n) \to (x, y).$$

Proof Let $\| \cdot \|$ be the norm induced by the inner product in V. Since x_n and y_n converge, $\|x_n\|$ and $\|y_n\|$ are bounded. Therefore

$$|(x_n, y_n) - (x, y)| = |(x_n - x, y_n) + (x, y_n - y)|$$
$$\leq \|x_n - x\|\|y_n\| + \|x\|\|y_n - y\|,$$

which implies that $(x_n, y_n) \to (x, y)$. $\qquad\qquad\square$

In particular, this lemma means that we can swap inner products and sums: if

$$x = \sum_{j=1}^{\infty} x_j, \qquad \text{i.e. if} \quad \sum_{j=1}^{n} x_j \to x,$$

then

$$\left(\sum_{j=1}^{\infty} x_j, y \right) = \left(\lim_{n \to \infty} \sum_{j=1}^{n} x_j, y \right) = \lim_{n \to \infty} \left(\sum_{j=1}^{n} x_j, y \right) = \lim_{n \to \infty} \sum_{j=1}^{n} (x_j, y)$$
$$= \sum_{j=1}^{\infty} (x_j, y). \qquad\qquad (8.10)$$

8.4 Hilbert Spaces

We are now in a position to introduce Hilbert spaces.

Definition 8.10 A *Hilbert space* is a complete inner-product space. (Here 'complete' is understood with respect to the norm induced by the inner product.)

For the inner-product spaces we have introduced above – $\mathbb{R}^n$, $\mathbb{C}^n$, ℓ^2, and $L^2(\Omega)$ – the norms induced by the inner products are the 'standard norms', with which they are complete. So these are all Hilbert spaces.

Since closed subspaces of Banach spaces are again Banach spaces (see Lemma 4.3) the same is true of Hilbert spaces.

Lemma 8.11 *Suppose that H is a Hilbert space and U is a closed subspace of H. Then U is a Hilbert space when equipped with the inner product from H.*

From now on if we have a Hilbert space H we will denote its inner product by $(\cdot, \cdot)$ and the induced norm by $\|\cdot\|$; unless stated otherwise we treat the case $\mathbb{K} = \mathbb{C}$, since taking $\mathbb{K} = \mathbb{R}$ only simplifies matters by removing the complex conjugates.

Exercises

8.1 Check that the expression

$$(f, g) = \int_\Omega f(x)\overline{g(x)}\,dx$$

(see Example 8.4) defines an inner product on $L^2(\Omega)$.

8.2 Show that if $|\Omega| = \int_\Omega 1\,dx < \infty$, then $L^\infty(\Omega) \subset L^2(\Omega) \subset L^1(\Omega)$ with

$$\|f\|_{L^1} \le |\Omega|^{1/2}\|f\|_{L^2} \le |\Omega|\|f\|_{L^\infty}. \qquad (8.11)$$

8.3 Show that if H and K are Hilbert spaces with inner products $(\cdot, \cdot)_H$ and $(\cdot, \cdot)_K$, respectively, then $H \times K$ is a Hilbert space with inner product

$$((x, \xi), (y, \eta))_{H \times K} := (x, y)_H + (\xi, \eta)_K.$$

8.4 Let $T: H \to K$ be a linear surjective mapping between two real Hilbert spaces. Use the polarisation identity (8.8) to show that

$$(Tx, Ty)_K = (x, y)_H \qquad \text{for every } x, y \in H \qquad (8.12)$$

if and only if $\|Tx\|_K = \|x\|_H$ for every $x \in H$. (In this case we say that T is *unitary*.) (Young, 1988)

8.5 Show that if $p \ne 2$ there is no inner product on ℓ^p that induces the ℓ^p norm.

8.6 Show that if $p \ne 2$ there is no inner product on $C([0, 1])$ that induces the L^p norm.

8.7 If $\|\cdot\|$ is a norm on a vector space X induced by an inner product show that it satisfies Apollonius's identity,

$$\|z - x\|^2 + \|z - y\|^2 = \frac{1}{2}\|x - y\|^2 + 2\left\|z - \frac{1}{2}(x + y)\right\|^2$$

for every $x, y, z \in X$. (One can argue directly, expanding out the inner products, but it is much easier to use the parallelogram law.)

8.8 Show that if $\|\cdot\|$ is a norm on a real vector space V that satisfies the parallelogram identity

$$\|x + y\|^2 + \|x - y\|^2 = 2(\|x\|^2 + \|y\|^2)$$

then

$$\langle x, y \rangle := \frac{1}{4}\left(\|x + y\|^2 - \|x - y\|^2\right)$$

defines an inner product on V. This result is due to Jordan and von Neumann (1935). [Hint: properties (i) and (iv) from Definition 8.1 are immediate. To prove that $\langle x, z \rangle + \langle y, z \rangle = \langle x + y, z \rangle$ for all $x, y, z \in V$, i.e. property (ii) from the definition, use the parallelogram identity. Finally, deduce that $\langle \alpha x, y \rangle = \alpha \langle x, y \rangle$ for every $\alpha \in \mathbb{Q}$, $x, y \in V$, and use this to prove the same identity for every $\alpha \in \mathbb{R}$ (which is property (iii) of an inner product).] (Yosida, 1980)

8.9 Show that if H is a Hilbert space and U a closed linear subspace of H, then H/U is also a Hilbert space. [Hint: Exercise 4.5 shows that H/U is a Banach space; show that the norm satisfies the parallelogram law and hence comes from an inner product.]

8.10 A Banach space is called *uniformly convex* if for every $\varepsilon > 0$ there exists $\delta > 0$ such that

$$\|x - y\| > \varepsilon, \ \|x\| = \|y\| = 1 \quad \Rightarrow \quad \left\| \frac{x + y}{2} \right\| < 1 - \delta.$$

Use the parallelogram identity to show that every Hilbert space is uniformly convex.

9

Orthonormal Sets and Orthonormal Bases for Hilbert Spaces

In this chapter we discuss how to define a basis for a general normed space. Since we are in a normed space and have a notion of convergence, we can now allow infinite linear combinations of basis elements (in contrast to the Hamel bases we considered in Chapter 1). We then consider orthonormal sets in inner-product spaces, and orthonormal bases for separable Hilbert spaces, which share much in common with the standard orthonormal basis in $\mathbb{R}^n$.

9.1 Schauder Bases in Normed Spaces

In an infinite-dimensional normed space X we cannot hope to find a finite basis, since then the space would by definition be finite-dimensional. Assuming that X is separable, the best that we can hope for is to find a countable basis $\{e_j\}_{j=1}^{\infty}$, in terms of which to expand any $x \in X$ as

$$x = \sum_{j=1}^{\infty} \alpha_j e_j,$$

where the sum converges in X in the sense discussed in Section 3.3. (If X is not separable it cannot have a countable basis; see Lemma 3.23 part (iii).) We now formalise the idea of a basis in a normed space, restricting to separable spaces for simplicity.

Definition 9.1 A countable set $\{e_j\}_{j=1}^{\infty}$ is a *Schauder basis* for a normed space X if every $x \in X$ can be written uniquely[1] as

$$x = \sum_{j=1}^{\infty} \alpha_j e_j \qquad \text{for some } \alpha_j \in \mathbb{K}. \tag{9.1}$$

---

[1] Equivalently, we could require that (i) every $x \in X$ can be written in the form $x = \sum_{j=1}^{\infty} \alpha_j e_j$ and (ii) if $\sum_{j=1}^{\infty} \alpha_j e_j = 0$, then $\alpha_j = 0$ for every j.

The equality here is to be understood as equality 'in X', i.e. the sum converges in X.

Note that if $\{e_j\}_{j=1}^{\infty}$ is a basis in the sense of Definition 9.1, then the uniqueness of the expansion implies that $\{e_j\}_{j=1}^{\infty}$ is a linearly independent set: if

$$0 = \sum_{j=1}^{n} \alpha_j e_j,$$

then as there is a unique expansion for zero we must have $\alpha_j = 0$ for all $j = 1, \ldots, n$.

Example 9.2 The collection $\{e^{(j)}\}_{j=1}^{\infty}$ is a Schauder basis for ℓ^p for every $1 \le p < \infty$, but is not a Schauder basis for ℓ^{∞}.

Proof We proved the first part of this in the course of Lemma 3.24: in (3.13) we showed if $1 \le p < \infty$ then for every $\varepsilon > 0$ there exists $N > 0$ such that

$$\left\| x - \sum_{j=1}^{n} x_j e^{(j)} \right\|_{\ell^p} < \varepsilon$$

for all $n \ge N$. So we can write $x = \sum_{j=1}^{\infty} x_j e^{(j)}$ as an equality in ℓ^p, in the sense that the sum converges in ℓ^p.

To see that these elements do not form a basis for ℓ^{∞}, we could either observe that if they did ℓ^{∞} would be separable (using Lemma 3.23), and we know that it is not (from Lemma 3.24); or more directly consider the element $x \in \ell^{\infty}$ with $x_j = 1$ for every j. The equality

$$\sum_{j=1}^{\infty} \alpha_j e^{(j)} = x$$

would mean that the partial sums converge to x in ℓ^{∞}; but for any finite n we have

$$\left\| \sum_{j=1}^{n} \alpha_j e^{(j)} - x \right\|_{\ell^{\infty}} = \|(\alpha_1 - 1, \alpha_2 - 1, \ldots, \alpha_n - 1, 1, 1, \ldots)\|_{\ell^{\infty}} \ge 1,$$

and so the partial sums cannot converge whatever our choice of coefficients $\{\alpha_j\}$. □

We will return briefly to the topic of Schauder bases in Banach spaces in Section 23.2, but now we turn our attention back to the particular case of inner-product spaces.

9.2 Orthonormal Sets

In $\mathbb{R}^n$ two vectors are orthogonal if $x \cdot y = 0$. We can make a corresponding definition in a general inner-product space V. (Throughout this section V will denote a general inner-product space.)

Definition 9.3 Two elements x and y of an inner-product space V are said to be *orthogonal* if $(x, y) = 0$.

If $(x, y) = 0$, then

$$\|x + y\|^2 = (x + y, x + y) = \|x\|^2 + (x, y) + (y, x) + \|y\|^2 = \|x\|^2 + \|y\|^2.$$

Sums of orthogonal vectors are therefore very useful in calculations, since all the cross terms in their norm vanish.

Definition 9.4 A set E in an inner-product space is *orthonormal* if $\|e\| = 1$ for every $e \in E$ and $(e_1, e_2) = 0$ for any $e_1, e_2 \in E$ with $e_1 \neq e_2$.

Note that this definition does not require E to be countable, although in most of what follows we will be primarily concerned with countable orthonormal sets $\{e_j\}_{j=1}^{\infty}$; in this case we require $\|e_j\| = 1$ for every $j \in \mathbb{N}$ and $(e_i, e_j) = 0$ if $i \neq j$, which we can write more compactly as

$$(e_i, e_j) = \delta_{ij}$$

(recall that δ_{ij} is the Kronecker delta, defined in (1.4)).

Note that any orthonormal set must be linearly independent, since if

$$\sum_{j=1}^{n} \alpha_j e_j = 0 \qquad n \in \mathbb{N}, \ \alpha_j \in \mathbb{K}, \ e_j \in E,$$

taking the inner product with each e_j in turn shows that $\alpha_j = 0$ for every $j = 1, \ldots, n$.

Example 9.5 The set $\{e^{(j)}\}_{j=1}^{\infty}$ from (1.3) is orthonormal in ℓ^2.

Example 9.6 The set

$$\left\{ e_k := \frac{1}{\sqrt{2\pi}} e^{ikx} \right\}_{k \in \mathbb{Z}}$$

is orthonormal in $L^2(-\pi, \pi)$.

Proof We have

$$(e_k, e_k) = \frac{1}{2\pi} \int_{-\pi}^{\pi} e^{ikx} e^{-ikx} \, dx = 1$$

and for $k \neq k'$

$$(e_k, e_{k'}) = \frac{1}{2\pi} \int_{-\pi}^{\pi} e^{ikx} e^{-ik'x} \, dx = \frac{1}{2\pi(k - k')} \left[e^{i(k-k')x} \right]_{-\pi}^{\pi} = 0. \quad \square$$

Example 9.7 The set

$$\{1\} \cup \{\sqrt{2} \cos k\pi x\}_{k=1}^{\infty}$$

is orthonormal in $L^2(0, 1)$.

Proof For $n \neq m$ we have

$$\int_0^1 (\cos n\pi x)(\cos m\pi x) \, dx = \frac{1}{2} \int_0^1 \cos(n + m)\pi x + \cos(n - m)\pi x \, dx$$

$$= \frac{1}{2} \left[\frac{\sin(n + m)\pi x}{(n + m)\pi} + \frac{\sin(n - m)\pi x}{(n - m)\pi} \right]_0^1 = 0,$$

while for $n = m$

$$\int_0^1 \cos^2 n\pi x \, dx = \frac{1}{2} \int_0^1 1 + \cos 2n\pi x \, dx = \frac{1}{2}. \quad \square$$

Expansions involving orthonormal elements of an inner-product space are easy to treat, essentially due to the following simple lemma.

Lemma 9.8 (Generalised Pythagoras) *If $\{e_1, \ldots, e_n\}$ is an orthonormal set in an inner-product space V, then for any $\{\alpha_j\}_{j=1}^n \in \mathbb{K}$*

$$\left\| \sum_{j=1}^n \alpha_j e_j \right\|^2 = \sum_{j=1}^n |\alpha_j|^2.$$

Proof By definition

$$\left\| \sum_{j=1}^n \alpha_j e_j \right\|^2 = \left(\sum_{i=1}^n \alpha_i e_i, \sum_{j=1}^n \alpha_j e_j \right)$$

$$= \sum_{i,j=1}^n (\alpha_i e_i, \alpha_j e_j)$$

$$= \sum_{i,j=1}^{n} \alpha_i \overline{\alpha_j}(e_i, e_j) = \sum_{i=1}^{n} |\alpha_i|^2,$$

since $(e_i, e_j) = \delta_{ij}$.   □

We now recall the Gram–Schmidt process for generating orthonormal sets in finite-dimensional inner-product spaces, and show that this generalises to infinite-dimensional spaces.

Proposition 9.9 (Gram–Schmidt orthonormalisation) *Suppose that V is an inner-product space and $E = (e_j)_{j \in \mathcal{J}} \in H$, with $\mathcal{J} = \{1, \ldots, n\}$ or $\mathcal{J} = \mathbb{N}$, is a linearly independent sequence. Then there exists an orthonormal sequence $\tilde{E} = (\tilde{e}_j)_{j \in \mathcal{J}}$ such that*

$$\mathrm{Span}(e_1, \ldots, e_k) = \mathrm{Span}(\tilde{e}_1, \ldots, \tilde{e}_k) \qquad (9.2)$$

for every $k \in \mathcal{J}$, and so $\mathrm{clin}(\tilde{E}) = \mathrm{clin}(E)$.

Proof We proceed by induction, starting with $\tilde{e}_1 = e_1 / \|e_1\|$.

Suppose that we already have an orthonormal set $(\tilde{e}_1, \ldots, \tilde{e}_n)$ whose linear span is the same as $(e_1, \ldots, e_n)$. Then we can define $\tilde{e}_{n+1}$ by setting

$$e'_{n+1} = e_{n+1} - \sum_{i=1}^{n} (e_{n+1}, \tilde{e}_i)\tilde{e}_i \qquad \text{and} \qquad \tilde{e}_{n+1} = \frac{e'_{n+1}}{\|e'_{n+1}\|}.$$

The span of $(\tilde{e}_1, \ldots, \tilde{e}_{n+1})$ is the same as the span of $(\tilde{e}_1, \ldots, \tilde{e}_n, e_{n+1})$, which is the same as the span of $(e_1, \ldots, e_n, e_{n+1})$ using the induction hypothesis. Clearly $\|\tilde{e}_{n+1}\| = 1$ and for $m \leq n$ we have

$$(\tilde{e}_{n+1}, \tilde{e}_m) = \frac{1}{\|e'_{n+1}\|}\left((e_{n+1}, \tilde{e}_m) - \sum_{i=1}^{n}(e_{n+1}, \tilde{e}_i)(\tilde{e}_i, \tilde{e}_m) \right) = 0$$

since $(\tilde{e}_1, \ldots, \tilde{e}_n)$ are orthonormal.

That the closed linear spans of $\tilde{E}$ and E coincide is a consequence of (9.2): any element in $\mathrm{clin}(E)$ can be approximated arbitrarily closed by finite linear combinations of the $\{e_j\}$, and hence by finite linear combinations of the $\{\tilde{e}_j\}$, so is an element of $\mathrm{clin}(\tilde{E})$. The same argument in reverse yields the equality of the closed linear spans.   □

In a finite-dimensional inner-product space this process guarantees the existence of an orthonormal basis, i.e. a basis of orthonormal elements: starting with any basis we use the Gram–Schmidt process to find an orthonormal basis. (A similar result holds in any Hilbert space; see Proposition 9.17 and Exercise 9.15.)

Lemma 9.10 *Let* $(\cdot, \cdot)$ *be any inner product on a vector space* V *of dimension* n. *Then there exists an orthonormal basis* $\{e_j\}_{j=1}^n$ *of* V.

It follows that in some sense the dot product (8.2) is the canonical inner product on any finite-dimensional space V. Indeed, with respect to any orthonormal basis $\{e_j\}_{j=1}^n$ of V the inner product $(\cdot, \cdot)$ has the form (8.2), i.e.

$$\left(\sum_{j=1}^n x_j e_j, \sum_{k=1}^n y_k e_k \right) = \sum_{j,k=1}^n x_j \overline{y_k}(e_j, e_k) = x_1 \overline{y_1} + \cdots + x_n \overline{y_n}.$$

9.3 Convergence of Orthogonal Series

Suppose that $\{e_j\}_{j=1}^\infty$ is an orthonormal set in an inner-product space V. We now investigate when the series

$$\sum_{j=1}^\infty \alpha_j e_j,$$

converges. If the series converges to some $x \in V$, then, taking the inner product with some e_k, we obtain

$$(x, e_k) = \left(\sum_{j=1}^\infty \alpha_j e_j, e_k \right) = \sum_{j=1}^\infty \alpha_j (e_j, e_k) = \alpha_k$$

(using (8.10) to change the order of the inner product and the sum), which shows that the coefficients (α_k) are completely determined, with $\alpha_k = (x, e_k)$.

We start with the following lemma.

Lemma 9.11 (Bessel's inequality) *Let* V *be an inner-product space and* $\{e_j\}_{j=1}^\infty$ *an orthonormal set in* V. *Then for any* $x \in V$ *we have*

$$\sum_{n=1}^\infty |(x, e_j)|^2 \leq \|x\|^2 \tag{9.3}$$

and in particular the left-hand side converges.

Proof Let us denote by x_k the partial sum

$$x_k = \sum_{j=1}^k (x, e_j) e_j.$$

By Lemma 9.8 we have

$$\|x_k\|^2 = \sum_{j=1}^{k} |(x, e_j)|^2$$

and so

$$
\begin{aligned}
\|x - x_k\|^2 &= (x - x_k, x - x_k) \\
&= \|x\|^2 - (x_k, x) - (x, x_k) + \|x_k\|^2 \\
&= \|x\|^2 - \sum_{j=1}^{k}(x, e_j)(e_j, x) - \sum_{j=1}^{k}\overline{(x, e_j)}(x, e_j) + \|x_k\|^2 \\
&= \|x\|^2 - \|x_k\|^2.
\end{aligned}
$$

It follows that

$$\sum_{j=1}^{k} |(x, e_j)|^2 = \|x_k\|^2 = \|x\|^2 - \|x - x_k\|^2 \le \|x\|^2.$$

Since this gives a bound uniform for all k, (9.3) now follows on letting $k \to \infty$.   □

We now use Bessel's inequality to give a simple criterion for the convergence of a sum $\sum_{j=1}^{\infty} \alpha_j e_j$ in a Hilbert space when the $\{e_j\}$ are orthonormal. Note that this will be the first time in this section that we have required our inner-product space to be complete.

Lemma 9.12 *Let H be a Hilbert space and $\{e_n\}_{n=1}^{\infty}$ an orthonormal set in H. The series $\sum_{n=1}^{\infty} \alpha_n e_n$ converges if and only if*

$$\sum_{n=1}^{\infty} |\alpha_n|^2 < \infty,$$

and then

$$\left\| \sum_{n=1}^{\infty} \alpha_n e_n \right\|^2 = \sum_{n=1}^{\infty} |\alpha_n|^2. \tag{9.4}$$

Equivalently, $\sum_{n=1}^{\infty} \alpha_n e_n$ converges if and only if $\boldsymbol{\alpha} := (\alpha_1, \alpha_2, \ldots) \in \ell^2$. The equality in (9.4) is one form of *Parseval's identity*.

Proof Suppose that $\sum_{j=1}^{n} \alpha_j e_j$ converges to some $x \in H$ as $n \to \infty$; then

$$\left\| \sum_{j=1}^{n} \alpha_j e_j \right\|^2 = \sum_{j=1}^{n} |\alpha_j|^2$$

converges to $\|x\|^2$ as $n \to \infty$ (see Lemma 3.14); so in particular we have $\sum_{j=1}^{\infty} |\alpha_j|^2 < \infty$, and (9.4) holds.

Conversely, if

$$\sum_{j=1}^{\infty} |\alpha_j|^2 < \infty,$$

then $(\sum_{j=1}^{n} |\alpha_j|^2)_{n=1}^{\infty}$ is a Cauchy sequence. Setting $x_n = \sum_{j=1}^{n} \alpha_j e_j$ we have, taking $m > n$,

$$\|x_m - x_n\|^2 = \left\| \sum_{j=n+1}^{m} \alpha_j e_j \right\|^2 = \sum_{j=n+1}^{m} |\alpha_j|^2.$$

It follows that (x_n) is a Cauchy sequence, and since H is complete it therefore converges to some $x \in H$. The equality in (9.4) now follows as above. $\square$

By combining this lemma with Bessel's inequality we obtain the following convergence result.

Corollary 9.13 *Let H be a Hilbert space and $\{e_n\}_{n=1}^{\infty}$ an orthonormal set in H. Then $\sum_{n=1}^{\infty} (x, e_n) e_n$ converges for every $x \in H$.*

Note that $\sum_{n=1}^{\infty} (x, e_n) e_n$ need not converge to x; we investigate how to ensure this in the next section.

9.4 Orthonormal Bases for Hilbert Spaces

If E is a Schauder basis for an inner-product space V and is also orthonormal, we refer to it as an *orthonormal basis* for V. Note that an orthonormal set $\{e_j\}_{j=1}^{\infty}$ is a basis provided that every x can be written in the form (9.1), since in this case the uniqueness of the expansion follows from the orthonormality: indeed, if

$$x = \sum_{j=1}^{\infty} \alpha_j e_j = \sum_{j=1}^{\infty} \beta_j e_j \quad \Rightarrow \quad \sum_{j=1}^{\infty} (\alpha_j - \beta_j) e_j = 0$$

then taking the inner product with e_i we have

$$0 = \left(\sum_{j=1}^{\infty} (\alpha_j - \beta_j) e_j, e_i \right) = \sum_{j=1}^{\infty} (\alpha_j - \beta_j)(e_j, e_i) = \alpha_i - \beta_i,$$

where we have used (8.10) to move the inner product inside the summation.

So if E is an orthonormal basis we can expect to have

$$x = \sum_{j=1}^{\infty} (x, e_j) e_j \qquad \text{for every } x \in H,$$

and in this case Parseval's identity (9.4) from Lemma 9.12 will guarantee that

$$\sum_{j=1}^{\infty} |(x, e_j)|^2 = \|x\|^2.$$

We now show that $E = \{e_j\}_{j=1}^{\infty}$ forms a basis for H if and only if this equality holds for every $x \in H$, and also provide some other equivalent conditions.

Proposition 9.14 *Let $E = \{e_j\}_{j=1}^{\infty}$ be an orthonormal set in a Hilbert space H. Then the following statements are equivalent:*

(a) *E is a basis for H;*
(b) *for any x we have*

$$x = \sum_{j=1}^{\infty} (x, e_j) e_j \qquad \text{for all} \qquad x \in H;$$

(c) *Parseval's identity holds:*

$$\|x\|^2 = \sum_{j=1}^{\infty} |(x, e_j)|^2 \qquad \text{for all} \qquad x \in H;$$

(d) *$(x, e_j) = 0$ for all j implies that $x = 0$; and*
(e) *$\operatorname{clin}(E) = H$.*

Part (e) means that the linear span of E is dense in H, i.e. for any $x \in H$ and any $\varepsilon > 0$ there exists an $n \in \mathbb{N}$ and $\alpha_j \in \mathbb{K}$ such that

$$\left| x - \sum_{j=1}^{n} \alpha_j e_j \right| < \varepsilon.$$

See Exercise 9.8 for an example showing that if E is linearly independent but not orthonormal, then $\operatorname{clin}(E) = H$ does not necessarily imply that E is a basis for H.

Proof First we show (a)$\Leftrightarrow$(b). If E is an orthonormal basis for H, then we can write

$$x = \sum_{j=1}^{\infty} \alpha_j e_j, \quad \text{i.e.} \quad x = \lim_{j \to \infty} \sum_{j=1}^{n} \alpha_j e_j.$$

Clearly if $k \leq n$ we have

$$\left(\sum_{j=1}^{n} \alpha_j e_j, e_k \right) = \alpha_k;$$

taking the limit $n \to \infty$ and using the compatibility of inner products and limits from (8.10) it follows that $\alpha_k = (x, e_k)$ and hence (a) holds. The same argument shows that if we assume (b), then this expansion is unique, and so E is a basis.

We show that (b) $\Rightarrow$ (c) $\Rightarrow$ (d) $\Rightarrow$ (b), and then that (b) $\Rightarrow$ (e) and (e) $\Rightarrow$ (d).

(b) $\Rightarrow$ (c) is immediate from (3.14).

(c) $\Rightarrow$ (d) is immediate since $\|x\| = 0$ implies that $x = 0$.

(d) $\Rightarrow$ (b) Take $x \in H$ and let

$$y = x - \sum_{j=1}^{\infty} (x, e_j) e_j.$$

For each $m \in \mathbb{N}$ we have, using Lemma 8.9 (continuity of the inner product),

$$(y, e_m) = (x, e_m) - \lim_{n \to \infty} \left(\sum_{j=1}^{n} (x, e_j) e_j, e_m \right)$$

$$= 0$$

since eventually $n \geq m$. It follows from (c) that $y = 0$, i.e. that

$$x = \sum_{j=1}^{\infty} (x, e_j) e_j$$

as required.

(b) $\Rightarrow$ (e) is clear, since given any x and $\varepsilon > 0$ there exists an n such that

$$\left| \sum_{j=1}^{n} (x, e_j) e_j - x \right| < \varepsilon.$$

(e) $\Rightarrow$ (d) Suppose that $x \in H$ with $(x, e_j) = 0$ for every j. Choose x_n contained in the linear span of E such that $x_n \to x$. Then

$$\|x\|^2 = (x, x) = \left(\lim_{n \to \infty} x_n, x \right) = \lim_{n \to \infty} (x_n, x) = 0,$$

since x_n is a (finite) linear combination of the e_j. So $x = 0$. $\qquad\square$

Example 9.15 The sequence $(e^{(j)})_{j=1}^{\infty}$ defined in (1.3) is an orthonormal basis for ℓ^2, since it is clear that if $(x, e^{(j)}) = x_j = 0$ for all j then $x = 0$.

We now use Proposition 9.14 to show that we can expand any function $f \in L^2(-\pi, \pi)$ as a Fourier series, which will converge to f in L^2; we use the orthonormality of the exponential functions in (9.5) to find an explicit expression for the Fourier coefficients.

Lemma 9.16 *The exponential functions*

$$\left\{ \frac{1}{\sqrt{2\pi}} e^{ikx} : k \in \mathbb{Z} \right\} \tag{9.5}$$

(as in Example 9.6) form an orthonormal basis for $L^2(-\pi, \pi)$. In particular, any $f \in L^2$ can be written as the Fourier series

$$f = \sum_{k=-\infty}^{\infty} c_k e^{ikx}, \tag{9.6}$$

where

$$c_n = \frac{1}{2\pi} \int_{-\pi}^{\pi} f(x) e^{-inx} \, dx \tag{9.7}$$

and the sum in (9.6) converges in L^2. Furthermore, we have

$$\int_{-\pi}^{\pi} |f(x)|^2 \, dx = 2\pi \sum_{k=-\infty}^{\infty} |c_k|^2. \tag{9.8}$$

Before we give the proof, note that we have already shown in Corollary 6.10 that any continuous function $f \in C([-\pi, \pi])$ with $f(\pi) = f(-\pi)$ can be approximated uniformly by expressions of the form

$$\sum_{k=-n}^{n} c_k e^{ikx}$$

and we will use this is the proof below. Note, however, that this requires a different set of coefficients for each approximation. In contrast, taking partial sums of the series on the right-hand side of the equality in (9.6) produces approximating expressions of this form, but in general these will only converge in the L^2 norm and not uniformly. Indeed, in Section 22.3 we will show that there exist 2π-periodic continuous functions for which the Fourier series in (9.6) fails to converge at $x = 0$ (and, in fact, at 'very many' points).

Proof We showed in Example 9.6 that these functions are orthonormal in $L^2(-\pi, \pi)$. To show that they form a basis for $L^2(-\pi, \pi)$ it is enough to show that their linear span is dense, by part (e) of Proposition 9.14.

Given any $f \in L^2(-\pi, \pi)$ and $\varepsilon > 0$, we first use the density of $C([-\pi, \pi])$ in $L^2(-\pi, \pi)$ (see Lemma 7.6) to find $g \in C([-\pi, \pi])$ such that

$$\|f - g\|_{L^2} < \varepsilon/3.$$

Now let

$$\tilde{g}(x) := \begin{cases} \frac{x+\pi}{\delta} g(-\pi + \delta) & -\pi \le x \le -\pi + \delta \\ g(x) & -\pi + \delta < x < \pi - \delta \\ \frac{\pi-x}{\delta} g(\pi - \delta), & \pi - \delta < x \le \pi, \end{cases}$$

with $\delta = \varepsilon/18\|g\|_\infty$, so that $\|\tilde{g} - g\|_{L^2} < \varepsilon/3$ and $g \in C([-\pi, \pi])$ with $g(\pi) = g(-\pi) = 0$.

Now we can use Corollary 6.10 to find coefficients α_j such that

$$\left\| \tilde{g} - \sum_{j=-k}^{k} \alpha_j e^{ijx} \right\|_\infty < \frac{\varepsilon}{3\sqrt{2\pi}}.$$

Then since for any $\phi \in C([-\pi, \pi])$ we have

$$\|\phi\|_{L^2} = \left(\int_{-\pi}^{\pi} |\phi(x)|^2 \, dx \right)^{1/2} \le \left(\int_{-\pi}^{\pi} \|\phi\|_\infty^2 \, dx \right)^{1/2} = \sqrt{2\pi} \|\phi\|_\infty$$

it follows that

$$\left\| \tilde{g} - \sum_{j=-k}^{k} \alpha_j e^{ijx} \right\|_{L^2} \le \sqrt{2\pi} \left\| \tilde{g} - \sum_{j=-k}^{k} \alpha_j e^{ijx} \right\|_\infty < \frac{\varepsilon}{3}.$$

Now using the triangle inequality we have

$$\left\| f - \sum_{j=-k}^{k} \alpha_k e^{ikx} \right\|_{L^2} \le \|f - g\|_{L^2} + \|g - \tilde{g}\|_{L^2} + \left\| \tilde{g} - \sum_{j=-k}^{k} \alpha_k e^{ikx} \right\|_{L^2}$$

$$< \frac{\varepsilon}{3} + \frac{\varepsilon}{3} + \frac{\varepsilon}{3} = \varepsilon,$$

as required.

The expression for c_k in (9.7) comes from taking the inner product of the expansion in (9.6) with $\frac{1}{2\pi} e^{inx}$:

$$\frac{1}{2\pi} \int_{-\pi}^{\pi} f(x) e^{-inx} \, dx = \left(f, \frac{1}{2\pi} e^{inx} \right) = \left(\sum_{k=-\infty}^{\infty} c_k \frac{1}{\sqrt{2\pi}} e^{ikx}, \frac{1}{\sqrt{2\pi}} e^{inx} \right) = c_n,$$

and the equality in (9.8) is Parseval's identity from part (c) of Proposition 9.14.
□

See Exercise 9.11 for a similar result for Fourier cosine series, and the book by Körner (1989) for much more on Fourier series.

9.5 Separable Hilbert Spaces

We end this chapter by showing that all separable Hilbert spaces are isometrically isomorphic to ℓ^2, so that in some sense ℓ^2 is the only (infinite-dimensional) separable Hilbert space.

We first show that every separable Hilbert space has a countable orthonormal basis. (In fact, every Hilbert space, separable or not, has an orthonormal basis; see Exercise 9.15.)

Proposition 9.17 *An infinite-dimensional Hilbert space is separable if and only if it has a countable orthonormal basis.*

Proof If a Hilbert space has a countable basis, then we can construct a countable dense set by taking finite linear combinations of the basis elements with rational coefficients, and so it is separable.

If H is separable, let $E' = (x_n)_{n=1}^{\infty}$ be a countable dense subset. In particular, the closed linear span of E' is the whole of H. Remove from E' any element x_n that can be written as a linear combination of $\{x_1, \ldots, x_{n-1}\}$, to give a new set E whose linear span is still dense but that is linearly independent. Now use the Gram–Schmidt process from Proposition 9.9 to obtain a countable orthonormal set whose closed linear span is all of H; Proposition 9.14 guarantees that E is therefore a countable orthonormal basis.   □

By using this countable orthonormal basis we can construct an isometric isomorphism between any separable Hilbert space and ℓ^2.

Theorem 9.18 *Any infinite-dimensional separable Hilbert space H over $\mathbb{K}$ is isometrically isomorphic to $\ell^2(\mathbb{K})$, i.e. $H \equiv \ell^2(\mathbb{K})$.*

Proof Since H is separable, Proposition 9.17 guarantees that it has a countable orthonormal basis $\{e_j\}_{j=1}^{\infty}$. Define a linear map $\varphi \colon H \to \ell^2$ by setting

$$\varphi(u) := \Big((u, e_1), (u, e_2), \ldots, (u, e_n), \ldots \Big);$$

clearly the inverse map, given by

$$\phi^{-1}(\alpha) = \sum_{j=1}^{\infty} \alpha_j e_j \qquad \text{where} \qquad \alpha = (\alpha_1, \alpha_2, \ldots, \alpha_n, \ldots),$$

is also linear. Lemma 9.12 guarantees that φ maps H onto ℓ^2, and also that φ^{-1} maps ℓ^2 onto H. The characterisation of a basis in Proposition 9.14 (c) shows that $\|u\|_H = \|\varphi(u)\|_{\ell^2}$, so φ is an isometry. □

Most Hilbert spaces that occur in applications are separable, but there are non-separable Hilbert spaces. For example, if Γ is uncountable, then the space $\ell^2(\Gamma)$ consisting of all functions $f : \Gamma \to \mathbb{R}$ such that f is non-zero for at most countably many $\gamma \in \Gamma$ and

$$\sum_{\gamma \in \Gamma} |f(\gamma)|^2 < \infty$$

is a Hilbert space but is not separable: the functions $\{1_\gamma : \gamma \in \Gamma\}$ that are zero everywhere except at x where they take the value 1 form an uncountable set in which any two distinct elements are a distance $\sqrt{2}$ apart.

Exercises

9.1 Show that the set

$$E = \frac{1}{\sqrt{2\pi}}, \ \frac{1}{\sqrt{\pi}} \cos t, \ \frac{1}{\sqrt{\pi}} \sin t, \ \frac{1}{\sqrt{\pi}} \cos 2t, \ \frac{1}{\sqrt{\pi}} \sin 2t, \ \ldots.$$

is orthonormal in $L^2(-\pi, \pi)$.

9.2 Show that $(x, y) = 0$ if and only if (i) $\|x + \alpha y\| \geq \|x\|$ for every $\alpha \in \mathbb{K}$ or (ii) $\|x + \alpha y\| = \|x - \alpha y\|$ for every $\alpha \in \mathbb{K}$. (Giles, 2000)

9.3 Use Bessel's inequality to show that if $\{e_j\}_{j=1}^{\infty}$ is an orthonormal set in an inner-product space V, then for any $x \in V$

$$|\{j : |(x, e_j)| > M\}| \leq \frac{\|x\|^2}{M^2}.$$

9.4 Extend the argument from the previous exercise to show that if E is an uncountable orthonormal set in an inner product space V, then for each $x \in V$,

$$\{e \in E : (x, e) \neq 0\}$$

is at most a countable set.

9.5   Show that if $\{e_j\}_{j=1}^{\infty}$ is an orthonormal basis for H then

$$(u, v) = \sum_{j=1}^{\infty}(u, e_j)(e_j, v)$$

for every $u, v \in H$. (This is a more general version of Parseval's identity.)

9.6   Suppose that $(e_j)_{j=1}^{\infty}$ is an orthonormal sequence that forms a basis for a Hilbert space H. Show that the 'Hilbert cube'

$$Q := \left\{ \sum_{j=1}^{\infty} \alpha_j e_j : |\alpha_j| \le \frac{1}{j} \right\}$$

is a compact subset of H.

9.7   Use Proposition 9.14 to deduce that if $E = \{e_j\}_{j=1}^{\infty}$ is an orthonormal set in a Hilbert space, then

$$\mathrm{clin}(E) = \left\{ x \in H : x = \sum_{j=1}^{\infty} \alpha_j e_j, \ (\alpha_j) \in \mathbb{K} \right\}.$$

9.8   Proposition 9.14 shows that if $\{e_j\}_{j=1}^{\infty}$ is an orthonormal set, then it is a basis if its linear span is dense in H. This exercise gives an example to show that this is not true without the assumption that $\{e_j\}$ is orthonormal.

Let $(e_j)_{j=1}^{\infty}$ be an orthonormal sequence that forms a basis for a Hilbert space H. Set

$$f_n = \sum_{j=1}^{n} \frac{1}{j} e_j.$$

Show that the linear span of $\{f_j\}$ is dense in H, but that $\{f_j\}$ is not a basis for H. (Show that $x = \sum_{j=1}^{\infty} j^{-1} e_j$ cannot be written in terms of the $\{f_j\}$.)

9.9   Suppose that H is a Hilbert space. Show that if $(x, z) = (y, z)$ for all z in a dense subset of H, then $x = y$.

9.10 Suppose that $f, g \in L^2(a, b)$ are such that

$$\int_a^b x^n f(x)\, dx = \int_a^b x^n g(x)\, dx$$

for every n. Use the result of the previous exercise along with Lemma 7.7 to show that $f = g$ in L^2. (Goffman and Pedrick, 1983; Pryce, 1973)

9.11 Use the results of Example 9.7 and Corollary 6.5 to show that any real $f \in L^2(0, 1)$ can be written as

$$f(x) = \sum_{k=0}^{\infty} a_k \cos k\pi x,$$

where the sum converges in $L^2(0, 1)$. Find an expression for the coefficients a_k.

9.12 Let $f(x) = x$ on $[-\pi, \pi]$. By finding the Fourier coefficients in the expansion $f(x) = \sum_{k=-\infty}^{\infty} c_k e^{ikx}$ and using the Parseval identity (9.8) show that

$$\sum_{k=1}^{\infty} \frac{1}{k^2} = \frac{\pi^2}{6}.$$

9.13 Show that any infinite-dimensional Hilbert H space contains a countably infinite orthonormal sequence.

9.14 Use the result of the previous exercise to show that a Hilbert space is finite-dimensional if and only if its closed unit ball is compact.

9.15 Show that any Hilbert space H has an orthonormal basis. (Use Zorn's Lemma to show that H has a maximal orthonormal subset $E = \{e_\alpha\}_{\alpha \in \mathbb{A}}$, and then show that every element of H can be written as $\sum_{j=1}^{\infty} a_j e_{\alpha_j}$ for some $a_j \in \mathbb{K}$ and $\alpha_j \in \mathbb{A}$.)

10

Closest Points and Approximation

In this chapter we consider the existence of 'closest points' in convex subsets of Hilbert spaces. In particular, this will enable us to define the orthogonal projection onto a closed linear subspace U of a Hilbert space H, and thereby decompose any $x \in H$ as $x = u + v$, where $u \in U$ and $v \in U^\perp$. Here $U^\perp$ is the 'orthogonal complement' of U,

$$U^\perp := \{y \in H : (y, u) = 0 \text{ for every } u \in U\}.$$

10.1 Closest Points in Convex Subsets of Hilbert Spaces

In a Hilbert space there is always a unique closest point in any closed convex set. (The same result is not true in a general Banach space without additional assumptions; see Exercises 10.2–10.7.)

Proposition 10.1 *Let A be a non-empty closed convex subset of a Hilbert space H and let $x \in H \setminus A$. Then there exists a unique $\hat{a} \in A$ such that*

$$\|x - \hat{a}\| = \text{dist}(x, A) := \inf\{\|x - a\| : a \in A\}.$$

Moreover, for every $a \in A$ we have

$$\text{Re}\,(x - \hat{a}, a - \hat{a}) \le 0. \tag{10.1}$$

See Figure 10.1.

Proof Set $\delta = \inf\{\|x - a\| : a \in A\} > 0$; that this is strictly positive follows from the facts that $x \notin A$ and A is closed (see the beginning of the proof of Lemma 5.4). Then we can find a sequence $(a_n) \in A$ such that

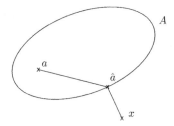

Figure 10.1 The point $\hat{a}$ is the unique closest point to x in the convex set A. The inequality in (10.1) means that in $\mathbb{R}^n$, the angle between $x - \hat{a}$ and $a - \hat{a}$ is always at least a right angle.

$$\|x - a_n\|^2 \leq \delta^2 + \frac{1}{n}. \tag{10.2}$$

We show that (a_n) is a Cauchy sequence by using the parallelogram law (Lemma 8.7) to write

$$\|(x - a_n) + (x - a_m)\|^2 + \|(x - a_n) - (x - a_m)\|^2 = 2\big[\|x - a_n\|^2 + \|x - a_m\|^2\big].$$

This gives

$$\|2x - (a_n + a_m)\|^2 + \|a_n - a_m\|^2 \leq 4\delta^2 + \frac{2}{m} + \frac{2}{n}$$

or

$$\|a_n - a_m\|^2 \leq 4\delta^2 + \frac{2}{m} + \frac{2}{n} - 4\left\|x - \tfrac{1}{2}(a_n + a_m)\right\|^2.$$

Since A is convex, $\frac{1}{2}(a_n + a_m) \in A$, and so $\|x - \frac{1}{2}(a_n + a_m)\|^2 \geq \delta^2$, which implies that

$$\|a_n - a_m\|^2 \leq \frac{2}{m} + \frac{2}{n}.$$

It follows that (a_n) is Cauchy; since H is a Hilbert space it is complete, and so $a_n \to \hat{a}$ for some $\hat{a} \in H$. Since A is closed, $\hat{a} \in A$, and taking limits in (10.2) shows that

$$\|x - \hat{a}\| = \delta$$

(since $\|x - a\| \geq \delta$ for every $a \in A$).

Finally, to prove (10.1), let a be any other point in A; then, since A is convex, $(1 - t)\hat{a} + ta \in A$ for all $t \in (0, 1)$. Since $\hat{a}$ is the closest point to x in A,

$$\begin{aligned}
\|x - \hat{a}\|^2 &\leq \|x - [(1 - t)\hat{a} + ta]\|^2 \\
&= \|(x - \hat{a}) - t(a - \hat{a})\|^2 \\
&= \|x - \hat{a}\|^2 - 2t \operatorname{Re}\,(x - \hat{a}, a - \hat{a}) + t^2\|a - \hat{a}\|^2.
\end{aligned}$$

It follows that $\text{Re}(a - \hat{a}, x - \hat{a}) \leq 0$; otherwise for t sufficiently small we obtain a contradiction.

Finally, if both $\hat{a}$ and a' satisfy (10.1), then

$$\text{Re}\,(x - \hat{a}, a' - \hat{a}) \leq 0 \qquad \text{and} \qquad \text{Re}\,(x - a', \hat{a} - a') \leq 0.$$

Rewriting the second of these as $\text{Re}\,(a' - x, a' - \hat{a}) \leq 0$ and adding yields

$$\text{Re}\,(a' - \hat{a}, a' - \hat{a}) \leq 0$$

which implies that $\|a' - \hat{a}\|^2 = 0$ so $a' = \hat{a}$. $\qquad\square$

The following corollary shows that we can 'separate' $x \notin A$ from A by looking at inner products with some $v \in H$. We will reinterpret this result later in Corollary 12.5.

Corollary 10.2 *Suppose that A is a non-empty closed convex subset of a Hilbert space, and $x \notin A$. Then there exists $v \in H$ such that*

$$\text{Re}\,(a, v) + d^2 \leq \text{Re}\,(x, v) \qquad \text{for every } a \in A,$$

where $d = \text{dist}(x, A)$; see Figure 10.2.

Proof Let $\hat{a}$ be the closest point to x in A, and set $v = x - \hat{a}$. The result now follows from (10.1) since

$$\text{Re}\,(x, v) = \text{Re}\,(\hat{a} + (x - \hat{a}), v) = \text{Re}\,(\hat{a}, v) + \|v\|^2$$

$$\geq \text{Re}\,(a, v) + d^2. \qquad\square$$

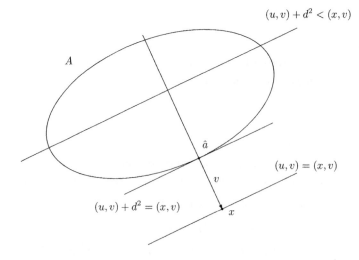

Figure 10.2 The sets of points $u \in H$ with (u, v) taking various constant values in the case of a real space.

10.2 Linear Subspaces and Orthogonal Complements

If X is a subset of a Hilbert space H, then the *orthogonal complement of X in H* is

$$X^{\perp} = \{u \in H : (u, x) = 0 \qquad \text{for all} \qquad x \in X\}.$$

Clearly, if $Y \subseteq X$, then $X^{\perp} \subseteq Y^{\perp}$. Note also that $X \cap X^{\perp} = \{0\}$ if $0 \in X$, and that this intersection is empty otherwise.

We have already remarked (see Section 4.1) that linear subspaces of infinite-dimensional spaces are not always closed; however, orthogonal complements are always closed.

Lemma 10.3 *If X is a subset of H, then* $X^{\perp}$ *is a closed linear subspace of H.*

Proof It is clear that $X^{\perp}$ is a linear subspace of H: indeed, if $u, v \in X^{\perp}$ and $\alpha \in \mathbb{K}$, then

$$(u + v, x) = (u, x) + (v, x) = 0 \qquad \text{and} \qquad (\alpha u, x) = \alpha(u, x) = 0$$

for every $x \in X$. To show that $X^{\perp}$ is closed, suppose that $u_n \in X^{\perp}$ and $u_n \to u$; then for every $x \in X$

$$(u, x) = \left(\lim_{n \to \infty} u_n, x \right) = \lim_{n \to \infty} (u_n, x) = 0$$

(using (8.10)) and so $X^{\perp}$ is closed. $\qquad \square$

Note that Proposition 9.14 shows that E is a basis for H if and only if $E^{\perp} = \{0\}$, since this is just a rephrasing of part (d) of that result: $(u, e_j) = 0$ for all j implies that $u = 0$.

We now show that given any closed linear subspace U of H, any $x \in H$ has a unique decomposition in the form $x = u + v$, where $u \in U$ and $v \in U^{\perp}$: we say that H is the direct sum of U and $U^{\perp}$ and write $H = U \oplus U^{\perp}$.

Proposition 10.4 *If U is a closed linear subspace of a Hilbert space H, then any $x \in H$ can be written uniquely as*

$$x = u + v \qquad \text{with} \qquad u \in U, \quad v \in U^{\perp},$$

i.e. $H = U \oplus U^{\perp}$. *The map* $P_U : H \to U$ *defined by*

$$P_U x := u$$

is called the orthogonal projection *of x onto U, and satisfies*

$$P_U^2 x = P_U x \qquad \text{and} \qquad \|P_U x\| \leq \|x\| \qquad \text{for all } x \in H.$$

See Figure 10.3.

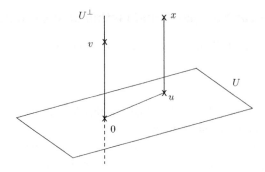

Figure 10.3 Decomposing $x = u + v$ with $u \in U$ and $v \in U^{\perp}$. The point u is the orthogonal projection of x onto U.

Proof If U is a closed linear subspace, then U is closed and convex, so the result of Proposition 10.1 shows that given $x \in H$ there is a unique closest point $u \in U$. It is now simple to show that $x - u \in U^{\perp}$ and then such a decomposition is unique.

To show that $x - u \in U^{\perp}$ we use (10.1). First, given any $v \in U$, we have $u \pm v \in U$, so (10.1) yields

$$\mathrm{Re}\,(x - u, \pm v) \leq 0,$$

which shows that $\mathrm{Re}\,(x - u, v) = 0$. Choosing instead $u \pm iv \in U$ we obtain $\mathrm{Im}\,(x - u, v) = 0$, and so $(x - u, v) = 0$ for every $v \in U$, i.e. $x - u \in U^{\perp}$.

Finally, the uniqueness follows easily: if $x = u_1 + v_1 = u_2 + v_2$, then $u_1 - u_2 = v_2 - v_1$, and so

$$\|u_1 - u_2\|^2 = (u_1 - u_2, u_1 - u_2) = (u_1 - u_2, v_2 - v_1) = 0,$$

since $u_1 - u_2 \in U$ and $v_2 - v_1 \in U^{\perp}$.

If $P_U x$ denotes the closest point to x in U, then clearly $P_U^2 = P_U$, and it follows from the fact that $(u, x - u) = 0$ that

$$\|x\|^2 = \|u\|^2 + \|x - u\|^2,$$

and so

$$\|P_U x\| \leq \|x\|,$$

i.e. the projection can only decrease the norm. $\square$

We will now show that in general $X \subseteq (X^{\perp})^{\perp}$; we can use the decomposition result we have just proved to show that this is an equality if X is a closed linear subspace.

Lemma 10.5 *If $X \subseteq H$, then $X \subseteq (X^{\perp})^{\perp}$ with equality if and only if X is a closed linear subspace of H.*

Proof Any $x \in X$ satisfies

$$(x, z) = 0 \qquad \text{for every } z \in X^\perp;$$

so $X \subseteq (X^\perp)^\perp$.

Now suppose that $z \in (X^\perp)^\perp$, so that $(z, y) = 0$ for every $y \in X^\perp$. If X is a closed linear subspace, then we can use Proposition 10.4 to write $z = x + \xi$, where $x \in X$ and $\xi \in X^\perp$. But then (since $z \in (X^\perp)^\perp$) we have

$$0 = (z, \xi) = (x + \xi, \xi) = \|\xi\|^2,$$

so in fact $\xi = 0$ and therefore $z \in X$.

Finally, it follows from Lemma 10.3 that if $X = (X^\perp)^\perp$, then X must be a closed linear subspace. $\qquad \square$

Exercise 10.9 shows that $E^\perp = [\text{clin}(E)]^\perp$. In particular, whenever X is a linear subspace of H we have $(X^\perp)^\perp = \overline{X}$.

10.3 Best Approximations

We now investigate the best approximation of elements of H using (possibly infinite) linear combinations of elements of an orthonormal set E. Exercise 9.7 shows that when E is an orthonormal sequence the set of all such linear combinations is[1] precisely $\text{clin}(E)$. Since this is a closed subspace of H, by Proposition 10.4 the closest point to any $x \in H$ in $\text{clin}(E)$ is the orthogonal projection of x onto $\text{clin}(E)$.

Theorem 10.6 *Let* $E = \{e_j\}_{j \in \mathfrak{I}}$ *be an orthonormal set, where* $\mathfrak{I} = \mathbb{N}$ *or* $(1, 2, \ldots, n)$. *Then for any* $x \in H$, *the orthogonal projection of* x *onto* $\text{clin}(E)$, *which is the closest point to* x *in* $\text{clin}(E)$, *is given by*

$$P_E x := \sum_{j \in \mathfrak{I}} (x, e_j) e_j.$$

(Of course, if E is a basis for H, then there is no approximation involved.)

Proof Consider $x - \sum_{j \in \mathfrak{I}} \alpha_j e_j$. Then

$$\left\| x - \sum_{j \in \mathfrak{I}} \alpha_j e_j \right\|^2 = \|x\|^2 - \sum_{j \in \mathfrak{I}} (x, \alpha_j e_j) - \sum_{j \in \mathfrak{I}} (\alpha_j e_j, x) + \sum_{j \in \mathfrak{I}} |\alpha_j|^2$$

$$= \|x\|^2 - \sum_{j \in \mathfrak{I}} \overline{\alpha_j}(x, e_j) - \sum_{j \in \mathfrak{I}} \alpha_j \overline{(x, e_j)} + \sum_{j \in \mathfrak{I}} |\alpha_j|^2$$

---

[1] Note that this is not necessarily the case when E is not orthonormal; see Exercise 9.8.

$$= \|x\|^2 - \sum_{j \in \mathcal{J}} |(x, e_j)|^2$$

$$+ \sum_{j \in \mathcal{J}} \left[|(x, e_j)|^2 - \overline{\alpha_j}(x, e_j) - \alpha_j \overline{(x, e_j)} + |\alpha_j|^2 \right]$$

$$= \|x\|^2 - \sum_{j \in \mathcal{J}} |(x, e_j)|^2 + \sum_{j \in \mathcal{J}} |(x, e_j) - \alpha_j|^2,$$

and so the minimum occurs when $\alpha_j = (x, e_j)$ for all $j \in \mathcal{J}$.   □

Example 10.7 If $E = \{e_j\}_{j=1}^{\infty}$ is an orthonormal basis in H, then the best approximation of an element of H in terms of $\{e_j\}_{j=1}^{n}$ is just given by the partial sum

$$\sum_{j=1}^{n} (x, e_j) e_j.$$

For example, the best approximation of an element $x \in \ell^2$ in terms of $\{e^{(j)}\}_{j=1}^{n}$ (elements of the standard basis) is simply

$$\sum_{j=1}^{n} (x, e^{(j)}) e^{(j)} = (x_1, x_2, \ldots, x_n, 0, 0, \ldots).$$

Now suppose that E is a finite or countable set that is not orthonormal. We can still find the best approximation to any $u \in H$ that lies in clin(E) by using the Gram–Schmidt orthonormalisation process from Proposition 9.9. First we find an orthonormal set $\tilde{E}$ whose linear span is the same as that of E, and then we use the result of Theorem 10.6.

Example 10.8 Consider approximation of functions in $L^2(-1, 1)$ by polynomials of degree up to n. We can start with the set $\{1, x, x^2, \ldots, x^n\}$ and then use the Gram–Schmidt process to construct polynomials that are orthonormal with respect to the $L^2(-1, 1)$ inner product. We do this here for polynomials up to degree 3.

We begin with $e_1 = 1/\sqrt{2}$ and then consider

$$e_2' = x - \left(x, \frac{1}{\sqrt{2}} \right) \frac{1}{\sqrt{2}} = x - \frac{1}{2} \int_{-1}^{1} t \, dt = x$$

so

$$\|e_2'\|^2 = \int_{-1}^{1} t^2 \, dt = \frac{2}{3}; \qquad e_2 = \sqrt{\frac{3}{2}} x.$$

Then

$$e_3' = x^2 - \left(x^2, \sqrt{\frac{3}{2}}x\right)\sqrt{\frac{3}{2}}x - \left(x^2, \frac{1}{\sqrt{2}}\right)\frac{1}{\sqrt{2}}$$

$$= x^2 - \frac{3x}{2}\int_{-1}^{1}\frac{3t^3}{2}\,dt - \frac{1}{2}\int_{-1}^{1}t^2\,dt$$

$$= x^2 - \frac{1}{3},$$

so

$$\|e_3'\|^2 = \int_{-1}^{1}\left(t^2 - \frac{1}{3}\right)^2 dt = \left[\frac{t^5}{5} - \frac{2t^3}{9} + \frac{t}{9}\right]_{-1}^{1} = \frac{8}{45}$$

which gives

$$e_3 = \sqrt{\frac{5}{8}}\left(3x^2 - 1\right).$$

Exercise 10.11 asks you to check that $e_4 = \sqrt{\frac{7}{8}}(5x^3 - 3x)$ and that this is orthogonal to e_1, e_2, and e_3.

Using these orthonormal functions we can find the best approximation of any function $f \in L^2(-1, 1)$ by a degree three polynomial:

$$\left(\frac{7}{8}\int_{-1}^{1} f(t)(5t^3 - 3t)\,dt\right)(5x^3 - 3x)$$

$$+ \left(\frac{5}{8}\int_{-1}^{1} f(t)(3t^2 - 1)\,dt\right)(3x^2 - 1)$$

$$+ \left(\frac{3}{2}\int_{-1}^{1} f(t)t\,dt\right)x + \frac{1}{2}\int_{-1}^{1} f(t)\,dt.$$

Example 10.9 The best approximation in $L^2(-1, 1)$ of $f(x) = |x|$ by a third degree polynomial is

$$f_3(x) = \frac{15x^2 + 3}{16},$$

with $\|f - f_3\|_{L^2} = \sqrt{3}/4$.

Proof Since $\int_{-1}^{1} |t|t^k\,dt = 0$ for k odd and $2\int_{0}^{1} t^{k+1}\,dt$ for k even, we only need to calculate

$$\left(\frac{5}{4}\int_{0}^{1} 3t^3 - t\,dt\right)(3x^2 - 1) + \int_{0}^{1} t\,dt = \frac{5}{16}(3x^2 - 1) + \frac{1}{2} = \frac{15x^2 + 3}{16}.$$

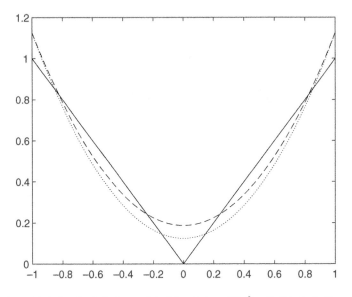

Figure 10.4 Graph of $f(x) = |x|$ (solid), $f(x) = (15x^2 + 3)/16$ (dashed), and $f(x) = x^2 + \frac{1}{8}$ (dotted).

The L^2 distance of f_3 from f can be found by integrating

$$\|f - f_3\|^2 = \frac{2}{16^2} \int_0^1 (15x^2 - 16x + 3)^2 \, dx = \frac{3}{16}. \qquad \square$$

Of course, the meaning of the 'best approximation' is that this choice minimises the L^2 norm of the difference. It is not the 'best approximation' in terms of the supremum norm: we have $f_3(0) = 3/16$, while

$$\sup_{x\in[-1,1]} \left| |x| - (x^2 + \frac{1}{8}) \right| = \frac{1}{8} < \frac{3}{16} \le \|f - f_3\|_\infty$$

(see Figure 10.4).

For other examples of orthonormal sets see Section 4.10 in Goffman and Pedrick (1983).

Exercises

10.1   Show that the result on the existence of a unique closest point in Proposition 10.1 is equivalent to the following statement: if K is a non-empty closed convex subset of a Hilbert space that does not contain zero, then K contains a unique element with minimum norm.

10.2 Let $X = C([-1, 1])$ with the supremum norm, and let

$$U := \left\{ f \in C([-1, 1]) : \int_{-1}^{0} f(t)\, dt = \int_{0}^{1} f(t)\, dt = 0 \right\},$$

which is a closed linear subspace of X. Let g be a function in X such that

$$\int_{-1}^{0} g(t)\, dt = 1 \qquad \text{and} \qquad \int_{0}^{1} g(t)\, dt = -1.$$

Show that $\operatorname{dist}(g, U) = 1$ but that $\operatorname{dist}(g, f) > 1$ for every $f \in U$, so that there is no closest point to g in U. (Lax, 2002)

10.3 A Banach space is called *strictly convex* if $x, y \in X$, $x \neq y$, with $\|x\| = \|y\| = 1$ implies that $\|x + y\| < 2$. Show that if X is strictly convex and U is a closed linear subspace of X, then given any $x \notin U$, any closest point to x in U (should one exist) is unique.

10.4 Show that a uniformly convex Banach space (see Exercise 8.10) is strictly convex.

10.5 Show that if $2 \le p < \infty$, then for any $\alpha, \beta \ge 0$

$$\alpha^p + \beta^p \le (\alpha^2 + \beta^2)^{p/2},$$

and deduce, using the fact that $t \mapsto |t|^{p/2}$ is convex, that

$$\left| \frac{a+b}{2} \right|^p + \left| \frac{a-b}{2} \right|^p \le \frac{1}{2}(|a|^p + |b|^p), \qquad a, b \in \mathbb{C}.$$

Hence obtain Clarkson's first inequality

$$\left\| \frac{f+g}{2} \right\|_{L^p}^p + \left\| \frac{f-g}{2} \right\|_{L^p}^p \le \frac{1}{2}(\|f\|_{L^p}^p + \|g\|_{L^p}^p), \qquad f, g \in L^p. \tag{10.3}$$

(The same inequality also holds in ℓ^p, $2 \le p < \infty$, by a similar argument.)

10.6 Use Clarkson's first inequality to show that L^p is uniformly convex for all $2 \le p < \infty$.

Clarkson's second equality is valid for the range $1 < p \le 2$: for all $f, g \in L^p$ we have

$$\left\| \frac{f+g}{2} \right\|_{L^p}^q + \left\| \frac{f-g}{2} \right\|_{L^p}^q \le \left(\frac{1}{2}\|f\|_{L^p}^p + \frac{1}{2}\|g\|_{L^p}^p \right)^{1/(p-1)},$$

where p and q are conjugate; this is much less straightforward to prove than Clarkson's first inequality (Clarkson, 1936). Use this inequality to show that L^p is uniformly convex for $1 < p \le 2$. (The same arguments work in the ℓ^p spaces for $1 < p < \infty$.)

10.7　In this exercise we show that if X is a uniformly convex Banach space and K is a closed convex subset of X that does not contain 0, then K has a unique element of minimum norm. Let $(k_n) \in K$ be a sequence such that $k_n \to \inf_{k \in K} \|k\|$.

(i) Set $x_n = k_n / \|k_n\|$ and use the convexity of K to show that

$$\|\tfrac{1}{2}(x_n + x_m)\| \geq \frac{d}{2}\left(\frac{1}{\|k_n\|} + \frac{1}{\|k_m\|}\right);$$

(ii) deduce that $\|\tfrac{1}{2}(x_n + x_m)\| \to 1$ as $\max(n, m) \to \infty$;

(iii) use the uniform convexity of X to show that (x_n) is a Cauchy sequence; and

(iv) finally, use the fact that $k_n = \|k_n\| x_n$ to show that (k_n) is also Cauchy.

This result implies that if K is a closed convex subset of a uniformly convex Banach space and $x \notin K$, then there exists a unique closest point to x in K (just consider $K' = K - x$; see the solution of Exercise 10.1). (Lax, 2002)

10.8　Show that $(X + Y)^\perp = X^\perp \cap Y^\perp$.

10.9　Show that $E^\perp = (\mathrm{clin}(E))^\perp$.

10.10　Suppose that M is a closed subspace of a Hilbert space H. Show that H/M (see Exercise 8.9) is isometrically isomorphic to $M^\perp$ via the mapping $T \colon H/M \to M^\perp$ given by $T([x]) = P^\perp x$, where $P^\perp$ is the orthogonal projection onto $M^\perp$.

10.11　Continuing the analysis in Example 10.8, show that

$$e_4 = \sqrt{\frac{7}{8}}\,(5x^3 - 3x).$$

10.12　The polynomials from Example 10.8 and the previous exercise are closely related to the Legendre polynomials (P_n). The nth Legendre polynomial is given by the formula

$$P_n(x) = \frac{1}{2^n n!}\frac{d^n}{dx^n}u_n(x), \qquad u_n(x) := (x^2 - 1)^n.$$

Noting that $u_n^{(j)}(\pm 1) = 0$ for all $j = 0, \ldots, n$, show that

$$\int_{-1}^{1} x^k u_n(x)\,dx = 0$$

whenever $0 \leq k < n$, and deduce that $(P_m, P_n) = 0$ for $m \neq n$. (We use $u_n^{(j)}$ to denote the jth derivative of u_n.) (Rynne and Youngson, 2008)

11

Linear Maps between Normed Spaces

We now consider linear maps between general normed spaces: throughout the chapter unless explicitly stated otherwise X and Y are normed spaces with norms $\| \cdot \|_X$ and $\| \cdot \|_Y$, respectively.

We say that a linear map $T : X \to Y$ is bounded if

$$\|Tx\|_Y \leq M \|x\|_X \qquad \text{for every } x \in X$$

for some $M > 0$; we will show that this is equivalent to continuity of T. After giving a number of examples we show that if Y is a Banach space, then the collection of all bounded linear maps from X into Y is a Banach space when equipped with the norm

$$\|T\|_{B(X,Y)} = \inf\{M : \|Tx\|_Y \leq M \|x\|_X\}.$$

We then discuss inverses of linear maps, pointing out that a bounded linear map can have an unbounded inverse; we reserve the term 'invertible' for those maps whose inverse is also bounded.

We will return to the particular case of linear maps between Hilbert spaces in the next chapter.

11.1 Bounded Linear Maps

Recall from Section 1.5 that if U and V are vector spaces over $\mathbb{K}$, then a map $T : U \to V$ is *linear* if

$$T(\alpha x + \beta y) = \alpha Tx + \beta Ty \qquad \text{for all} \qquad \alpha, \beta \in \mathbb{K}, \ x, y \in U;$$

the collection $L(U, V)$ of all linear maps from U into V is a vector space; and we write $L(U)$ for $L(U, U)$.

137

Definition 11.1 A linear map $T : (X, \| \cdot \|_X) \to (Y, \| \cdot \|_Y)$ is *bounded* if there exists a constant M such that

$$\|Tx\|_Y \leq M \|x\|_X \qquad \text{for all} \qquad x \in X. \qquad (11.1)$$

Linear maps defined on finite-dimensional spaces are automatically bounded.

Lemma 11.2 *If X is a finite-dimensional vector space, then any linear map $T : (X, \| \cdot \|_X) \to (Y, \| \cdot \|_Y)$ is bounded.*

Proof If $E = \{e_j\}_{j=1}^n$ is a basis for X, then recall that

$$\left\| \sum_{j=1}^n \alpha_j e_j \right\|_E = \left(\sum_{j=1}^n |\alpha_j|^2 \right)^{1/2}$$

defines a norm on X (Lemma 3.7). For any $x \in X$ with $x = \sum_{j=1}^n \alpha_j e_j$ we have

$$
\begin{aligned}
\|Tx\|_Y &= \left\| T \left(\sum_{j=1}^n \alpha_j e_j \right) \right\|_Y \\
&= \left\| \sum_{j=1}^n \alpha_j T e_j \right\|_Y \leq \sum_{j=1}^n |\alpha_j| \|T e_j\| \\
&\leq \left(\sum_{j=1}^n \|T e_j\|_Y^2 \right)^{1/2} \left(\sum_{j=1}^n |\alpha_j|^2 \right)^{1/2} \\
&= C \|x\|_E,
\end{aligned}
$$

where $C := (\sum_{j=1}^n \|T e_j\|_Y^2)^{1/2}$. Since X is finite-dimensional, all norms on X are equivalent (Theorem 5.1); in particular, we have $\|x\|_E \leq C' \|x\|$ for some $C' > 0$. It follows that

$$\|Tx\|_Y \leq CC' \|x\|_X,$$

so T is bounded from $(X, \| \cdot \|_X)$ into $(Y, \| \cdot \|_Y)$ as claimed. $\qquad \square$

However, linear operators on infinite-dimensional spaces need not be bounded. For example,[1] the linear map $T : (c_{00}, \ell^2) \to \ell^2$ that maps $e^{(j)}$ to $j e^{(j)}$ is not bounded; nor is the linear map from

---

[1] Recall that c_{00} is the space of all sequences with only a finite number of non-zero terms; see Example 1.8.

$$(C^1([0, 1]), \| \cdot \|_\infty) \to (C([0, 1]), \| \cdot \|_\infty)$$

given by $f \mapsto f'$, since $x^n \mapsto nx^{n-1}$ for any $n \in \mathbb{N}$.

Lemma 11.3 *A linear map* $T: X \to Y$ *is continuous if and only if it is bounded.*

Proof Suppose that T is bounded; then for some $M > 0$

$$\|Tx_n - Tx\|_Y = \|T(x_n - x)\|_Y \le M\|x_n - x\|_X,$$

and so T is continuous. Now suppose that T is continuous; then in particular it is continuous at zero, and so, taking $\varepsilon = 1$ in the definition of continuity, there exists a $\delta > 0$ such that

$$\|Tx\| \le 1 \qquad \text{for all} \qquad \|x\| \le \delta.$$

It follows that for $z \ne 0$

$$\|Tz\| = \left\| T\left(\frac{\|z\|}{\delta} \frac{\delta z}{\|z\|} \right) \right\| = \frac{\|z\|}{\delta} \left\| T\left(\frac{\delta z}{\|z\|} \right) \right\| \le \frac{1}{\delta}\|z\|,$$

and so T is bounded. □

The space of all bounded linear maps from X into Y is denoted by $B(X, Y)$; we write $B(X)$ for the space $B(X, X)$ of all bounded linear maps from X into itself.

Definition 11.4 The norm in $B(X, Y)$ or *operator norm* of a linear map $T: X \to Y$ is the smallest value of M such that (11.1) holds,

$$\|T\|_{B(X,Y)} := \inf \{M : \|Tx\|_Y \le M\|x\|_X \text{ for all } x \in X\}. \qquad (11.2)$$

The infimum in (11.2) is attained: since for each $x \in X$, $\|Tx\|_Y \le M\|x\|_X$ for every $M > \|T\|_{B(X,Y)}$, it follows that

$$\|Tx\|_Y \le \|T\|_{B(X,Y)}\|x\|_X \qquad \text{for all} \qquad x \in X.$$

Note that from now on we will use the terms 'linear map' and 'linear operator' interchangeably. The former is perhaps more common when considering maps between vector spaces, but as many of the examples in applications are differential operators, the latter terminology is more common when considering spectral theory, for example.

We now show that (11.2) really does define a norm on $B(X, Y)$.

Lemma 11.5 *As defined in (11.2)* $\| \cdot \|_{B(X,Y)}$ *is a norm on* $B(X, Y)$.

Proof If $\|T\|_{B(X,Y)} = 0$, then for every $x \in X$ we have $\|Tx\|_Y = 0$, which shows that $Tx = 0$ and so $T = 0$. Since, by definition, $(\lambda T)x = \lambda Tx$, the homogeneity property $\|\lambda T\|_{B(X,Y)} = |\lambda| \|T\|_{B(X,Y)}$ is immediate, and for the triangle inequality observe that

$$
\begin{aligned}
\|(T + S)x\|_Y = \|Tx + Sx\|_Y &\leq \|Tx\|_Y + \|Sx\|_Y \\
&\leq \|T\|_{B(X,Y)}\|x\|_X + \|S\|_{B(X,Y)}\|x\|_X \\
&= \left[\|T\|_{B(X,Y)} + \|S\|_{B(X,Y)}\right]\|x\|_X,
\end{aligned}
$$

from which it follows that

$$
\|T + S\|_{B(X,Y)} \leq \|T\|_{B(X,Y)} + \|S\|_{B(X,Y)}
$$

as required.   □

Lemma 11.6 *The norm in $B(X, Y)$ is also given by*

$$
\|T\|_{B(X,Y)} = \sup_{\|x\|_X = 1} \|Tx\|_Y. \tag{11.3}
$$

Proof Let us denote by $\|T\|_1$ the value defined in (11.2), and by $\|T\|_2$ the value defined in (11.3). Then, given $x \neq 0$, we have

$$
\left\|T\frac{x}{\|x\|_X}\right\|_Y \leq \|T\|_2 \qquad \text{i.e.} \quad \|Tx\|_Y \leq \|T\|_2\|x\|_X,
$$

and so $\|T\|_1 \leq \|T\|_2$. It is also clear that if $\|x\|_X = 1$ then

$$
\|Tx\|_Y \leq \|T\|_1\|x\|_X = \|T\|_1,
$$

and so $\|T\|_2 \leq \|T\|_1$. It follows that $\|T\|_1 = \|T\|_2$.   □

We also have

$$
\|T\|_{B(X,Y)} = \sup_{\|x\|_X \leq 1} \|Tx\|_Y = \sup_{x \neq 0} \frac{\|Tx\|_Y}{\|x\|_X}; \tag{11.4}
$$

see Exercise 11.1.

We remarked in Section 1.5 that the composition of two linear maps is linear. In a similar way, the composition of two bounded linear maps is another bounded linear map:

$$
T \in B(X, Y), \ S \in B(Y, Z) \qquad \Rightarrow \qquad S \circ T \in B(X, Z), \tag{11.5}
$$

since

$$
\|(S \circ T)x\|_Z \leq \|S\|_{B(Y,Z)}\|Tx\|_Y \leq \|S\|_{B(Y,Z)}\|T\|_{B(X,Y)}\|x\|_X,
$$

and so

$$\|S \circ T\|_{B(X,Z)} \leq \|S\|_{B(Y,Z)} \|T\|_{B(X,Y)}. \tag{11.6}$$

In particular, it follows that if $T \in B(X)$, then $T^n \in B(X)$, where T^n is T composed with itself n times, and $\|T^n\|_{B(X)} \leq \|T\|_{B(X)}^n$.

11.2 Some Examples of Bounded Linear Maps

When there is no room for confusion we will omit the $B(X, Y)$ subscript on the norm of a linear map.

If $T: X \to Y$, then in order to find $\|T\|$ one can try the following: first show that

$$\|Tx\|_Y \leq M \|x\|_X \tag{11.7}$$

for some $M > 0$, i.e. show that T is bounded. It follows that $\|T\| \leq M$ (since $\|T\|$ is the infimum of all M such that (11.7) holds). Then, in order to show that in fact $\|T\| = M$, find an example of a particular $z \in X$ such that

$$\|Tz\|_Y = M \|z\|_X.$$

This shows from the definition in (11.4) that $\|T\| \geq M$ and hence that in fact $\|T\| = M$.

Example 11.7 Consider the right- and left- shift operators $s_r: \ell^2 \to \ell^2$ and $s_l: \ell^2 \to \ell^2$, given by

$$s_r(x) = (0, x_1, x_2, \ldots) \qquad \text{and} \qquad s_l(x) = (x_2, x_3, x_4, \ldots).$$

Both operators are linear with $\|s_r\| = \|s_l\| = 1$.

Proof It is clear that the operators are linear. We have

$$\|s_r(x)\|_{\ell^2}^2 = \sum_{i=1}^{\infty} |x_i|^2 = \|x\|_{\ell^2}^2,$$

so that $\|s_r\| = 1$, and

$$\|s_l(x)\|_{\ell^2}^2 = \sum_{i=2}^{\infty} |x_i|^2 \leq \|x\|_{\ell^2}^2,$$

so that $\|s_l\| \leq 1$. However, if we choose an x with

$$x = (0, x_2, x_3, \ldots)$$

then we have

$$\|s_l(x)\|_{\ell^2}^2 = \sum_{j=2}^{\infty} |x_j|^2 = \|x\|_{\ell^2}^2,$$

and so we must have $\|s_l\| = 1$. ☐

In other cases one may need to do a little more; for example, given the bound $\|Tx\|_Y \le M\|x\|_X$ find a sequence $(z_n) \in X$ such that

$$\frac{\|Tz_n\|_Y}{\|z_n\|_X} \to M$$

as $n \to \infty$, which shows, using (11.4), that $\|T\| \ge M$ and hence that $\|T\| = M$.

Example 11.8 Take $X = L^2(a, b)$ and, for some $g \in C([a, b])$, define the multiplication operator T from $L^2(a, b)$ into itself by

$$[Tf](x) := f(x)g(x) \qquad x \in [a, b].$$

Then T is linear and $\|T\|_{B(X)} = \|g\|_\infty$.

Proof It is clear that T is linear. For the upper bound on $\|T\|_{B(X)}$ observe that

$$\|Tf\|_{L^2}^2 = \int_a^b |f(x)g(x)|^2 \, dx$$

$$= \int_a^b |f(x)|^2 |g(x)|^2 \, dx$$

$$\le \left(\max_{a \le x \le b} |g(x)|^2 \right) \int_a^b |f(x)|^2 \, dx;$$

so

$$\|Tf\|_{L^2} \le \|g\|_\infty \|f\|_{L^2},$$

i.e. $\|T\|_{B(X)} \le \|g\|_\infty$.

Now let s be a point at which $|g|$ attains its maximum. Assume for simplicity that $s \in (a, b)$, and for each $\varepsilon > 0$ consider

$$f_\varepsilon(x) = \begin{cases} 1 & |x - s| < \varepsilon \\ 0 & \text{otherwise}, \end{cases}$$

then

$$\frac{\|Tf_\varepsilon\|^2}{\|f_\varepsilon\|^2} = \frac{1}{2\varepsilon} \int_{s-\varepsilon}^{s+\varepsilon} |g(x)|^2 \, dx \to |g(s)|^2 \qquad \text{as} \qquad \varepsilon \to 0$$

since g is continuous. Therefore in fact

$$\|T\|_{B(X)} = \|g\|_\infty.$$

If $s = a$, then we replace $|x - s| < \varepsilon$ in the definition of f_ε by $a \le x < a + \varepsilon$, and if $s = b$ we replace it by $b - \varepsilon < x \le b$; the rest of the argument is identical. $\qquad\qquad\square$

Example 11.9 Consider the map from $X = C([a, b])$ to $\mathbb{R}$ given by

$$Tf = \int_a^b \phi(x) f(x) \, dx,$$

where $\phi \in C([a, b])$. Then T is linear with $\|T\|_{B(X;\mathbb{R})} = \|\phi\|_{L^1}$.

Proof Linearity is clear. For the upper bound we have

$$|Tf| \le \int_a^b \|f\|_\infty |\phi(x)| \, dx = \|f\|_\infty \|\phi\|_{L^1}.$$

The lower bound is a little more involved. Ideally we would choose

$$f(x) = \text{sign}(\phi(x)) = \begin{cases} +1 & \phi(x) > 0 \\ 0 & \phi(x) = 0 \\ -1 & \phi(x) < 0, \end{cases}$$

as for such an f we have $\|f\|_\infty = 1$ and

$$\int_a^b \phi(x) f(x) \, dx = \int_a^b |\phi(x)| \, dx = \|\phi\|_{L^1} = \|\phi\|_{L^1} \|f\|_\infty;$$

however, the function f will not be continuous if ϕ changes sign. We therefore consider a sequence of continuous functions f_ε that approximate this choice of f, setting

$$f_\varepsilon(x) = \frac{\phi(x)}{|\phi(x)| + \varepsilon}$$

for $\varepsilon > 0$. Since ϕ attains its maximum on $[a, b]$, we have

$$\frac{\|\phi\|_\infty}{\|\phi\|_\infty + \varepsilon} \le \|f_\varepsilon\|_\infty \le 1.$$

Then

$$\int_a^b |\phi(x)| \, dx - \int_a^b \phi(x) f_\varepsilon(x) \, dx = \int_a^b |\phi(x)| - \frac{|\phi(x)|^2}{|\phi(x)| + \varepsilon} \, dx$$

$$= \int_a^b \frac{\varepsilon |\phi(x)|}{|\phi(x)| + \varepsilon}$$

$$\le 2(b - a)\varepsilon.$$

This shows that

$$\left| \int_a^b \phi(x) f_\varepsilon(x) \, dx \right| \geq \left(\|\phi\|_{L^1} - 2(b-a)\varepsilon \right)$$

$$\geq \left(\|\phi\|_{L^1} - 2(b-a)\varepsilon \right) \|f_\varepsilon\|_\infty;$$

letting $\varepsilon \to 0$ guarantees that $\|T\| \geq \|\phi\|_{L^1}$, yielding the required equality. $\qquad\square$

We now consider another example to which we will return a number of times in what follows.

Example 11.10 Take $X := L^2(a,b)$ and $K \in C([a,b] \times [a,b])$. Define $T : X \to X$ as the integral operator

$$(Tf)(x) = \int_a^b K(x,y) f(y) \, ds \qquad \text{for all} \qquad x \in [a,b].$$

Then T is a bounded linear map with

$$\|T\|_{B(X)}^2 \leq \int_a^b \int_a^b |K(x,y)|^2 \, dx \, dy. \tag{11.8}$$

Proof The operator T is clearly linear, and

$$\|Tf\|_{L^2}^2 = \int_a^b \left| \int_a^b K(x,y) f(y) \, dy \right|^2 dx$$

$$\leq \int_a^b \left[\int_a^b |K(x,y)|^2 \, dy \int_a^b |f(y)|^2 \, dy \right] dx$$

$$= \left(\int_a^b \int_a^b |K(x,y)|^2 \, dy \, dx \right) \|f\|_{L^2}^2,$$

where we have used the Cauchy–Schwarz inequality to move from the first to the second line. This yields (11.8).

This example can be extended to treat the case $K \in L^2((a,b) \times (a,b))$; we now need to appeal to the Fubini–Tonelli Theorem (Theorem B.9) to justify the integration steps. $\qquad\square$

Note that this upper bound on the operator norm can be strict; see Exercise 11.10.

11.3 Completeness of $B(X, Y)$ When Y Is Complete

The space $B(X, Y)$ (with the norm defined in (11.2)) is a Banach space whenever Y is a Banach space. Remarkably this does not depend on whether the space X is complete or not.

Theorem 11.11 *If X is a normed space and Y is a Banach space, then $B(X, Y)$ is a Banach space.*

Proof Given any Cauchy sequence (T_n) in $B(X, Y)$ we need to show that $T_n \to T$ for some $T \in B(X, Y)$. Since (T_n) is Cauchy, given $\varepsilon > 0$ there exists an N_ε such that

$$\|T_n - T_m\|_{B(X,Y)} \le \varepsilon \qquad \text{for all} \quad n, m \ge N_\varepsilon. \qquad (11.9)$$

We now show that for every fixed $x \in X$ the sequence $(T_n x)$ is Cauchy in Y. This follows since

$$\|T_n x - T_m x\|_Y = \|(T_n - T_m)x\|_Y \le \|T_n - T_m\|_{B(X,Y)} \|x\|_X, \qquad (11.10)$$

and (T_n) is Cauchy in $B(X, Y)$. Since Y is complete, it follows that

$$T_n x \to z$$

for some $z \in Y$, which depends on x. We can therefore define a mapping $T : X \to Y$ by setting $Tx = z$.

Now that we have identified our expected limit we need to make sure that $T \in B(X, Y)$ and that $T_n \to T$ in $B(X, Y)$.

First, T is linear since for any $x, y \in X, \alpha, \beta \in \mathbb{K}$,

$$T(\alpha x + \beta y) = \lim_{n \to \infty} T_n(\alpha x + \beta y) = \alpha \lim_{n \to \infty} T_n x + \beta \lim_{n \to \infty} T_n y$$
$$= \alpha T x + \beta T y.$$

To show that T is bounded take $n, m \ge N_\varepsilon$ (from (11.9)) in (11.10), and let $m \to \infty$. Since $T_m x \to T x$ this limiting process shows that

$$\|T_n x - T x\|_Y \le \varepsilon \|x\|_X. \qquad (11.11)$$

Since (11.11) holds for every x, it follows that

$$\|T_n - T\|_{B(X,Y)} \le \varepsilon \qquad \text{for } n \ge N_\varepsilon. \qquad (11.12)$$

In particular, $T_{N_\varepsilon} - T \in B(X, Y)$, and since $B(X, Y)$ is a vector space and we have $T_{N_\varepsilon} \in B(X, Y)$, it follows that $T \in B(X, Y)$. Finally, (11.12) also shows that $T_n \to T$ in $B(X, Y)$. $\qquad \square$

11.4 Kernel and Range

Given a linear map $T : X \to Y$, recall (see Definition 1.20) that we define its kernel to be

$$\mathrm{Ker}(T) := \{x \in X : Tx = 0\}$$

and its range to be

$$\mathrm{Range}(T) := \{y \in Y : y = Tx \text{ for some } x \in X\}.$$

Lemma 11.12 *If $T \in B(X, Y)$, then* $\mathrm{Ker}\, T$ *is a closed linear subspace of* X.

Proof Given any $x, y \in \mathrm{Ker}(T)$ we have

$$T(\alpha x + \beta y) = \alpha Tx + \beta Ty = 0.$$

Furthermore, if $x_n \to x$ and $Tx_n = 0$ (i.e. $x_n \in \mathrm{Ker}(T)$), then since T is continuous $Tx = \lim_{n \to \infty} Tx_n = 0$, so $\mathrm{Ker}(T)$ is closed.

For a more topological proof, we could simply note that $\mathrm{Ker}(T) = T^{-1}(\{0\})$ and so is the preimage under T of $\{0\}$, which is a closed subset of Y, and so $\mathrm{Ker}(T)$ is closed in X since T is continuous (Lemma 2.13). □

While $T \in B(X, Y)$ implies that $\mathrm{Ker}(T)$ is closed, the same is not true for the range of T: it need not be closed. Indeed, consider the map from ℓ^2 into itself given by

$$Tx = \left(x_1, \frac{x_2}{2}, \frac{x_3}{3}, \frac{x_4}{4}, \ldots \right). \tag{11.13}$$

Since

$$\|Tx\|_{\ell^2}^2 = \sum_{j=1}^{\infty} \frac{1}{j^2} |x_j|^2 \le \sum_{j=1}^{\infty} |x_j|^2 = \|x\|_{\ell^2}^2,$$

we have $\|T\| \le 1$ and T is bounded. Now consider $y^{(n)} \in \mathrm{Range}(T)$, where

$$y^{(n)} = T(\underbrace{1, 1, 1, \ldots, 1}_{\text{first } n \text{ terms}}, 0 \ldots) = \left(1, \frac{1}{2}, \frac{1}{3}, \ldots, \frac{1}{n}, 0, \ldots \right).$$

We have $y^{(n)} \to y$, where y is the element of ℓ^2 with $y_j = j^{-1}$ (observe that $y \in \ell^2$ since $\sum_{j=1}^{\infty} j^{-2} < \infty$). However, there is no $x \in \ell^2$ such that $T(x) = y$: the only candidate is $x = (1, 1, 1, \ldots)$, but this is not in ℓ^2 since its ℓ^2 norm is not finite, so $y \notin \mathrm{Range}(T)$.

11.5 Inverses and Invertibility

We have already discussed the existence of inverses for linear maps between general vector spaces in Section 1.5, and we showed that a bijective linear map $T: X \to Y$ has an inverse $T^{-1}: Y \to X$ that is also linear (Lemma 1.24).

However, in the context of bounded linear maps between normed spaces there is no guarantee that the inverse is also bounded. As a somewhat artificial example, consider the subspace c_{00} of ℓ^∞ consisting of sequences with only a finite number of non-zero terms. Then the linear map $T: c_{00} \to c_{00}$ defined by setting

$$Te^{(n)} = \frac{1}{n} e^{(n)}$$

is bijective and bounded (it has norm 1). However, the inverse map is not bounded, since it maps $e^{(n)}$ to $ne^{(n)}$. To avoid this sort of pathology we incorporate boundedness into our requirement of invertibility. This gives rise to the somewhat strange situation in which $T \in B(X, Y)$ can 'have an inverse' but not be 'invertible'.

Definition 11.13 An operator $T \in B(X, Y)$ is *invertible* if there exists an $S \in B(Y, X)$ such that[2] $ST = I_X$ and $TS = I_Y$, and then $T^{-1} = S$ is the inverse of T.

Before continuing we make the (almost trivial but very useful) observation that for any non-zero $\alpha \in \mathbb{K}$, T is invertible if and only if αT is invertible.

We now relate the concept of invertibility from this definition and the usual notion of an inverse when $T: X \to Y$ is a bijection. We will see later in Theorem 23.2 that if X and Y *are both Banach spaces* and $T: X \to Y$ is a bijection, then T^{-1} is automatically bounded.

Lemma 11.14 *Suppose that X and Y are both normed spaces. Then for any $T \in B(X, Y)$, the following are equivalent:*

(i) *T is invertible;*
(ii) *T is a bijection and $T^{-1} \in B(Y, X)$;*
(iii) *T is onto and for some $c > 0$*

$$\|Tx\|_Y \geq c\|x\|_X \qquad \text{for every } x \in X. \tag{11.14}$$

---

[2] As before, we denote by $I_X: X \to X$ the identity map from X to itself.

Proof We showed in Lemma 1.24 that (i) implies (ii) apart from the bounded-ness of T^{-1}, but this is part of the definition of invertibility in Definition 11.13. That (ii) implies (i) is clear.

We now show that (i) $\Rightarrow$ (iii). If T is invertible, then it is onto (by (i) $\Rightarrow$ (ii)), and since $T^{-1} \in B(Y, X)$ we have $\|T^{-1}y\|_X \leq M\|y\|_Y$ for every $y \in Y$, for some $M > 0$; choosing $y = Tx$ we obtain $\|x\|_X \leq M\|Tx\|_Y$, which yields (iii) with $c = 1/M$.

Finally, we show that (iii) $\Rightarrow$ (ii). Note first that the lower bound on Tx in (iii) implies that T is one-to-one, since if $Tx = Tx'$, then

$$0 = \|T(x - x')\|_Y \geq c\|x - x'\|_X \quad \Rightarrow \quad x = y.$$

Since T is assumed to be onto, it is therefore a bijection. That the resulting inverse map $T^{-1} \colon Y \to X$ is bounded follows if we set $x = T^{-1}y$ in (11.14), since this yields $\|y\|_Y \geq c\|T^{-1}y\|_X$. $\qquad\square$

Corollary 11.15 *If X is finite-dimensional, then a linear operator $T \colon X \to X$ is invertible if and only if* $\mathrm{Ker}(T) = \{0\}$.

Proof Since X is finite-dimensional, we can use Lemma 1.22, which guaran-tees that T is injective if and only if it is surjective; injectivity is equivalent to bijectivity in this case, and T is injective if and only if $\mathrm{Ker}(T) = \{0\}$ (Lemma 1.21). Under this condition T^{-1} exists, and is linear by Lemma 1.24. Lemma 11.2 guarantees that T^{-1} is bounded, and so T is invertible. $\qquad\square$

We can use the equivalence between (i) and (iii) in Lemma 11.14 to prove the following useful result.

Lemma 11.16 *If X and Y are Banach spaces and $T \in B(X, Y)$ is invertible, then so is $T + S$ for any $S \in B(X, Y)$ with $\|S\|\|T^{-1}\| < 1$. Consequently, the subset of $B(X, Y)$ consisting of invertible operators is open.*

For an alternative (and more 'traditional') proof see Exercises 11.5 and 11.6.

Proof Suppose that $T \in B(X, Y)$ is invertible; then by (i)$\Rightarrow$(iii) we know that T is onto and that

$$\|Tx\|_Y \geq \frac{1}{\|T^{-1}\|}\|x\|_X.$$

We will show that for any $S \in B(X, Y)$ with $\|S\|\|T^{-1}\| = \alpha < 1$, $T + S$ is invertible.

First we show that $T + S$ is onto: given $y \in Y$, we want to ensure that there is an $x \in X$ such that

$$(T + S)x = y.$$

Consider the map $\mathfrak{I}\colon X \to X$ defined by setting

$$x \mapsto \mathfrak{I}(x) := T^{-1}(y - Sx).$$

Then

$$
\begin{aligned}
\|\mathfrak{I}(x) - \mathfrak{I}(x')\|_X &= \left\| T^{-1}(y - Sx) - T^{-1}(y - Sx') \right\|_X \\
&= \left\| T^{-1}S(x - x') \right\|_X \\
&\le \|T^{-1}\| \|S\| \|x - x'\|_X \\
&= \alpha \|x - x'\|_X,
\end{aligned}
$$

where $\alpha < 1$ by assumption. Since X is a Banach space, we can use the Contraction Mapping Theorem (Theorem 4.15) to ensure that there is a unique $x \in X$ such that $x = \mathfrak{I}(x)$, i.e. such that $x = T^{-1}(y - Sx)$. Applying T to both sides guarantees that $y = (T + S)x$ and so $T + S$ is onto.

We now just have to check that $T + S$ is bounded below in the sense of (iii). Note that since

$$\|S\| \|T^{-1}\| < 1$$

we have

$$\frac{1}{\|T^{-1}\|} - \|S\| = c > 0.$$

Therefore

$$
\begin{aligned}
\|(T + S)x\|_Y &\ge \|Tx\|_Y - \|Sx\|_Y \\
&\ge \frac{1}{\|T^{-1}\|} \|x\|_X - \|S\| \|x\|_X = c\|x\|_X.
\end{aligned}
$$

Now using (iii)$\Rightarrow$(i) we deduce that $(T + S)$ is invertible. $\qquad\square$

We finish this section by considering inverses of products.

Lemma 11.17 *If $T \in B(X, Y)$ and $S \in B(Y, Z)$ are invertible, then so is $ST \in B(X, Z)$, and $(ST)^{-1} = T^{-1}S^{-1}$.*

Proof We have $T^{-1} \in B(Y, X)$ and $S^{-1} \in B(Z, Y)$, so (see (11.6)) $T^{-1}S^{-1} \in B(Z, X)$, and

$$T^{-1}S^{-1}ST = I_X \qquad \text{and} \qquad STT^{-1}S^{-1} = I_Z. \qquad\square$$

Two operators $T, S \in L(X, X)$ commute if $TS = ST$. We make the simple observation that if S and T commute and T is invertible, then S commutes with T^{-1}; indeed, since $ST = TS$, we have

$$T^{-1}[ST]T^{-1} = T^{-1}[TS]T^{-1} \qquad \Rightarrow \qquad T^{-1}S = ST^{-1}.$$

Proposition 11.18 *If* $\{T_1, \ldots, T_n\}$ *are commuting operators in* $B(X)$, *then*

$$T_1 \cdots T_n$$

is invertible if and only if every T_j, $j = 1, \ldots, n$, *is invertible.*

Proof One direction follows from Lemma 11.17 and induction. For the other direction, suppose that $\mathcal{T} = T_1 \cdots T_n$ is invertible; since T_1 commutes with $\mathcal{T}$ it also commutes with $\mathcal{T}^{-1}$, and so

$$T_1[\mathcal{T}^{-1}T_2 \cdots T_n] = \mathcal{T}^{-1}T_1 T_2 \cdots T_n = \mathcal{T}^{-1}\mathcal{T} = I.$$

Since $\{T_1, \ldots, T_n\}$ commute we have

$$[\mathcal{T}^{-1}T_2 \cdots T_n]T_1 = \mathcal{T}^{-1}\mathcal{T} = I$$

as well. All the operators are bounded, so T_1 is invertible.

For other values of j we can use the fact that the $\{T_j\}$ commute to reorder the factors of $\mathcal{T}$ so that the first is T_j. We can now apply the same argument to show that T_j is invertible. $\qquad\square$

No such result is true if the operators do not commute: if we consider the left and right shifts $\mathfrak{s}_l$ and $\mathfrak{s}_r$ on ℓ^2, then $\mathfrak{s}_l\mathfrak{s}_r$ is the identity (so clearly invertible), but neither $\mathfrak{s}_l$ nor $\mathfrak{s}_r$ are invertible. (The left shift $\mathfrak{s}_l$ is not injective, and the right shift $\mathfrak{s}_r$ is not surjective.)

Exercises

11.1   Show that

$$\|T\|_{B(X,Y)} = \sup_{\|x\|_X \leq 1} \|Tx\|_Y$$

and

$$\|T\|_{B(X,Y)} = \sup_{x \neq 0} \frac{\|Tx\|_Y}{\|x\|_X}.$$

11.2   Let $X = C_b([0, \infty))$ with the supremum norm. Show that the map $T \colon X \to X$ defined by setting $[Tf](0) = f(0)$ and

$$[Tf](x) = \frac{1}{x} \int_0^x f(s)\, ds$$

is linear and bounded with $\|T\|_{B(X)} = 1$.

11.3 Show that $T \in L(X, Y)$ is bounded if and only if

$$\sum_{j=1}^{\infty} Tx_j = T\left(\sum_{j=1}^{\infty} x_j\right)$$

whenever the sum on the right-hand side converges. (Pryce, 1973)

11.4 Suppose that $(T_n) \in B(X, Y)$ and $(S_n) \in B(Y, Z)$ are such that $T_n \to T$ and $S_n \to S$. Show that $S_n T_n \to ST$ in $B(X, Z)$.

11.5 Suppose that X is a Banach space and $T \in B(X)$ is such that

$$\sum_{j=1}^{\infty} \|T^n\|_{B(X)} < \infty.$$

Show that

$$(I - T)^{-1} = I + T + T^2 + \cdots = \sum_{j=0}^{\infty} T^j.$$

(This is known as the Neumann series for $(I - T)^{-1}$.) In the case that $\|T\| < 1$ deduce that

$$\|(I - T)^{-1}\| \le (1 - \|T\|)^{-1}.$$

11.6 Use the result of the previous exercise to show that if X and Y are Banach spaces and $T \in B(X, Y)$ is invertible, then so is $T + S$ for any $S \in B(X, Y)$ with $\|S\|\|T^{-1}\| < 1$, and then

$$\|(T + S)^{-1}\| \le \frac{\|T^{-1}\|}{1 - \|S\|\|T^{-1}\|}. \qquad (11.15)$$

(This is the usual way to prove Lemma 11.16.)

11.7 Suppose that $K \in C([a, b] \times [a, b])$ with $\|K\|_\infty \le M$. Show that the operator T defined on $X := C([a, b])$ by setting

$$[Tf](x) = \int_a^x K(x, y) f(y)\, dy$$

is a bounded linear operator from X into itself. Show by induction that

$$|T^n f(x)| \le M^n \|f\|_\infty \frac{(x - a)^n}{n!}$$

and use the result of Exercise 11.5 to deduce that the equation

$$f(x) = g(x) + \lambda \int_a^x K(x, y) f(y) \, dy \qquad (11.16)$$

has a unique solution $f \in X$ for any $g \in X$ and any $\lambda \in \mathbb{R}$.

11.8   Show that if $T \in B(X, Y)$ is a bijection, then T is an isometry if and only if $\|T\|_{B(X,Y)} = \|T^{-1}\|_{B(Y,X)} = 1$.

11.9   Suppose that X is a Banach space, Y a normed space, and take some $T \in B(X, Y)$. Show that if there exists $\alpha > 0$ such that $\|Tx\| \geq \alpha \|x\|$, then Range(T) is closed. (Rynne and Youngson, 2008)

11.10 Let $e_1(x) = 1/\sqrt{2}$ and $e_2(x) = \sqrt{3/2}x$, which are orthonormal functions in $L^2(-1, 1)$, and set

$$K(x, y) = 1 + 6xy = 2e_1(x)e_1(y) + 4e_2(x)e_2(y).$$

Show that the norm of the operator $T : L^2(-1, 1) \to L^2(-1, 1)$ defined by setting

$$Tf(t) := \int_{-1}^1 K(x, y) f(y) \, dy$$

is strictly less than $\|K\|_{L^2((-1,1)\times(-1,1))}$.

11.11 Show that if X is a Banach space and $T \in B(X)$, then

$$\exp(T) := \sum_{k=0}^{\infty} \frac{T^k}{k!}$$

defines an element of $B(X)$.

12

Dual Spaces and the Riesz Representation Theorem

If X is a normed space over $\mathbb{K}$, then a linear map from X into $\mathbb{K}$ is called a *linear functional* on X. Linear functionals therefore take elements of X (which could be a very abstract space) and return a number. It is one of the central observations in functional analysis that understanding all of these linear functionals on X (the 'dual space' X^*) gives us a good understanding of the space X itself.

In this chapter we concentrate on linear functionals on Hilbert spaces and show that any bounded linear functional $f \colon H \to \mathbb{K}$ must actually be of the form

$$f(x) = (x, y)$$

for some $y \in H$ (this is the Riesz Representation Theorem). We also discuss a more geometric interpretation of this result and show how linear functionals are closely related to hyperplanes (sets of codimension one) in H.

12.1 The Dual Space

We denote by X^* the collection of all *bounded* linear functionals on X, i.e. $X^* = B(X, \mathbb{K})$; we equip X^* with the norm

$$\|f\|_{X^*} = \sup_{\|x\|=1} |f(x)| \qquad \text{for each } f \in X^*,$$

i.e. the standard norm in $B(X, \mathbb{K})$. The space X^* is called[1] the *dual (space) of X*.

---

[1] Strictly speaking there is a distinction to be made between the 'algebraic dual' of X, which is the collection of all linear functionals on X, and the 'normed dual' of X, which is the normed space formed by this collection of all *bounded* linear functionals, equipped with the $B(X, \mathbb{K})$ norm.

Example 12.1 Take $X = \mathbb{R}^n$. Then if $e^{(j)}$ is the jth coordinate vector, we have $x = \sum_{j=1}^n x_j e^{(j)}$, and so if $f : \mathbb{R}^n \to \mathbb{R}$ is linear, then

$$f(x) = f\left(\sum_{j=1}^n x_j e^{(j)}\right) = \sum_{j=1}^n x_j f(e^{(j)});$$

if we write y for the element of $\mathbb{R}^n$ with $y_j = f(e^{(j)})$, then we can write this as

$$f(x) = \sum_{j=1}^n x_j y_j = (x, y). \tag{12.1}$$

So with any $f \in (\mathbb{R}^n)^*$ we can associate some $y \in \mathbb{R}^n$ such that (12.1) holds; since

$$|f(x)| \le \|y\|_{\ell^2} \|x\|_{\ell^2} \qquad \text{and} \qquad |f(y)| = \|y\|_{\ell^2}^2,$$

it follows that

$$\|f\|_{(\mathbb{R}^n)^*} = \|y\|_{\ell^2}.$$

In this way $(\mathbb{R}^n)^* \equiv \mathbb{R}^n$.

Example 12.2 Let X be $L^2(a, b)$, take any $\phi \in L^2(a, b)$, and consider the map $f : L^2(a, b) \to \mathbb{R}$ defined by setting

$$f(u) = \int_a^b \phi(t) u(t) \, dt.$$

Then

$$|f(u)| = \left| \int_a^b \phi(t) u(t) \, dt \right| = |(\phi, u)_{L^2}| \le \|\phi\|_{L^2} \|u\|_{L^2},$$

using the Cauchy–Schwarz inequality, and so $f \in X^*$ with

$$\|f\|_{X^*} \le \|\phi\|_{L^2}.$$

If we choose $u = \phi / \|\phi\|_{L^2}$, then $\|u\|_{L^2} = 1$ and

$$|f(u)| = \int_a^b \frac{|\phi(t)|^2}{\|\phi\|_{L^2}} \, dt = \|\phi\|_{L^2},$$

and so $\|f\|_{X^*} = \|\phi\|_{L^2}$.

This example shows that any element u of L^2 gives rise to a bounded linear functional on L^2, which is defined by taking the inner product with u. It is natural to ask whether any bounded linear functional on L^2 can be obtained in this way, and remarkably this is true, not only for L^2 but for any Hilbert space. We will prove this result, the Riesz Representation Theorem, in the next section: it is one of the most useful fundamental properties of Hilbert spaces.

12.2 The Riesz Representation Theorem

In an abstract Hilbert space H we can generalise Examples 12.1 and 12.2 to give a very important example of a linear functional on H (i.e. an element of H^*). In fact we will prove that any element of H^* must be of this particular form.

Lemma 12.3 *If H is a Hilbert space over $\mathbb{K}$ and $y \in H$, then the map $f_y \colon H \to \mathbb{K}$ defined by setting*

$$f_y(x) = (x, y) \tag{12.2}$$

is an element of H^* with $\|f_y\|_{H^*} = \|y\|_H$.

Note that this shows in particular that $\|x\| = \max_{\|y\|=1} |(x, y)|$.

Proof The map f_y is linear since the inner product is always linear in its first argument (although conjugate-linear in its second argument when $\mathbb{K} = \mathbb{C}$). Using the Cauchy–Schwarz inequality we have

$$|f_y(x)| = |(x, y)| \le \|x\| \|y\|$$

and so it follows that $f_y \in H^*$ with $\|f_y\|_{H^*} \le \|y\|$.
Choosing $x = y$ in (12.2) shows that

$$|f_y(y)| = |(y, y)| = \|y\|^2$$

and hence $\|f_y\|_{H^*} = \|y\|$. □

The *Riesz map* $R \colon H \to H^*$ given by setting $R(y) = f_y$ is therefore an isometry from H into H^*; it is linear when H is real, and conjugate-linear when H is complex (because in this case $y \mapsto (x, y)$ is conjugate-linear).

The Riesz Representation Theorem shows that the map R is onto, so that this example can be 'reversed', i.e. every linear functional on H can be realised as an inner product with some element $y \in H$.

Theorem 12.4 (Riesz Representation Theorem) *If H is a Hilbert space, then for every $f \in H^*$ there exists a unique element $y \in H$ such that*

$$f(x) = (x, y) \qquad \text{for all} \qquad x \in H; \tag{12.3}$$

and $\|y\|_H = \|f\|_{H^*}$. In particular, the Riesz map $R \colon H \to H^*$ defined via (12.2) by setting $R(y) = f_y$ maps H onto H^*.

Note if H is real, then R is a bijective linear isometry and $H \equiv H^*$.

Proof Let $K = \operatorname{Ker} f$; since f is bounded this is a closed linear subspace of H (Lemma 11.12). We claim that $K^\perp$ is a one-dimensional linear subspace of H. Indeed, given $u, v \in K^\perp$ we have

$$f\left(f(u)v - f(v)u \right) = f(u)f(v) - f(v)f(u) = 0, \qquad (12.4)$$

since f is linear. Since $u, v \in K^\perp$, it follows that $f(u)v - f(v)u \in K^\perp$, while (12.4) shows that $f(u)v - f(v)u \in K$. Since $K \cap K^\perp = \{0\}$, it follows that

$$f(u)v - f(v)u = 0,$$

and so u and v are linearly dependent.

Therefore we can choose $z \in K^\perp$ such that $\|z\| = 1$, and use Proposition 10.4 to decompose any $x \in H$ as

$$x = (x, z)z + w \qquad \text{with} \qquad w \in (K^\perp)^\perp = K,$$

where we have used Lemma 10.5 and the fact that K is closed to guarantee that $(K^\perp)^\perp = K$. Thus

$$f(x) = (x, z)f(z) = (x, \overline{f(z)}z),$$

and setting $y = \overline{f(z)}z$ we obtain (12.3).

To show that this choice of y is unique, suppose that

$$(x, y) = (x, \hat{y}) \qquad \text{for all} \qquad x \in H.$$

Then $(x, y - \hat{y}) = 0$ for all $x \in H$; taking $x = y - \hat{y}$ gives $\|y - \hat{y}\|^2 = 0$.

Finally, Lemma 12.3 shows that $\|y\|_H = \|f\|_{H^*}$. $\square$

This gives a way to rephrase the result of Corollary 10.2 somewhat more elegantly, by letting $f(u) = (u, v)$. (We will obtain a similar result later in a more general context as Theorem 21.2.)

Corollary 12.5 *Suppose that A is a non-empty closed convex subset of a real Hilbert space, and $x \notin A$. Then there exists $f \in H^*$ such that*

$$f(a) + d^2 \leq f(x) \qquad \text{for every } a \in A,$$

where $d = \operatorname{dist}(x, A)$; see Figure 12.1.

The Riesz Representation Theorem also allows us to form a geometric picture of the action of a linear functional on a Hilbert space. Given any $f \in H^*$, the proof of Theorem 12.4 shows that the value of f at any point in $y \in H$ is determined by its projection (y, z) onto any normal z to the set $\operatorname{Ker}(f)$.

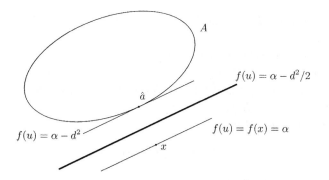

Figure 12.1 Separating x from A using a linear functional. In the figure, $\hat{a}$ is the closest point to x in A; cf. Figure 10.2.

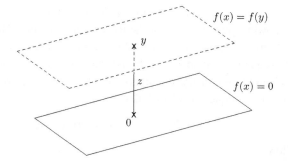

Figure 12.2 Illustration of the foliation of a Hilbert space by the sets $f(x) = c$, which are translated copies of Ker(f). In the Riesz Representation Theorem the linear functional f is reconstructed as the inner product with an element z in the direction normal to Ker(f) = $\{x : f(x) = 0\}$.

In this way the space H is 'foliated' by translated copies of Ker(f), i.e. $\{x : f(x) = c, c \in \mathbb{K}\}$; see Figure 12.2.

The space Ker(f) is an example of a hyperplane: a linear subspace that has 'codimension 1'. We will make this more precise when we consider the situation in a more general setting in Chapter 21.

Exercises

12.1 Show that there are discontinuous linear functionals on any infinite-dimensional normed space $(X, \| \cdot \|)$.

12.2 Show that if V is a finite-dimensional space, then dim $V^* = $ dim V. (Given a basis $\{e_j\}_{j=1}^n$ for V show that the set of linear functionals $\{\phi_i\}_{i=1}^n$ with $\phi_i(e_j) = \delta_{ij}$ form a basis for V^*.) (Kreyszig, 1978)

12.3 Suppose that H is a real Hilbert space and that $B: H \times H \to \mathbb{R}$ is such
that
  (i) $B(x, y)$ is linear in x and in y;
  (ii) $|B(x, y)| \leq c\|x\|\|y\|$ for some $c > 0$, for all $x, y \in H$;
  (iii) $|B(x, x)| \geq b\|x\|^2$ for some $b \in \mathbb{R}$, for all $x \in H$; and
  (iv) $B(x, y) = B(y, x)$ for every $x, y \in H$.
  Show that $u \in H$ minimises

$$F(u) := \frac{1}{2}B(u, u) - f(u),$$

where $f \in H^*$, if and only if

$$B(u, v) = f(v) \qquad \text{for every } v \in H.$$

[Hint: consider $\phi(t) := F(u + tv)$.] (Zeidler, 1995)

12.4 The following generalisation of the Riesz Representation Theorem, the
Lax–Milgram Lemma, is very useful in the analysis of linear partial differential equations. Suppose that H and B are as in the previous exercise,
but without the symmetry assumption (iv). Show that for every $f \in H^*$
there exists a unique $y \in H$ such that

$$f(x) = B(x, y) \qquad \text{for every } x \in H \tag{12.5}$$

as follows:
  (i) show that for each fixed $y \in H$ the map $x \mapsto B(x, y)$ is a bounded
  linear functional on H, so that $B(x, y) = (x, w)$ for some $w \in H$
  by the Riesz Representation Theorem;
  (ii) define $A: H \to H$ by setting $Ay = w$ and show that $A \in B(H)$;
  (iii) given $f \in H$, use the Riesz Representation Theorem to find $z \in H$
  such that $f(x) = (x, z)$ for every $x \in H$ and rewrite (12.5) as

$$(x, z) = (x, Ay),$$

  i.e. $Ay = z$. For any choice of $\varrho \in \mathbb{R}$ this equality holds if and
  only if $y = y - \varrho(Ay - z)$. Use the Contraction Mapping Theorem
  applied to the map $T: H \to H$ defined by setting

$$Ty := y - \varrho(Ay - z)$$

  to show that if ϱ is sufficiently small T has a fixed point, which is
  the required solution y of our original equation. (To show that T is
  a contraction consider $\|Ty - Ty'\|^2$.)

13

The Hilbert Adjoint of a Linear Operator

We now use the Riesz Representation Theorem to define the adjoint of a linear operator $T \colon H \to K$, where H and K are Hilbert spaces; this is a linear operator $T^* \colon K \to H$ such that

$$(Tx, y)_K = (x, T^*y)_H \qquad x \in H, \ y \in K.$$

Properties of operators and their adjoints are closely related, and we will be able to develop a good spectral theory for operators $T \colon H \to H$ that are 'self-adjoint', i.e. for which $T = T^*$.

13.1 Existence of the Hilbert Adjoint

We let H and K be Hilbert spaces with inner products $(\cdot, \cdot)_H$ and $(\cdot, \cdot)_K$ respectively; these induce corresponding norms $\| \cdot \|_H$ and $\| \cdot \|_K$.

Theorem 13.1 *Let H and K be Hilbert spaces and $T \in B(H, K)$. Then there exists a unique operator $T^* \in B(K, H)$, which we call the* (Hilbert) adjoint *of T, such that*

$$(Tx, y)_K = (x, T^*y)_H \tag{13.1}$$

for all $x \in H$, $y \in K$. Furthermore, $T^{**} := (T^*)^* = T$ and

$$\|T^*\|_{B(K,H)} = \|T\|_{B(H,K)}.$$

Proof Let $y \in K$ and consider $f \colon H \to \mathbb{K}$ defined by $f(x) := (Tx, y)_K$. Then clearly f is linear and

$$
\begin{aligned}
|f(x)| &= |(Tx, y)_K| \\
&\le \|Tx\|_K \|y\|_K \\
&\le \|T\|_{B(H,K)} \|x\|_H \|y\|_K.
\end{aligned}
$$

159

It follows that $f \in H^*$, and so by the Riesz Representation Theorem there exists a unique $z \in H$ such that

$$(Tx, y)_K = (x, z)_H \qquad \text{for all} \qquad x \in H.$$

We now define $T^* \colon K \to H$ by setting $T^*y = z$. By definition we have

$$(Tx, y)_K = (x, T^*y)_H \qquad \text{for all} \qquad x \in H, \ y \in K,$$

i.e. (13.1). However, it remains to show that $T^* \in B(K, H)$. First, T^* is linear since for all $\alpha, \beta \in \mathbb{K}$, $y_1, y_2 \in Y$,

$$
\begin{aligned}
(x, T^*(\alpha y_1 + \beta y_2))_H &= (Tx, \alpha y_1 + \beta y_2)_K \\
&= \overline{\alpha}(Tx, y_1)_K + \overline{\beta}(Tx, y_2)_K \\
&= \overline{\alpha}(x, T^*y_1)_H + \overline{\beta}(x, T^*y_2)_H \\
&= (x, \alpha T^*y_1 + \beta T^*y_2)_H,
\end{aligned}
$$

i.e. $T^*(\alpha y_1 + \beta y_2) = \alpha T^*y_1 + \beta T^*y_2$. To show that T^* is bounded, we can write

$$
\begin{aligned}
\|T^*y\|_H^2 &= (T^*y, T^*y)_H \\
&= (TT^*y, y)_K \\
&\leq \|TT^*y\|_K \|y\|_K \\
&\leq \|T\|_{B(H,K)} \|T^*y\|_H \|y\|_K.
\end{aligned}
$$

If $\|T^*y\|_H \neq 0$, then we can divide both sides by $\|T^*y\|_H$ to obtain

$$\|T^*y\|_H \leq \|T\|_{B(H,K)} \|y\|_K,$$

while this final inequality is trivially true if $\|T^*y\|_H = 0$. Thus $T^* \in B(K, H)$ with $\|T^*\|_{B(K,H)} \leq \|T\|_{B(H,K)}$.

We now show that $T^{**} := (T^*)^* = T$, from which can obtain equality of the norms of T and T^*. Indeed, if we have $T^{**} = T$, then it follows that

$$\|T\|_{B(H,K)} = \|(T^*)^*\|_{B(H,K)} \leq \|T^*\|_{B(K,H)},$$

which combined with $\|T^*\|_{B(K,H)} \leq \|T\|_{B(H,K)}$ shows that

$$\|T^*\|_{B(K,H)} = \|T\|_{B(H,K)}.$$

To prove that $T^{**} = T$, note that since $T^* \in B(K, H)$ it follows that $(T^*)^* \in B(H, K)$, and by definition for all $x \in K$, $y \in H$ we have

$$
\begin{aligned}
(x, (T^*)^*y)_K &= (T^*x, y)_H \\
&= \overline{(y, T^*x)_H}
\end{aligned}
$$

$$= \overline{(Ty, x)_K}$$
$$= (x, Ty)_K,$$

i.e. $(T^*)^* y = Ty$ for all $y \in H$, which is exactly $(T^*)^* = T$.

Finally, we show that the requirement that (13.1) holds defines T^* uniquely. Suppose that $T^*, \hat{T} : K \to H$ are such that

$$(x, T^* y)_H = (x, \hat{T} y)_H \qquad \text{for all} \quad x \in H, \; y \in K.$$

Then for each $y \in K$ we have

$$(x, (T^* - \hat{T})y)_H = 0 \qquad \text{for every } x \in H;$$

this shows that $(T^* - \hat{T})y = 0$ for each $y \in K$, i.e. that $\hat{T} = T^*$. $\qquad \square$

Before we give some examples we first prove some simple properties of the adjoint operation, and give an important definition.

Lemma 13.2 *Let* H, K, *and* J *be Hilbert spaces,* $R, S \in B(H, K)$, *and* $T \in B(K, J)$; *then*

(a) $(\alpha R + \beta S)^* = \overline{\alpha} R^* + \overline{\beta} S^*$ *and*
(b) $(TR)^* = R^* T^*$.

Proof (a) For any $x \in H$, $y \in K$ we have

$$\begin{aligned}
(x, (\alpha R + \beta S)^* y)_H &= ((\alpha R + \beta S)x, y)_K \\
&= \alpha (Rx, y)_K + \beta (Sx, y)_K \\
&= \alpha (x, R^* y)_H + \beta (x, S^* y)_H \\
&= (x, \overline{\alpha} R^* y + \overline{\beta} S^* y)_H = (x, (\overline{\alpha} R^* + \overline{\beta} S^*)y)_H;
\end{aligned}$$

the uniqueness argument from Theorem 13.1 now guarantees that (a) holds.

(b) We have

$$(x, (TR)^* y)_H = (TRx, y)_J = (Rx, T^* y)_K = (x, R^* T^* y)_H,$$

and again we use the uniqueness argument from Theorem 13.1. $\qquad \square$

The following definition should seem natural.

Definition 13.3 If H is a Hilbert space and $T \in B(H)$, then T is *self-adjoint* if $T = T^*$.

Equivalently $T \in B(H)$ is self-adjoint if and only if it is *symmetric*, i.e.

$$(x, Ty) = (Tx, y) \qquad \text{for all} \qquad x, y \in H. \tag{13.2}$$

Note that this means that for operators $T \in B(H)$ we do not actually need the definition of the adjoint T^* in order to define what it means to be self-adjoint. (We will see later in Chapter 25 that self-adjointness of unbounded operators requires more than just symmetry.)

Note that it is a consequence of part (b) of Lemma 13.2 that if T and R are both self-adjoint, then $(TR)^* = R^*T^* = RT$, and so TR is self-adjoint if and only if T and R commute.

In the next chapter we will introduce the notion of the spectrum of a linear operator. In Chapter 16 we will be able to give a full analysis of the spectrum of compact self-adjoint operators on Hilbert spaces; we will define what it means for an operator to be compact in Chapter 15.

13.2 Some Examples of the Hilbert Adjoint

We now give three examples of operators and their adjoints, and (in some cases) conditions under which they are self-adjoint.

Example 13.4 Let $H = K = \mathbb{K}^n$ with its standard inner product. Then any matrix $A = (a_{ij}) \in \mathbb{K}^{n \times n}$ defines a linear map T_A on $\mathbb{K}^n$ by mapping x to Ax, where

$$(Ax)_i = \sum_{j=1}^{n} a_{ij} x_j.$$

Then we have

$$(T_A x, y) = \sum_{i=1}^{n} \left(\sum_{j=1}^{n} a_{ij} x_j \right) \overline{y_i}$$

$$= \sum_{j=1}^{n} x_j \sum_{i=1}^{n} \overline{(\overline{a_{ij}} y_i)} = (x, T_{A^*} y),$$

where A^* is the Hermitian conjugate of A, i.e. $A^* = \overline{A}^T$.

If $\mathbb{K} = \mathbb{R}$, then T_A is self-adjoint if and only if $A^T = A$, i.e. if A is symmetric. If $\mathbf{K} = \mathbf{C}$, then T_A is self-adjoint if and only if $\overline{A}^T = A$, i.e. if A is Hermitian.

Example 13.5 Let $H = K = \ell^2$ and consider the shift operators from Example 11.7. If we start with the right-shift operator $s_r x = (0, x_1, x_2, \ldots)$ we have

$$(\mathsf{s}_r\boldsymbol{x}, \boldsymbol{y}) = x_1 y_2 + x_2 y_3 + x_3 y_4 + \cdots = (\boldsymbol{x}, \mathsf{s}_r^* \boldsymbol{y});$$

so $\mathsf{s}_r^* \boldsymbol{y} = (y_2, y_3, y_4, \ldots)$, i.e. $\mathsf{s}_r^* = \mathsf{s}_l$.

Similarly for the left shift $\mathsf{s}_l \boldsymbol{x} = (x_2, x_3, x_4, \ldots)$ we have

$$(\mathsf{s}_l\boldsymbol{x}, \boldsymbol{y}) = x_2 y_1 + x_3 y_2 + x_4 y_3 + \cdots = (\boldsymbol{x}, \mathsf{s}_l^* \boldsymbol{y});$$

so $\mathsf{s}_l^* \boldsymbol{y} = (0, y_1, y_2, \ldots)$, i.e. $\mathsf{s}_l^* = \mathsf{s}_r$.

These maps are not self-adjoint, but we do have $\mathsf{s}_l^{**} = \mathsf{s}_l$ and $\mathsf{s}_r^{**} = \mathsf{s}_r$ (as is guaranteed by Theorem 13.1).

We will return to our next example again later.

Lemma 13.6 *For $K \in C((a, b) \times (a, b))$ define $T : L^2(a, b) \to L^2(a, b)$ by setting*

$$(Tf)(x) := \int_a^b K(x, y) f(y) \, \mathrm{d}y$$

(see Example 11.10). Then

$$T^* g(x) = \int_a^b K(y, x) g(y) \, \mathrm{d}y, \qquad (13.3)$$

and T is self-adjoint if $K(x, y) = K(y, x)$.

Proof For $f, g \in L^2(a, b)$ we have

$$
\begin{aligned}
(Tf, g)_H &= \int_a^b \int_a^b K(x, y) f(y) \, \mathrm{d}y \, \overline{g(x)} \, \mathrm{d}x \\
&= \int_a^b \int_a^b K(x, y) f(y) \overline{g(x)} \, \mathrm{d}y \, \mathrm{d}x \\
&= \int_a^b f(y) \overline{\left(\int_a^b K(x, y) g(x) \, \mathrm{d}x \right)} \, \mathrm{d}y = (f, T^* g)_H,
\end{aligned}
$$

with $T^* g$ defined as in (13.3). In order to justify the change in the order of integration in this calculation we can either appeal to Fubini's Theorem (Theorem B.9) or, without recourse to measure theoretic results, use the fact that $C([a, b])$ is dense in L^2: given $f, g \in L^2(a, b)$, find sequences (f_n) and (g_n) in $C([a, b])$ such that $f_n \to f$ and $g_n \to g$ in L^2. Then, with f_n and g_n replacing f and g, the above calculation is valid (using the result of Exercise 6.8, for example), yielding

$$(Tf_n, g_n) = (f_n, T^* g_n).$$

Since $f_n \to f$, $g_n \to g$, and T and T^* are continuous from $L^2(a, b)$ into $L^2(a, b)$, we can take $n \to \infty$ and deduce that $(Tf, g) = (f, T^*g)$ for every $f, g \in L^2(a, b)$. $\square$

Exercises

13.1 Show that the adjoint of the operator $T : L^2(0, 1) \to L^2(0, 1)$ defined by setting

$$(Tf)(x) := \int_0^x K(x, y) f(y) \, dy$$

is given by

$$(T^*g)(x) = \int_x^1 K(y, x) g(y) \, dy.$$

13.2 Show that if $(T_n) \in B(H)$ is a sequence of self-adjoint operators such that $T_n \to T$ in $B(H)$, then T is also self-adjoint.

13.3 Show that if $T \in B(H, K)$, then $\mathrm{Ker}(T) = (\mathrm{Range}(T^*))^\perp$. (It then follows from the fact that $T^{**} = T$ that $\mathrm{Ker}(T^*) = (\mathrm{Range}(T))^\perp$.)

13.4 Show that if $T \in B(H, K)$, then $T^*T \in B(H, H)$ with

$$\|T^*T\|_{B(H,H)} = \|T\|_{B(H,K)}^2.$$

13.5 Show that if $T \in B(H, K)$ is invertible, then $T^* \in B(K, H)$ is invertible with $(T^*)^{-1} = (T^{-1})^*$. (In particular, this shows that if $T \in B(H)$ is self-adjoint and invertible, then T^{-1} is also self-adjoint.)

14

The Spectrum of a Bounded Linear Operator

In the theory of linear operators on finite-dimensional spaces, the eigenvalues play a prominent role. In this case the eigenvalues form the entire 'spectrum' of the operator; but we will see that in the case of infinite-dimensional spaces the situation is somewhat more subtle.

If $T : X \to X$ then λ is an eigenvalue of T if there exists a non-zero $x \in X$ such that $Tx = \lambda x$. Since this implies that

$$(T - \lambda I)x = 0,$$

for λ to be an eigenvalue it must be the case that $T - \lambda I$ is not invertible; otherwise multiplying on the left by $(T - \lambda I)^{-1}$ would show that $x = 0$.

When discussing the spectral properties of operators it is convenient to treat Banach spaces over $\mathbb{C}$, but this is no restriction, since we can always consider the 'complexification' of a Banach space over $\mathbb{R}$; see Exercises 14.1, 14.2, and 16.9.

14.1 The Resolvent and Spectrum

We have already remarked that for linear operators between infinite-dimensional spaces there is a distinction to be made between 'having an inverse' and 'being invertible', with the latter requiring the inverse to be bounded (see Section 11.5). We incorporate the requirement of invertibility into the following definition of the resolvent set and its complement, the spectrum.

Definition 14.1 Let X be a complex Banach space and $T \in B(X)$. The *resolvent set of T*, $\rho(T)$, is

$$\rho(T) = \{\lambda \in \mathbb{C} : T - \lambda I \text{ is invertible}\}.$$

165

The *spectrum of T*, $\sigma(T)$, is the complement of $\rho(T)$,

$$\sigma(T) = \mathbb{C} \setminus \rho(T)$$
$$= \{\lambda \in \mathbb{C} : \ T - \lambda I \text{ is not invertible}\}.$$

In a finite-dimensional space the spectrum consists entirely of eigenvalues, since Lemma 11.15 guarantees that the resolvent set $\rho(T)$ is exactly

$$\{\lambda \in \mathbb{C} : \ \mathrm{Ker}(T - \lambda I) = \{0\}\},$$

and hence its complement is the set where $(T - \lambda I)x = 0$ for some non-zero $x \in X$, i.e. the eigenvalues.

In an infinite-dimensional space the spectrum can be strictly larger than the set of eigenvalues, which we term the 'point spectrum':

$$\sigma_{\mathrm{p}}(T) = \{\lambda \in \mathbb{C} : \ (T - \lambda I)x = 0 \text{ for some non-zero } x \in X\}.$$

If $\lambda \in \sigma_{\mathrm{p}}(T)$, then λ is an *eigenvalue* of T, $E_\lambda := \mathrm{Ker}(T - \lambda I)$ is the *eigenspace* corresponding to λ, and any non-zero $x \in E_\lambda$ is one of the corresponding *eigenvectors* (if $x \in E_\lambda$, then $Tx = \lambda x$); the dimension of E_λ is the *multiplicity* of λ.

To begin with we prove two simple results about the eigenvalues (and corresponding eigenvectors) of any bounded linear operator. First, we observe that any $\lambda \in \sigma_{\mathrm{p}}(T)$ satisfies $|\lambda| \le \|T\|$: if there exists $x \ne 0$ such that $Tx = \lambda x$, then

$$|\lambda|\|x\| = \|\lambda x\| = \|Tx\| \le \|T\|\|x\|, \tag{14.1}$$

which shows that $|\lambda| < \|T\|$. We now show that eigenvectors corresponding to distinct eigenvalues are linearly independent.

Lemma 14.2 *Suppose that $T \in B(X)$ and that $\{\lambda_j\}_{j=1}^n$ are distinct eigenvalues of T. Then any set $\{e_j\}_{j=1}^n$ of corresponding eigenvectors (i.e. $Te_j = \lambda_j e_j$) is linearly independent.*

Proof We argue by induction. Suppose that $\{e_1, \ldots, e_k\}$ are linearly independent and that

$$\sum_{j=1}^{k+1} \alpha_j e_j = 0, \qquad \{\alpha_j\}_{j=1}^k \in \mathbb{K}. \tag{14.2}$$

By (i) applying T to both sides and (ii) multiplying both sides by λ_{k+1} we obtain

$$\sum_{j=1}^{k+1} \lambda_j \alpha_j e_j = 0 = \sum_{j=1}^{k+1} \lambda_{k+1} \alpha_j e_j.$$

It follows that

$$\sum_{j=1}^{k}(\lambda_{k+1} - \lambda_j)\alpha_j e_j = 0.$$

Since $\lambda_j \neq \lambda_{k+1}$ and $\{e_1, \ldots, e_k\}$ are linearly independent the preceding equation implies that $\alpha_j = 0$ for $j = 1, \ldots, k$; then $\alpha_{k+1} = 0$ from (14.2). It follows that $\{e_1, \ldots, e_{k+1}\}$ are linearly independent. $\qquad\square$

We remarked above that any $\lambda \in \sigma_p(T)$ satisfies $|\lambda| \leq \|T\|$. We now show, using Lemma 11.16, that the same bound holds for any $\lambda \in \sigma(T)$.

Lemma 14.3 *If $T \in B(X)$, then $\sigma(T)$ is a closed subset of*

$$\sigma(T) \subseteq \{\lambda \in \mathbb{C} : |\lambda| \leq \|T\|\}. \tag{14.3}$$

Proof First we prove the inclusion in (14.3). To do this, note that for any $\lambda \neq 0$ we can write

$$T - \lambda I = \lambda\left(\frac{1}{\lambda}T - I\right),$$

so if $I - \frac{1}{\lambda}T$ is invertible, $\lambda \notin \sigma(T)$. But for $|\lambda| > \|T\|$ we have

$$\left\|\frac{1}{\lambda}T\right\| \|I\| < 1,$$

and then Lemma 11.16 guarantees that $I - \frac{1}{\lambda}T$ is invertible, i.e. $\lambda \in \rho(T)$, and the result follows.

To show that the spectrum is closed we show that the resolvent set is open. If $\lambda \in \rho(T)$, then $T - \lambda I$ is invertible and Lemma 11.16 shows that $(T - \lambda I) - \delta I$ is invertible provided that

$$\|\delta T\|\|(T - \lambda I)^{-1}\| < 1,$$

i.e. $T - (\lambda + \delta)I$ is invertible for all δ with $|\delta| < \|(T - \lambda I)^{-1}\|^{-1}$, and so $\rho(T)$ is open. $\qquad\square$

We now, following Rynne and Youngson (2008), consider the illustrative examples of the shift operators from Example 11.7. These allow us to show that the spectrum can be significantly larger than the point spectrum.

Example 14.4 The right-shift operator s_r on ℓ^2 from Example 11.7 has no eigenvalues.

Proof Observe that $s_r x = \lambda x$ implies that

$$(0, x_1, x_2, \ldots) = \lambda(x_1, x_2, x_3, \ldots)$$

and so

$$\lambda x_1 = 0, \quad \lambda x_2 = x_1, \quad \lambda x_3 = x_2, \ldots.$$

If $\lambda \neq 0$, then this implies that $x_1 = 0$, and then $x_2 = x_3 = x_4 = \ldots = 0$, and so λ is not an eigenvalue. If $\lambda = 0$, then we also obtain $x = 0$, and so there are no eigenvalues, i.e. $\sigma_p(s_r) = \varnothing$.   □

Example 14.5 For the left-shift operator s_l on ℓ^2 every $\lambda \in \mathbb{C}$ with $|\lambda| < 1$ is an eigenvalue.

Proof Observe that $\lambda \in \mathbb{C}$ is an eigenvalue if $s_l x = \lambda x$, i.e. if

$$(x_2, x_3, x_4 \ldots) = \lambda(x_1, x_2, x_3, \ldots),$$

i.e. if

$$x_2 = \lambda x_1, \quad x_3 = \lambda x_2, \quad x_4 = \lambda x_3, \quad \cdots$$

Given $\lambda \neq 0$ this gives a candidate eigenvector

$$x = (1, \lambda, \lambda^2, \lambda^3, \ldots),$$

which is an element of ℓ^2 (and so is an actual eigenvector) provided that

$$\sum_{n=1}^{\infty} |\lambda|^{2n} = \frac{1}{1 - |\lambda|^2} < \infty,$$

which is the case for any λ with $|\lambda| < 1$. It follows that

$$\{\lambda \in \mathbb{C} : |\lambda| < 1\} \subseteq \sigma_p(s_l).$$ □

We showed in Example 13.5 that $s_r^* = s_l$ and $s_l^* = s_r$. The following result about the spectrum of the Hilbert adjoint will therefore allow us to relate the spectra of these two operators.

Lemma 14.6 *If H is a Hilbert space and $T \in B(H)$, then*

$$\sigma(T^*) = \{\bar{\lambda} : \lambda \in \sigma(T)\}.$$

Proof If $\lambda \notin \sigma(T)$, then $T - \lambda I$ has a bounded inverse,

$$(T - \lambda I)(T - \lambda I)^{-1} = I = (T - \lambda I)^{-1}(T - \lambda I).$$

Taking adjoints and using Lemma 13.2 we obtain

$$[(T - \lambda I)^{-1}]^*(T^* - \bar{\lambda} I) = I = (T^* - \bar{\lambda} I)[(T - \lambda I)^{-1}]^*,$$

and so $T^* - \bar{\lambda} I$ has a bounded inverse, i.e. $\bar{\lambda} \notin \sigma(T^*)$. Starting instead with T^* we deduce that $\lambda \notin \sigma(T^*) \Rightarrow \bar{\lambda} \notin \sigma(T)$, which completes the proof. $\square$

Example 14.7 Let $\mathfrak{s}_r$ be the right-shift operator on ℓ^2. We saw above that $\mathfrak{s}_r$ has no eigenvalues (Example 14.4), but that for $\mathfrak{s}_r^* = \mathfrak{s}_l$ the interior of the unit disc is contained in the point spectrum (Example 14.5). It follows from Lemma 14.6 that

$$\{\lambda \in \mathbb{C} : |\lambda| < 1\} \subseteq \sigma(\mathfrak{s}_r)$$

even though $\sigma_p(\mathfrak{s}_r) = \varnothing$.

Combining the above argument with the fact that $\sigma(T)$ is a compact subset of $\{\lambda : |\lambda| \le \|T\|\}$ (Lemma 14.3) allows us to determine the spectrum of these two shift operators.

Example 14.8 The spectrum of $\mathfrak{s}_l$ and of $\mathfrak{s}_r$ (as operators on ℓ^2) are both equal to the unit disc in the complex plane:

$$\sigma(\mathfrak{s}_l) = \sigma(\mathfrak{s}_r) = \{\lambda \in \mathbb{C} : |\lambda| \le 1\}.$$

Proof We showed earlier that for the shift operators $\mathfrak{s}_r$ and $\mathfrak{s}_l$ on ℓ^2,

$$\sigma(\mathfrak{s}_l) = \sigma(\mathfrak{s}_r) \supseteq \{\lambda \in \mathbb{C} : |\lambda| < 1\}.$$

Since the spectrum is closed and $\|\mathfrak{s}_r\| = \|\mathfrak{s}_l\| = 1$, it follows from Lemma 14.3 that

$$\sigma(\mathfrak{s}_l) = \sigma(\mathfrak{s}_r) = \{\lambda \in \mathbb{C} : |\lambda| \le 1\}. \qquad \square$$

14.2 The Spectral Mapping Theorem for Polynomials

We end this chapter with a relatively simple version of the 'spectral mapping theorem', which in full generality guarantees that the spectrum of $f(T)$ consists of $\{f(\lambda) : \lambda \in \sigma(T)\}$. Here we prove a simpler result where we restrict to the case that f is a polynomial; this does not require the theory of operator-valued complex functions used in the more general case (see Kreyszig, 1978, for example).

If $P(x) = \sum_{k=0}^{n} a_k x^k$ is a polynomial and $T \in L(X)$, then we can consider the operator

$$P(T) = \sum_{k=0}^{n} a_k T^k,$$

which is another linear operator from X into itself.

Theorem 14.9 *If $T \in B(X)$ and P is a polynomial, then*

$$\sigma(P(T)) = P(\sigma(T)) := \{P(\lambda) : \lambda \in \sigma(T)\}.$$

Proof If P has degree n, then for each fixed $\lambda \in \mathbb{C}$ we can write

$$\lambda - P(z) = a(\beta_1 - z) \cdots (\beta_n - z),$$

where $a \in \mathbb{K}$ and the β_j are the roots of the polynomial $\lambda - P(z)$. Note that the values of β_j depend on the choice of λ, and that setting $z = \beta_j$ shows that $\lambda = P(\beta_j)$ for any $j = 1, \ldots, n$. It follows that

$$\lambda I - P(T) = a(\beta_1 I - T) \cdots (\beta_n I - T). \qquad (14.4)$$

Note that all the factors on the right-hand side commute; this allows us to use Proposition 11.18 to deduce that $\lambda I - P(T)$ is invertible if and only if $\beta_j I - T$ is invertible for every $j = 1, \ldots, n$.

In particular, this means that when $\lambda \in \sigma(P(T))$, i.e. $\lambda I - P(T)$ is not invertible, it follows that $\beta_j I - T$ is not invertible for some j and so $\beta_j \in \sigma(T)$. We already observed that $\lambda = P(\beta_j)$, so $\sigma(P(T)) \subseteq P(\sigma(T))$.

Now suppose that $\lambda \notin \sigma(P(T))$, in which case $\lambda I - P(T)$ is invertible, with inverse S, say. For each $i = 1, \ldots, n$ we can write

$$\lambda I - P(T) = (\beta_i I - T) Q_i(T)$$

where Q_i is a polynomial in T of degree $n - 1$. Since all the factors on the right-hand side of (14.4) commute, $\beta_i I - T$ commutes with $Q_i(T)$; moreover, since $\beta_i I - T$ commutes with $\lambda I - P(T)$, it commutes with $(\lambda I - P(T))^{-1} = S$ (see comment immediately before Proposition 11.18). Therefore we have

$$I = S(\lambda I - P(T)) = S(\beta_i I - T) Q_i(T) = (\beta_i I - T) S Q_i(T)$$
$$= S Q_i(T)(\beta_i I - T).$$

It follows that $S Q_i(T)$ (which is bounded) is the inverse of $\beta_i I - T$, and so $\beta_i \notin \sigma(T)$.

By (14.4) the only possible choices of z such that $\lambda = P(z)$ are the $\{\beta_j\}$, and we have just shown that none of these are in $\sigma(T)$. It follows that $\lambda \neq P(z)$

for any $z \in \sigma(T)$, i.e. $\lambda \notin P(\sigma(T))$. Thus $\lambda \in P(\sigma(T)) \Rightarrow \lambda \in \sigma(P(T))$, i.e. $P(\sigma(T)) \subseteq \sigma(P(T))$, and we obtain the required equality. $\qquad \square$

Exercises

14.1 Let X be a real Banach space. We define its complexification $X_{\mathbb{C}}$ as the vector space

$$X_{\mathbb{C}} := \{(x, y) : x, y \in X\},$$

equipped with operations of addition and multiplication by complex numbers defined via

$$(x, y) + (x', y') = (x + x', y + y'), \qquad x, y, x', y' \in H$$

and

$$(a + ib)(x, y) = (ax - by, bx + ay) \qquad a, b \in \mathbb{R}, \ x, y \in H. \quad (14.5)$$

It is natural to denote (x, y) by $x + iy$, but this is purely 'notational', since multiplication by i in the original space X has no meaning; then it is easy to see that (14.5) corresponds to the usual rule of multiplication for complex numbers. (Zeidler, 1995)

When H is a Hilbert space, show that

$$(x + iy, x' + iy')_{H_{\mathbb{C}}} := (x, x') + i(y, x') - i(x, y') + (y, y')$$

is an inner product on $H_{\mathbb{C}}$, and that this makes $H_{\mathbb{C}}$ a Hilbert space. (Note that the induced norm on $H_{\mathbb{C}}$ is $\|(x, y)\|_{H_{\mathbb{C}}}^2 = \|x\|^2 + \|y\|^2$.)

14.2 Let H be a real Hilbert space and $H_{\mathbb{C}}$ its complexification. Given any $T \in L(H)$, define the complexification of T, $T_{\mathbb{C}} \in L(H_{\mathbb{C}})$, by setting

$$T_{\mathbb{C}}(x + iy) := Tx + iTy \qquad x, y \in H.$$

Show that

(i) if $T \in B(H)$, then $T_{\mathbb{C}} \in B(H_{\mathbb{C}})$ with $\|T_{\mathbb{C}}\|_{B(H_{\mathbb{C}})} = \|T\|_{B(H)}$; and

(ii) any eigenvalue of T is an eigenvalue of $T_{\mathbb{C}}$, and any real eigenvalue of $T_{\mathbb{C}}$ is an eigenvalue of T.

14.3 For any $\boldsymbol{\alpha} \in \ell^{\infty}(\mathbb{C})$ consider the operator $D_{\boldsymbol{\alpha}} : \ell^2(\mathbb{C}) \to \ell^2(\mathbb{C})$ given by

$$(x_1, x_2, x_3, \cdots) \mapsto (\alpha_1 x_1, \alpha_2 x_2, \alpha_3 x_3, \ldots),$$

i.e. $(D_{\boldsymbol{\alpha}} x)_j = \alpha_j x_j$. Show that

(i) $\sigma_{\mathrm{p}}(D_{\boldsymbol{\alpha}}) = \{\alpha_j\}_{j=1}^{\infty}$;

(ii) $\sigma(D_{\boldsymbol{\alpha}}) = \overline{\sigma_{\mathrm{p}}(D_{\boldsymbol{\alpha}})}$; and

(iii) any compact subset of $\mathbb{C}$ is the spectrum of an operator of this form. (Giles, 2000)

14.4 If X is a complex Banach space and $T \in B(X)$, then the *spectral radius* of T, $r_\sigma(T)$, is defined as

$$r_\sigma(T) = \sup\{|\lambda| : \lambda \in \sigma(T)\}.$$

Show that $r_\sigma(T) \le \liminf_{n \to \infty} \|T^n\|^{1/n}$.

14.5 Let $X = C([0, 1])$. Use the result of the previous exercise and Exercise 11.7 to show that every $\lambda \ne 0$ is in the resolvent set of the operator $T \in B(X)$ defined by setting

$$[Tf](x) = \int_0^x f(s)\,ds.$$

Show that $0 \in \sigma(T)$ but is not an eigenvalue of T. (Giles, 2000; Lax, 2002)

14.6 The Fourier transform of a function $f \in L^1$ is defined by

$$(\tilde{f})(k) = [\mathcal{F}f](k) := \frac{1}{\sqrt{2\pi}} \int_{-\infty}^{\infty} e^{ikx} f(x)\,dx,$$

and if $\tilde{f} \in L^1$, then we can recover f using the fact that

$$f(-x) = [\mathcal{F}\tilde{f}](x). \tag{14.6}$$

We can extend the definition of $\mathcal{F}$ to a map from $L^2(\mathbb{R})$ into itself using a density argument: we approximate f by smooth functions that decay rapidly at infinity and use the fact that $\|f\|_{L^2} = \|\tilde{f}\|_{L^2}$ for such functions. In this way we also preserve the relationship $[\mathcal{F}^2 f](x) = f(-x)$ from (14.6).

Show that if $\mathcal{F}$ is viewed as an operator from $L^2(\mathbb{R})$ into itself, then $\sigma(\mathcal{F}) \subseteq \{\pm 1, \pm i\}$. (Lax, 2002)

15

Compact Linear Operators

If a linear operator is not only bounded but compact (we define this below), then we can obtain more information about its spectrum. We prove results for compact self-adjoint operators on Hilbert spaces in the next chapter, and for compact operators on Banach spaces in Chapter 24.

15.1 Compact Operators

A linear operator is compact if it maps bounded sequences into sequences that have a convergent subsequence.

Definition 15.1 Let X and Y be normed spaces. A linear operator $T: X \to Y$ is *compact* if for any bounded sequence $(x_n) \in X$, the sequence $(Tx_n) \in Y$ has a convergent subsequence (whose limit lies in Y).

Alternatively T is compact if $T\mathbb{B}_X$ is a precompact subset of Y, i.e. if $\overline{T\mathbb{B}_X}$ is a compact subset of Y. The equivalence of these two definitions follows since we showed in Lemma 6.11 that a set A is precompact if and only if any sequence in A has a Cauchy subsequence. (See Exercise 15.1 for more details.)

Note that a compact operator must be bounded, since otherwise there exists a sequence $(x_n) \in X$ with $\|x_n\| = 1$ but $Tx_n \to \infty$, and clearly (Tx_n) cannot have a convergent subsequence.

Example 15.2 Take $T \in B(X, Y)$ with finite-dimensional range. Then T is compact, since any bounded sequence in a finite-dimensional space has a convergent subsequence.

Noting that if $T, S: X \to Y$ are both compact, then $T + S$ is also compact, and that λT is compact for any $\lambda \in \mathbb{K}$, we can define the space $K(X, Y)$ of all compact linear operators from X into Y, and this is then a vector space. Our next result shows that this is a closed subspace of $B(X, Y)$, and so is complete (by Lemma 4.3).

Theorem 15.3 *Suppose that X is a normed space and Y is a Banach space. If $(K_n)_{n=1}^{\infty}$ is a sequence of compact (linear) operators in $K(X, Y)$ that converges to some $K \in B(X, Y)$, i.e.*

$$\|K_n - K\|_{B(X,Y)} \to 0 \qquad as \qquad n \to \infty,$$

then $K \in K(X, Y)$. In particular, $K(X, Y)$ is complete.

Proof Let $(x_n)_n$ be a bounded sequence in X with $\|x_n\| \le M$ for all n. Then, since K_1 is compact, $(K_1(x_n))_n$ has a convergent subsequence, $(K_1(x_{n_{1,j}}))_j$. Since $(x_{n_{1,j}})_j$ is bounded, $(K_2(x_{n_{1,j}}))_j$ has a convergent subsequence, $(K_2(x_{n_{2,j}}))_j$. Repeat this process to get a family of nested subsequences, $(x_{n_{k,j}})_j$, with $(K_l(x_{n_{k,j}}))_j$ convergent for all $l \le k$.

As in the proof of the Arzelà–Ascoli Theorem we now consider the diagonal sequence $y_j = x_{n_{j,j}}$. Since (y_j) is a subsequence of $(x_{n_{k,i}})_i$ for $j \ge k$, it follows that $K_n(y_j)$ converges (as $j \to \infty$) for every n.

We now show that $(K(y_j))_{j=1}^{\infty}$ is Cauchy, and hence convergent, to complete the proof. Choose $\varepsilon > 0$, and use the triangle inequality to write

$$\|K(y_i) - K(y_j)\|_Y$$
$$\le \|K(y_i) - K_n(y_i)\|_Y + \|K_n(y_i) - K_n(y_j)\|_Y + \|K_n(y_j) - K(y_j)\|_Y.$$
$$(15.1)$$

Since (y_j) is bounded, with $\|y_j\| \le M$ for all j, and $K_n \to K$, pick n large enough that

$$\|K - K_n\|_{B(X,Y)} < \frac{\varepsilon}{3M};$$

then

$$\|K(y_j) - K_n(y_j)\|_Y \le \varepsilon/3 \qquad \text{for every } j.$$

For such an n, the sequence $(K_n(y_j))_{j=1}^{\infty}$ is Cauchy, and so there exists an N such that for $i, j \ge N$ we can guarantee that

$$\|K_n(y_i) - K_n(y_j)\|_Y \le \varepsilon/3.$$

So now from (15.1)

$$\|K(y_i) - K(y_j)\|_Y \le \varepsilon \qquad \text{for all} \qquad i, j \ge N,$$

and $(K(y_n))$ is a Cauchy sequence. Since Y is complete, it follows that $(K(y_n))$ converges, and so K is compact. $\qquad\square$

15.2 Examples of Compact Operators

For the remainder of this chapter we concentrate on operators on Hilbert spaces.

As a first example we use Theorem 15.3 to show that the integral operator from Example 11.10 is compact.

Proposition 15.4 *Suppose that* $K \in C([a, b] \times [a, b])$. *Then the integral operator* $T : L^2(a, b) \to L^2(a, b)$ *given by*

$$[Tu](x) := \int_a^b K(x, y)u(y)\,dy$$

is compact.

For another proof using the Arzelà–Ascoli Theorem see Exercise 15.4.

Proof First note that if

$$K_n(x, y) = \sum_{j=1}^n f_j(x)g_j(y), \tag{15.2}$$

with $f_j, g_j \in C([a, b])$, then the integral operator $T_n : L^2 \to L^2$ defined by setting

$$(T_n u)(x) := \int_a^b \sum_{j=1}^n \left[f_j(x)g_j(y) \right] u(y)\,dy = \sum_{j=1}^n \left[\int_a^b g_j(y)u(y)\,dy \right] f_j(x)$$

has finite-dimensional range, namely the linear span of $\{f_j(x)\}_{j=1}^n$, and so is compact.

Now use the result of Exercise 6.6 to approximate K uniformly on the set $[a, b] \times [a, b]$ by a sequence of (K_n) of the form in (15.2). Then

$$|Tu(x) - T_n u(x)| = \left| \int_a^b \{K(x, y) - K_n(x, y)\} u(y)\,dy \right|$$

$$\leq \int_a^b |K(x, y) - K_n(x, y)||u(y)|\,dy$$

$$\leq \left(\int_a^b \|K - K_n\|_\infty^2 \right)^{1/2} \left(\int_a^b |u(y)|^2\,dy \right)^{1/2}$$

$$= (b - a)^{1/2} \|K - K_n\|_\infty \|u\|_{L^2},$$

which shows that

$$\|T - T_n\|_{B(H)} \leq (b - a)^{1/2} \|K - K_n\|_\infty.$$

It follows that T is the limit in $B(H)$ of compact operators, and so T is compact by Theorem 15.3. □

Another class of compact operators on Hilbert spaces are the so-called Hilbert–Schmidt operators.

Definition 15.5 An operator $T \in B(H)$ is *Hilbert–Schmidt* if for some orthonormal basis $\{e_j\}_{j=1}^{\infty}$ of H

$$\|T\|_{\mathrm{HS}}^2 := \sum_{j=1}^{\infty} \|Te_j\|^2 < \infty.$$

This is a meaningful definition, since the quantity $\|T\|_{\mathrm{HS}}$ is independent of the orthonormal basis we choose. To see this, suppose that $\{e_j\}_{j=1}^{\infty}$ and $\{f_j\}_{j=1}^{\infty}$ are two orthonormal bases for H; then

$$\sum_{j=1}^{\infty} \|Te_j\|^2 = \sum_{j,k=1}^{\infty} |(Te_j, f_k)|^2 = \sum_{j,k=1}^{\infty} |(e_j, T^*f_k)|^2 = \sum_{k=1}^{\infty} \|T^*f_k\|^2.$$

Applying the resulting equality with $e_j = f_j$ shows that

$$\sum_{j=1}^{\infty} \|Tf_j\|^2 = \sum_{k=1}^{\infty} \|T^*f_k\|^2 = \sum_{j=1}^{\infty} \|Te_j\|^2,$$

and so $\|T\|_{\mathrm{HS}}$ is indeed independent of the choice of basis. Exercise 15.5 shows that $\|T\|_{B(H)} \le \|T\|_{\mathrm{HS}}$.

Proposition 15.6 *Any Hilbert–Schmidt operator T is compact.*

Proof Choose some orthonormal basis $\{e_j\}_{j=1}^{\infty}$ for H, and observe that since T is linear and continuous we can write

$$Tu = T\left(\sum_{j=1}^{\infty} (u, e_j)e_j\right) = \sum_{j=1}^{\infty} (u, e_j)Te_j.$$

Now for each n let $T_n: H \to H$ be defined by setting

$$T_n u := \sum_{j=1}^{n} (u, e_j)Te_j.$$

This operator is clearly linear, and its range is finite-dimensional since it is the linear span of $\{Te_j\}_{j=1}^{n}$. It follows that T_n is a compact operator for each n.

Now we have

$$\|(T_n - T)u\| = \left\| \sum_{j=n+1}^{\infty} (u, e_j) T e_j \right\|$$

$$\leq \sum_{j=n+1}^{\infty} |(u, e_j)| \|T e_j\|$$

$$\leq \left(\sum_{j=n+1}^{\infty} |(u, e_j)|^2 \right)^{1/2} \left(\sum_{j=n+1}^{\infty} \|T e_j\|^2 \right)^{1/2}$$

$$\leq \|u\| \left(\sum_{j=n+1}^{\infty} \|T e_j\|^2 \right)^{1/2},$$

which shows that

$$\|T_n - T\|_{B(H)} \leq \left(\sum_{j=n+1}^{\infty} \|T e_j\|^2 \right)^{1/2}.$$

Since T is Hilbert–Schmidt, $\sum_{j=1}^{\infty} \|T e_j\|^2 < \infty$, and so the right-hand side tends to zero as $n \to \infty$. It follows from Theorem 15.3 that T is compact. $\square$

In fact the operator T from Proposition 15.4 is Hilbert–Schmidt, which provides another proof that it is compact; see Exercise 15.6.

15.3 Two Results for Compact Operators

We end with two results for compact operators: we show that if T is a compact operator on a Hilbert space, then T^* is also compact; and that the spectrum of a compact operator on an infinite-dimensional Banach space always contains zero.

Lemma 15.7 *If H is a Hilbert space and $T \in K(H)$, then $T^* \in K(H)$.*

Proof Since T is compact and T^* is bounded (Theorem 13.1), it follows (see Exercise 15.2) that $T T^*$ is compact. So given any bounded sequence $(x_n) \in H$, $T T^* x_n$ has a convergent subsequence (which we relabel). Therefore

$$|(T T^*(x_n - x_m), x_n - x_m)| \leq \|T T^*(x_n - x_m)\| \|x_n - x_m\| \to 0$$

as $\min(m, n) \to \infty$. But the left-hand side of this expression is

$$|(T^*(x_n - x_m), T^*(x_n - x_m))| = \|T^*(x_n - x_m)\|^2,$$

which shows that (T^*x_n) is Cauchy and thus convergent, showing that T^* is compact. $\qquad\square$

We will investigate the spectrum of compact operators on Banach spaces in detail in Chapter 24, but for now we show that $0 \in \sigma(T)$ if T acts on an infinite-dimensional Banach space.

Theorem 15.8 *Suppose that X is an infinite-dimensional Banach space and $T \in K(X)$. Then $0 \in \sigma(T)$.*

Proof If $0 \notin \sigma(T)$, then T is invertible and so T^{-1} is bounded. Since the composition of a compact operator and a bounded operator is compact (see Exercise 15.2), it follows that $I = TT^{-1}$ is compact. But this implies that the unit ball in X is compact, so by Theorem 5.5 X is finite-dimensional, a contradiction. $\qquad\square$

While this result shows that the spectrum of a compact operator on an infinite-dimensional space is always non-empty, such an operator can have no eigenvalues (see Exercise 15.9).

Exercises

15.1   Show that $T \in B(X, Y)$ is compact if and only if $T\mathbb{B}_X$ is a precompact subset of Y, i.e. any sequence in $T\mathbb{B}_X$ has a convergent subsequence.

15.2   Suppose that $T \in B(X, Y)$ and $S \in B(Y, Z)$. Show that if either of S or T are compact, then $S \circ T$ is compact.

15.3   Show that the operator $T : \ell^2 \to \ell^2$ given by

$$(x_1, x_2, x_3, x_4, \ldots) \mapsto (x_1, \frac{x_2}{2}, \frac{x_3}{3}, \cdots)$$

is compact.

15.4   Suppose that $K \in C([a, b] \times [a, b])$. Use the Arzelà–Ascoli Theorem (Theorem 6.13) to show that the operator $T : C([a, b]) \to C([a, b])$ defined by

$$Tu(x) = \int_a^b K(x, y)u(y)\, dy$$

is compact. (Bollobás, 1990; Pryce, 1973)

15.5   Show that if T is a Hilbert–Schmidt operator, then

$$\|T\|_{B(H)} \le \|T\|_{HS}.$$

15.6 Show that the operator from Proposition 15.4 is a Hilbert–Schmidt operator on $L^2(a, b)$ as follows. Take any orthonormal basis $\{e_j\}_{j=1}^\infty$ for $L^2(a, b)$ and for each fixed $x \in (a, b)$ consider the function $\kappa_x \in C^0(a, b)$ given by $\kappa_x(y) = K(x, y)$. Then

$$(Te_j)(x) = \int_a^b K(x, y)e_j(y)\,dy = \int_a^b \kappa_x(y)e_j(y)\,dy = (\kappa_x, e_j).$$

Use this to show that

$$\sum_{j=1}^\infty \|Te_j\|^2 = \int_a^b \int_a^b |K(x, y)|^2\,dx\,dy < \infty,$$

and hence that T is Hilbert–Schmidt. (Young, 1988)

15.7 Suppose that $\{K_{ij}\}_{i,j=1}^\infty \in \mathbb{K}$ with

$$\sum_{i,j=1}^\infty |K_{ij}|^2 < \infty.$$

Show that the operator $S\colon \ell^2 \to \ell^2$ defined by setting

$$(Sx)_i = \sum_{i=1}^\infty K_{ij}x_j$$

is compact. (Show that S is Hilbert–Schmidt.)

15.8 Let X be an infinite-dimensional normed space and $T\colon X \to Y$ a compact linear operator. Show that there exists $(x_n) \in S_X$ such that $Tx_n \to 0$. Show by example that there need not exist $x \in S_X$ with $Tx = 0$. (Giles, 2000)

15.9 Show that the operator $T'\colon \ell^2 \to \ell^2$ defined by

$$(x_1, x_2, x_3, \cdots) \mapsto (0, x_1, \frac{x_2}{2}, \frac{x_3}{3}, \cdots)$$

is compact and has no eigenvalues. (Note that $T' = s_r \circ T$, where T is the compact operator from Exercise 15.3 and s_r is the right-shift operator from Example 11.7.) (Kreyszig, 1978)

15.10 Show that the operator $T\colon \ell^2 \to \ell^2$ that maps $e^{(j)}$ to $e^{(j+1)}$ when j is odd and $e^{(j)}$ to zero when j is even, i.e.

$$(x_1, x_2, x_3, \cdots) \mapsto (0, x_1, 0, x_3, 0, x_5, \cdots)$$

is not compact. (Since $T^2 = 0$, this gives an example of a non-compact operator whose square is compact.)

16

The Hilbert–Schmidt Theorem

It is one of the major results of finite-dimensional linear algebra that all eigen-values of real symmetric matrices are real and that the eigenvectors of distinct eigenvalues are orthogonal. In this chapter we prove similar results for compact self-adjoint operators on infinite-dimensional Hilbert spaces: we show that the spectrum consists entirely of real eigenvalues (except perhaps zero), that the multiplicity of every non-zero eigenvalue is finite, and that the eigenvectors form an orthonormal basis for H.

16.1 Eigenvalues of Self-Adjoint Operators

If T is self-adjoint, then the *numerical range* of T, $V(T)$, is the set

$$V(T) := \{(Tx, x) : x \in H, \|x\| = 1\}. \tag{16.1}$$

It is possible to deduce various facts about the spectrum of T from its numerical range (see Exercises 16.1, 16.3, and 16.4), but we will mainly use it for the following result.

Theorem 16.1 *Let H be a Hilbert space and $T \in B(H)$ a self-adjoint operator. Then $V(T) \subset \mathbb{R}$ and*

$$\|T\|_{B(H)} = \sup\{|\lambda| : \lambda \in V(T)\}. \tag{16.2}$$

(In fact when H is a complex Hilbert space an operator $T \in B(H)$ is self-adjoint if and only if $V(T) \subset \mathbb{R}$; see Exercise 16.1.)

Proof We have

$$(Tx, x) = (x, Tx) = \overline{(Tx, x)},$$

and so (Tx, x) is real for every $x \in H$.

To prove (16.2) we let $M = \sup\{|(Tx, x)| : x \in H, \|x\| = 1\}$. Clearly

$$|(Tx, x)| \le \|Tx\|\|x\| \le \|T\|\|x\|^2 = \|T\|$$

when $\|x\| = 1$, and so $M \le \|T\|$.

Now observe that for any $u, v \in H$ we have

$$(T(u + v), u + v) - (T(u - v), u - v) = 2[(Tu, v) + (Tv, u)]$$

$$= 2[(Tu, v) + (v, Tu)]$$

$$= 4\operatorname{Re}(Tu, v),$$

using the fact that $(Tv, u) = (v, Tu) = \overline{(Tu, v)}$ since T is self-adjoint. Therefore

$$4\operatorname{Re}(Tu, v) = (T(u + v), u + v) - (T(u - v), u - v)$$

$$\le M(\|u + v\|^2 + \|u - v\|^2)$$

$$= 2M(\|u\|^2 + \|v\|^2)$$

using the Parallelogram Law (Lemma 8.7). If $Tu \ne 0$ choose

$$v = \frac{\|u\|}{\|Tu\|} Tu$$

to obtain, since $\|v\| = \|u\|$, that

$$4\|u\|\|Tu\| \le 4M\|u\|^2,$$

i.e. $\|Tu\| \le M\|u\|$ if $Tu \ne 0$. The same inequality is trivial if $Tu = 0$, and so it follows that $\|T\| \le M$ and therefore we obtain $\|T\| = M$, as required. $\qquad \square$

Corollary 16.2 *If $T \in B(H)$ is self-adjoint, then*

(i) *all of its eigenvalues are real, and*
(ii) *if $Tx_1 = \lambda_1 x_1$ and $Tx_2 = \lambda_2 x_2$ with $\lambda_1 \ne \lambda_2$, then $(x_1, x_2) = 0$.*

Proof If $x \ne 0$ and $Tx = \lambda x$, then

$$(Tx, x) = (\lambda x, x) = \lambda\|x\|^2$$

and $(Tx, x) \in \mathbb{R}$ by the previous theorem.

If x_1 or x_2 is zero, then the result is immediate. Otherwise λ_1 and λ_2 are eigenvalues of T, and so by part (i) we must have $\lambda_1, \lambda_2 \in \mathbb{R}$; now simply note that

$$\lambda_1(x_1, x_2) = (Tx_1, x_2) = (x_1, Tx_2) = \lambda_2(x_1, x_2)$$

and so $(\lambda_1 - \lambda_2)(x_1, x_2) = 0$, which implies that $(x_1, x_2) = 0$ since $\lambda_1 \neq \lambda_2$. □

16.2 Eigenvalues of Compact Self-Adjoint Operators

We now show that any compact self-adjoint operator has at least one eigenvalue.

Theorem 16.3 *Let H be a Hilbert space and $T \in B(H)$ a compact self-adjoint operator. Then at least one of $\pm \|T\|$ is an eigenvalue of T, and so in particular*

$$\|T\| = \max\{|\lambda| : \lambda \in \sigma_p(T)\}. \tag{16.3}$$

Proof We assume that $T \neq 0$, otherwise the result is trivial. From Theorem 16.1 we have

$$\|T\| = \sup_{\|x\|=1} |(Tx, x)|,$$

so there exists a sequence (x_n) of unit vectors in H such that

$$(Tx_n, x_n) \to \alpha, \tag{16.4}$$

where α is either $\|T\|$ or $-\|T\|$. Since T is compact, there is a subsequence x_{n_j} such that Tx_{n_j} is convergent to some $y \in H$. Relabel x_{n_j} as x_n again, so that $Tx_n \to y$ and (16.4) still holds.

Now consider

$$\|Tx_n - \alpha x_n\|^2 = \|Tx_n\|^2 + \alpha^2 - 2\alpha(Tx_n, x_n)$$
$$\leq 2\alpha^2 - 2\alpha(Tx_n, x_n);$$

by our choice of x_n, the right-hand side tends to zero as $n \to \infty$. It follows, since $Tx_n \to y$, that

$$\alpha x_n \to y,$$

and since $\alpha \neq 0$ is fixed we have $x_n \to x := y/\alpha$; note that $\|x\| = 1$ since it is the limit of the x_n and $\|x_n\| = 1$ for every n. Since T is bounded, it is continuous, so therefore

$$Tx = \lim_{n \to \infty} Tx_n = y = \alpha x.$$

We have found $x \in H$ with $\|x\| = 1$ such that $Tx = \alpha x$, so $\alpha \in \sigma_p(T)$.

We already showed that any eigenvalue λ must satisfy $|\lambda| \leq \|T\|$ (see (14.1)) and so (16.3) follows. □

We now start to investigate the spectrum of self-adjoint compact operators. We show that the eigenvalues can only accumulate at 0, and that every eigenspace is finite-dimensional.

Proposition 16.4 *Let T be a compact self-adjoint operator on a separable Hilbert space H. Then $\sigma_p(T)$ is either finite or consists of a countable sequence $(\lambda_n)_{n=1}^{\infty}$ with $\lambda_n \to 0$ as $n \to \infty$. Furthermore, every distinct nonzero eigenvalue corresponds to only a finite number of linearly independent eigenvectors.*

Proof Suppose that T has infinitely many eigenvalues that do not form a sequence tending to zero. Then for some $\varepsilon > 0$ there exists a sequence (λ_n) of distinct eigenvalues with $|\lambda_n| > \varepsilon$ for every n. Let (x_n) be a corresponding sequence of eigenvectors (i.e. $Tx_n = \lambda_n x_n$) with $\|x_n\| = 1$; then

$$\|Tx_n - Tx_m\|^2 = (Tx_n - Tx_m, Tx_n - Tx_m)$$
$$= (\lambda_n x_n - \lambda_m x_m, \lambda_n x_n - \lambda_m x_m) = |\lambda_n|^2 + |\lambda_m|^2 \geq 2\varepsilon^2$$

since $(x_n, x_m) = 0$ (as we are assuming that T is self-adjoint we can use Corollary 16.2). It follows that (Tx_n) can have no convergent subsequence, which contradicts the compactness of T.

Now suppose that for some eigenvalue λ there exist an infinite number of linearly independent eigenvectors $\{e_n\}_{n=1}^{\infty}$. Using the Gram–Schmidt process from Proposition 9.9 we can find a countably infinite orthonormal set of eigenvectors $\{\hat{e}_j\}$, since any linear combination of the $\{e_j\}$ is still an eigenvector:

$$T\left(\sum_{j=1}^{n} \alpha_j e_j\right) = \sum_{j=1}^{n} \alpha_j T e_j = \lambda \left(\sum_{j=1}^{n} \alpha_j e_j\right).$$

So we have

$$\|T\hat{e}_n - T\hat{e}_m\| = \|\lambda \hat{e}_n - \lambda \hat{e}_m\| = |\lambda| \|\hat{e}_n - \hat{e}_m\| = \sqrt{2}|\lambda|.$$

It follows that $(T\hat{e}_n)$ can have no convergent subsequence, again contradicting the compactness of T. (Note that this second part does not use the fact that T is self-adjoint.) $\qquad\square$

Note that if T^n rather than T is compact, then we can apply the above result to T^n: the Spectral Mapping Theorem for polynomials (Theorem 14.9) tells us that $\sigma(T^n) = [\sigma(T)]^n$, and so in this case too the point spectrum of T is either finite or consists of a countable sequence tending to zero.

16.3 The Hilbert–Schmidt Theorem

We now prove our main result about compact self-adjoint operators, the Hilbert–Schmidt Theorem. We show that any such operator can be expressed in terms of its eigenvalues and eigenvectors, and that when augmented by a basis for the kernel of T the eigenvectors form an orthonormal basis for H.

We will need the following simple lemma.

Lemma 16.5 *If $T \in B(H)$ and Y is a closed linear subspace of H such that $TY \subseteq Y$, then $T^*Y^\perp \subseteq Y^\perp$. In particular, if $T \in B(H)$ is self-adjoint and Y is a closed linear subspace of H, then*

$$TY \subseteq Y \quad \Rightarrow \quad TY^\perp \subseteq Y^\perp.$$

Proof Let $x \in Y^\perp$ and $y \in Y$. Then $Ty \in Y$ and so

$$0 = (Ty, x) = (y, T^*x) \qquad \text{for all } y \in Y,$$

i.e. $T^*x \in Y^\perp$. $\square$

In the proof of the Hilbert–Schmidt Theorem we find successive eigenvalues of T, $\lambda_1, \lambda_2, \ldots$ by using Theorem 16.3 repeatedly. Each time we find a new eigenvector $w_1, w_2, \ldots$, and 'remove' these directions from H by considering $H_{n+1} = \mathrm{Span}(w_1, \ldots, w_n)^\perp$, which will be invariant under T due to Lemma 16.5.

Theorem 16.6 *(Hilbert–Schmidt Theorem). Let H be a Hilbert space and $T \in B(H)$ a compact self-adjoint operator. Then there exists a finite or countably infinite orthonormal sequence (w_j) consisting of eigenvectors of T, with corresponding non-zero real eigenvalues (λ_j), such that for all $x \in H$*

$$Tx = \sum_j \lambda_j (x, w_j) w_j. \tag{16.5}$$

Proof By Theorem 16.3 there exists $w_1 \in H$ such that $Tw_1 = \pm\|T\|w_1$ and $\|w_1\| = 1$.

Consider the subspace of H perpendicular to w_1,

$$H_2 = w_1^\perp.$$

Since $H_2 \subset H$ is closed, it is a Hilbert space (Lemma 8.11). Then since T is self-adjoint, Lemma 16.5 shows that T leaves H_2 invariant. If we consider $T_2 = T|_{H_2}$, then we have $T_2 \in B(H_2, H_2)$ with T_2 compact; this operator is still self-adjoint, since for all $x, y \in H_2$

$$(x, T_2 y) = (x, Ty) = (Tx, y) = (T_2 x, y).$$

Now apply Theorem 16.3 to the operator T_2 on the Hilbert space H_2 find an eigenvalue $\lambda_2 = \pm \|T_2\|$ and an eigenvector $w_2 \in H_2$ with $\|w_2\| = 1$.

Now if we let $H_3 = \{w_1, w_2\}^{\perp}$, then H_3 is a closed subspace of H_2 and $T_3 = T|_{H_3}$ is compact and self-adjoint. We can once more apply Theorem 16.3 to find an eigenvalue $\lambda_3 = \pm \|T_3\|$ and a corresponding eigenvector $w_3 \in H_3$ with $\|w_3\| = 1$. We continue this process as long as $T_n \neq 0$.

If $T_n = 0$ for some n, then, for any given $x \in H$, if we set

$$y := x - \sum_{j=1}^{n-1}(x, w_j)w_j \in H_n,$$

we have

$$0 = T_n y = Ty = Tx - \sum_{j=1}^{n-1}(x, w_j)Tw_j = Tx - \sum_{j=1}^{n-1}\lambda_j(x, w_j)w_j,$$

which is (16.5).

If T_n is never zero, then, given $x \in H$, consider

$$y_n := x - \sum_{j=1}^{n-1}(x, w_j)w_j \in H_n$$

(for $n \geq 2$). We have

$$\|x\|^2 = \|y_n\|^2 + \sum_{j=1}^{n-1}|(x, w_j)|^2,$$

and so $\|y_n\| \leq \|x\|$. It follows, since $T_n = T|_{H_n}$, that

$$\left\| Tx - \sum_{j=1}^{n-1}\lambda_j(x, w_j)w_j \right\| = \|Ty_n\| \leq \|T_n\|\|y_n\| = |\lambda_n|\|x\|,$$

and since $|\lambda_n| \to 0$ as $n \to \infty$ (Proposition 16.4) we obtain (16.5). $\qquad \square$

There is a partial converse to this theorem; see Exercise 16.5.

The orthonormal sequence constructed in this theorem is only a basis for the range of T; however, we can augment this by a basis for the kernel of T to obtain the following result.

Corollary 16.7 *Let H be an infinite-dimensional separable Hilbert space and $T \in B(H)$ a compact self-adjoint operator. Then there exists a countable* *orthonormal basis* $E = \{e_j\}_{j=1}^{\infty}$ *of* H *consisting of eigenvectors of* T, *and for any* $x \in H$

$$Tx = \sum_{j=1}^{\infty} \lambda_j (x, e_j) e_j, \qquad (16.6)$$

where $Te_j = \lambda_j e_j$. *In particular, if* $\mathrm{Ker}(T) = \{0\}$, *then* H *has an orthonormal basis consisting of (suitably normalised) eigenvectors of* T *corresponding to non-zero eigenvalues.*

Proof Theorem 16.6 gives a finite or countable sequence $W = (w_k)$ of eigenvectors of T such that

$$Tx = \sum_{k} \lambda_k (x, w_k) w_k. \qquad (16.7)$$

Since $\mathrm{Ker}(T)$ is a closed subspace of H, it is a Hilbert space in its own right (Lemma 8.11); since H is separable so is $\mathrm{Ker}(T)$, and as Proposition 9.17 shows that every separable Hilbert space has a countable orthonormal basis, it follows that $\mathrm{Ker}(T)$ has a finite or countable orthonormal basis F.

Note that each $f \in F$ is an eigenvector of T with eigenvalue zero, and since $Tf = 0$ but $Tw_k = \lambda_k w_k$ with $\lambda_k \neq 0$, we know that $(f, w_k) = 0$ for every $f \in F$, $w_k \in W$. So $F \cup W$ is a countable orthonormal set in H.

Now, (16.7) implies that

$$T\left[x - \sum_{j=1}^{\infty} (x, w_j) w_j \right] = 0,$$

i.e. that $x - \sum_{j=1}^{\infty} (x, w_j) w_j \in \mathrm{Ker}\, T$. It follows that $F \cup W$ is an orthonormal basis for H. Since it is countable, we can relabel it to yield $E = \{e_j\}_{j=1}^{\infty}$, and the expression in (16.6) follows directly from (16.7).   □

Finally, we use this result to show that for compact self-adjoint operators, any non-zero element of the spectrum must be an eigenvalue. (In fact self-adjointness is not necessary: we will see that the same result holds for compact operators on a Banach space in Theorem 24.7.)

Theorem 16.8 *If* T *is a compact self-adjoint operator on a separable Hilbert space* H, *then* $\sigma(T) = \overline{\sigma_p(T)}$. *Every non-zero* $\lambda \in \sigma(T)$ *is an eigenvalue, and either* $\sigma(T) = \sigma_p(T)$ *or* $\sigma(T) = \sigma_p(T) \cup \{0\}$.

Proof By Corollary 16.7 we have

$$Tx = \sum_{j=1}^{\infty} \lambda_j (x, e_j) e_j$$

for some orthonormal basis $\{e_j\}_{j=1}^{\infty}$ of H.

Now take $\lambda \notin \overline{\sigma_p(T)}$. For such λ, it follows that there exists a $\delta > 0$ such that

$$\sup_{j \in \mathbb{N}} |\lambda - \lambda_j| \geq \delta > 0, \tag{16.8}$$

for otherwise $\lambda \in \overline{\sigma_p(T)}$. We use this to show that $T - \lambda I$ is invertible, i.e. that $\lambda \notin \sigma(T)$.

Indeed,

$$(T - \lambda I)x = y \quad \Leftrightarrow \quad \sum_{j=1}^{\infty} (\lambda_j - \lambda)(x, e_j)e_j = \sum_{j=1}^{\infty} (y, e_j)e_j.$$

Taking the inner product of both sides with each e_i in turn we have

$$(T - \lambda I)x = y \quad \Leftrightarrow \quad (\lambda_i - \lambda)(x, e_i) = (y, e_i) \text{ for every } i \in \mathbb{N}.$$

Since $\lambda \neq \lambda_i$ for every i, we can solve $(T - \lambda I)x = y$ by setting

$$(x, e_i) = \frac{(y, e_i)}{\lambda_i - \lambda} \qquad \text{for every } i \in \mathbb{N}.$$

Then, using (16.8),

$$\|x\|^2 = \sum_{j=1}^{\infty} |(x, e_i)|^2 = \sum_{j=1}^{\infty} \frac{|(y, e_i)|^2}{|\lambda_i - \lambda|^2} \leq \frac{1}{\delta^2} \sum_{j=1}^{\infty} |(y, e_i)|^2 = \frac{\|y\|^2}{\delta^2}.$$

Therefore $(T - \lambda I)^{-1}$ exists and is bounded: thus $\lambda \in \rho(T)$, which shows that $\lambda \notin \sigma(T)$.

We have shown that if $\lambda \notin \overline{\sigma_p(T)}$, then $\lambda \notin \sigma(T)$, which implies that $\sigma(T) \subseteq \overline{\sigma_p(T)}$; since $\sigma_p(T) \subseteq \sigma(T)$ and $\sigma(T)$ is closed (Lemma 14.3) we also have $\overline{\sigma_p(T)} \subseteq \sigma(T)$, so we can conclude that $\sigma(T) = \overline{\sigma_p(T)}$, as claimed.

The proof of Proposition 16.4 shows that the only possible limit point of $\sigma_p(T)$ is 0. Since $\lambda \in \sigma(T) = \overline{\sigma_p(T)}$, either $\lambda \in \sigma_p(T)$ or $\lambda = 0$; in particular, every non-zero $\lambda \in \sigma(T)$ is an eigenvalue.

If H is infinite-dimensional, then we always have $0 \in \sigma(T)$ since T is compact (Theorem 15.8). It follows, since the only possible limit point of $\sigma_p(T)$ is 0, that either $\sigma(T) = \sigma_p(T)$ (if 0 is an eigenvalue of T) or $\sigma(T) = \sigma_p(T) \cup \{0\}$ (if 0 is not an eigenvalue of T). $\qquad \square$

Exercises

16.1 Suppose that H is a complex Hilbert space and $T \in B(H)$. Show that if $(Tx, x) \in \mathbb{R}$ for every $x \in H$, then T is self-adjoint. [Hint: consider the two expressions $(T(x + y), x + y)$ and $(T(x + \mathrm{i}y), x + \mathrm{i}y)$.]

16.2 Show that if $T \in B(H)$ and $(Tx, y) = 0$ for every $x, y \in H$ then $T = 0$. Show that if H is a complex Hilbert space then $(Tx, x) = 0$ for every $x \in H$ implies that $T = 0$. (Kreyszig, 1978)

16.3 Suppose that H is a Hilbert space, $T \in B(H)$ is self-adjoint, and $V(T) \subseteq [0, \beta]$ for some $\beta > 0$, where $V(T)$ is the numerical range of T defined in (16.1). Show that

$$\|Tx\|^2 \le \beta(Tx, x) \qquad x \in X.$$

[Hint: rewrite this inequality as $((\beta I - T)Tx, x) \ge 0$.] (Pryce, 1973)

16.4 Using the result of the previous exercise and adapting the argument of Theorem 16.3 show that $\alpha := \inf V(T)$ and $\beta := \sup V(T)$ are both eigenvalues of T. (Note that Theorem 16.1 shows that either $\alpha = -\|T\|$ or $\beta = \|T\|$ (or perhaps both).)

16.5 Let H be a Hilbert space, $(e_j)_{j=1}^\infty$ an orthonormal sequence in H, and define $T \in L(H)$ by setting

$$Tu := \sum_{j=1}^\infty \lambda_j(u, e_j)e_j,$$

where $(\lambda_j) \in \mathbb{R}$. Show that every λ_j is an eigenvalue of T and that there are no other eigenvalues. Show also that

  (i) T is bounded if and only if (λ_j) is bounded;
  (ii) if (λ_j) is bounded, then T is self-adjoint;
  (iii) T is compact if and only if $\lambda_j \to 0$ as $j \to \infty$;
  (iv) T is Hilbert–Schmidt if and only if $\sum |\lambda_j|^2 < \infty$.
  (Rynne and Youngson, 2008)

16.6 Suppose that $\{e_j(x)\}$ is an orthonormal set in $L^2((a, b); \mathbb{R})$ and that

$$K(x, y) = \sum_{j=1}^\infty \lambda_j e_j(x)e_j(y) \qquad \text{with} \qquad \sum_{j=1}^\infty |\lambda_j|^2 < \infty.$$

Show that the $\{e_j(x)\}$ are eigenvectors of

$$(Tu)(x) = \int_a^b K(x, y)u(y)\,\mathrm{d}y$$

with corresponding eigenvalues λ_j and that there are no other eigenvectors corresponding to non-zero eigenvalues.

16.7 In the setting of the previous exercise, consider
  (i) $(a, b) = (-\pi, \pi)$ and $K(t, s) = \cos(t - s)$; show that the eigenvalues of the corresponding operator T are $\pm 2\pi$ (use Exercise 9.1);
  (ii) $(a, b) = (-1, 1)$ and $K(t, s) = 1 - 3(t - s)^2 + 9t^2s^2$; show that the eigenvalues of the corresponding operator T are 4 and 8/5 (use Example 10.8).

16.8 If T is a compact self-adjoint operator on a Hilbert space H and $(\lambda_j)_{j=1}^{\infty}$, ordered so that $\lambda_{n+1} \leq \lambda_n$, are the eigenvalues of T, then

$$\lambda_{n+1} = \min_{V_n} \max_{x \in V_n^{\perp}, \, x \neq 0} \frac{(Tx, x)}{\|x\|^2}, \qquad (16.9)$$

where the minimum is taken over all n-dimensional subspaces V_n of H. To show this, prove that
  (i) if V is an n-dimensional subspace V, any $(n + 1)$-dimensional subspace W of H contains a vector orthogonal to V; and
  (ii) if $x \in \mathrm{Span}(e_1, \ldots, e_n)$ where (e_j) are orthonormal eigenvectors corresponding the to (λ_j), then

$$\frac{(Tx, x)}{\|x\|^2} \geq \lambda_n.$$

Deduce that (16.9) holds. (Lax, 2002)

16.9 Show that if $T : H \to H$ is compact, then $T_{\mathbb{C}} : H_C \to H_C$ is compact, where $H_{\mathbb{C}}$ and $T_{\mathbb{C}}$ are the complexifications of H and T from Exercises 14.1 and 14.2. Show similarly that if T is self-adjoint, then $T_{\mathbb{C}}$ is self-adjoint.

17

Application: Sturm–Liouville Problems

In this chapter we consider the Sturm–Liouville eigenvalue problem

$$-\frac{d}{dx}\left(p(x)\frac{du}{dx}\right) + q(x)u = \lambda u \qquad \text{with} \qquad u(a) = u(b) = 0. \quad (17.1)$$

Sturm–Liouville problems arise naturally in many situations via the technique of separation of variables.

Throughout this chapter we will assume that $p \in C^1([a, b])$ with $p(x) > 0$ on $[a, b]$ and $q \in C([a, b])$ with $q(x) \geq 0$ on $[a, b]$.

Any $\lambda \in \mathbb{R}$ for which there exists a non-zero function $u \in C^2([a, b])$ such that (17.1) holds is called an *eigenvalue* of this problem, and u is then the corresponding *eigenfunction*.

As a shorthand, we write $L[u]$ for the left-hand side of (17.1), i.e.

$$L[u] = -(p(x)u')' + q(x)u,$$

and then λ is an eigenvalue if $L[u] = \lambda u$ for some non-zero $u \in \mathcal{D}$, where

$$\mathcal{D} := \{u \in C^2([a, b]) : u(a) = u(b) = 0\}.$$

Our aim in this chapter is to show that the eigenfunctions of the Sturm–Liouville problem form an orthonormal basis for $L^2(a, b)$, and to prove some other properties of the eigenvalues and eigenfunctions. Some of these we will be able to prove relatively easily, while others will require the theory developed in the previous chapters. We will sum up all our results in Theorem 17.8 at the end of the chapter.

17.1 Symmetry of L and the Wronskian

Let us note first that L has a useful symmetry property. We use $(\cdot, \cdot)$ for the inner product in $L^2(a, b)$.

Lemma 17.1 *If $u, v \in \mathcal{D}$, then*

$$(L[u], v) = (u, L[v]).$$

For all $u \in \mathcal{D}$ we have $(L[u], u) \geq 0$ with $(L[u], u) = 0$ if and only if $u = 0$.

Proof We integrate by parts and use the boundary conditions:

$$
\begin{aligned}
(L[u], v) &= \int_a^b (-(pu')' + qu)v = -\int_a^b (pu')'v + \int_a^b quv \\
&= \int_a^b pu'v' - \left[pu'v\right]_a^b + \int_a^b quv \quad\quad (17.2) \\
&= -\int_a^b (pv')'u + \left[pv'u\right]_a^b + \int_a^b quv \\
&= -\int_a^b (pv')'u + \int_a^b quv = (u, L[v]).
\end{aligned}
$$

Putting $v = u$ it follows from (17.2) that we have

$$(L[u], u) = \int_a^b p(u')^2 + qu^2,$$

which is non-negative since $p > 0$ and $q \geq 0$ on $[a, b]$. If $(L[u], u) = 0$, then, since $p > 0$ on $[a, b]$, it follows that $u' = 0$ on $[a, b]$, and so u must be constant on $[a, b]$. Since $u(a) = 0$, it follows that $u \equiv 0$. $\qquad\square$

Using this result we can show that eigenvalues are positive and eigenfunctions with different eigenvalues are orthogonal.

Corollary 17.2 *If λ is an eigenvalue of (17.1), then $\lambda > 0$, and if u and v are eigenfunctions corresponding to distinct eigenvalues, then they are orthogonal in $L^2(a, b)$.*

Proof If $L[u] = \lambda u$ and $u \in \mathcal{D}$ is non-zero, then

$$\lambda \|u\|^2 = \lambda(u, u) = (L[u], u) > 0.$$

If $\lambda_1, \neq \lambda_2$ and u_1 and u_2 are non-zero with $L[u_i] = \lambda_i u_i, i = 1, 2$, then

$$\lambda_1(u_1, u_2) = (L[u_1], u_2) = (u_1, L[u_2]) = \lambda_2(u_1, u_2),$$

which implies that $(u_1, u_2) = 0$. $\qquad\square$

The following lemma will play an important role in our subsequent analysis.

Lemma 17.3 *Suppose that $u_1, u_2 \in C^2([a, b])$ are two non-zero solutions of*

$$-\frac{d}{dx}\left(p(x)\frac{du}{dx}\right) + w(x)u = 0,$$

where $p \in C^1([a, b])$ with $p(x) > 0$ and $w \in C([a, b])$. Then the 'Wronskian'

$$W_p(u_1, u_2)(x) := p(x)[u_1'(x)u_2(x) - u_2'(x)u_1(x)]$$

is constant, and W_p is non-zero if and only if u_1 and u_2 are linearly independent.

Note that we do not require $w(x) \geq 0$ for this result. When $w(x) = q(x)$ the functions u_1 and u_2 are not required to satisfy the boundary conditions from (17.1), so they are not eigenfunctions of L.

Proof Differentiate W_p with respect to x, then use that $L[u_1] = L[u_2] = 0$ to substitute for $pu_i'' = qu_i - p'u_i'$ to give

$$\begin{aligned}
W_p' &= p'u_1'u_2 + pu_1''u_2 + pu_1'u_2' - p'u_1u_2' - pu_1'u_2' - pu_1u_2'' \\
&= p'(u_1'u_2 - u_2'u_1) + p(u_1''u_2 - u_2''u_1) \\
&= p'(u_1'u_2 - u_2'u_1) + u_2(wu_1 - p'u_1') - u_1(wu_2 - p'u_2') \\
&= 0.
\end{aligned}$$

For the link between linear independence of u_1, u_2 and W_p, first note that since $p > 0$ and W_p is constant, either $u_1(x)u_2'(x) - u_2(x)u_1(x)' = 0$ for every $x \in [a, b]$ or $u_1(x)u_2(x)' - u_2(x)u_1'(x) \neq 0$ for every $x \in [a, b]$.

Suppose that $\alpha u_1(x) + \beta u_2(x) = 0$ for every $x \in [a, b]$. Then, differentiating this equality, we obtain $\alpha u_1'(x) + \beta u_2'(x) = 0$ for every $x \in [a, b]$, and we can combine these two equations as

$$\begin{pmatrix} u_1(x) & u_2(x) \\ u_1'(x) & u_2'(x) \end{pmatrix}\begin{pmatrix} \alpha \\ \beta \end{pmatrix} = 0.$$

If the determinant of this matrix is non-zero, i.e. if $W_p \neq 0$, then it is invertible and so we must have $\alpha = \beta = 0$, i.e. u_1 and u_2 are linearly independent. On the other hand, if the determinant is zero, then the matrix is not invertible; by Lemma 1.22 this means that its kernel is not trivial, so there exists a non-zero solution (α, β), which implies that u_1 and u_2 are linearly dependent. $\square$

We can now show that all eigenvalues of (17.1) are simple.

Corollary 17.4 *All eigenvalues λ of (17.1) are simple, i.e. the space*

$$E_\lambda := \{u \in \mathcal{D} : L[u] = \lambda u\}$$

has dimension one.

Proof Suppose that $L[u] = \lambda u$ and $L[v] = \lambda v$. Then both u and v are solutions of the problem

$$-(pu')' + (q - \lambda)u = 0, \qquad u(a) = u(b) = 0.$$

It follows from Lemma 17.3 that $W_p(u, v)$ is constant on $[a, b]$; since $W_p(u, v)(a) = 0$, $W_p \equiv 0$ and so u and v are linearly dependent, i.e. E_λ has dimension one. □

17.2 The Green's Function

To go further we will need to use the theory of self-adjoint compact operators developed in the previous chapter. In order to do this, we first turn the differential equation (17.1) into an integral equation by finding the Green's function for the problem: this is a function G such that the solution of $L[u] = f$ can be written as

$$u(x) = \int_a^b G(x, y) f(y) \, dy.$$

Theorem 17.5 *Suppose that $u_1, u_2 \in \mathcal{D}$ are linearly independent solutions of*

$$-\frac{d}{dx}\left(p(x)\frac{dy}{dx}\right) + q(x)y = 0, \tag{17.3}$$

with $u_1(a) = 0$ and $u_2(b) = 0$; set $W := W_p(u_1, u_2)$ and define

$$G(x, y) = \begin{cases} W^{-1}u_1(x)u_2(y) & a \le x < y \\ W^{-1}u_2(x)u_1(y) & y \le x \le b. \end{cases}$$

Then for any $f \in C([a, b])$ the function u given by

$$u(x) = \int_a^b G(x, y) f(y) \, dy \tag{17.4}$$

is an element of $\mathcal{D}$ and $L[u] = f$.

Proof Writing (17.4) out in full we have

$$u(x) = \frac{u_2(x)}{W} \int_a^x u_1(y) f(y) \, dy + \frac{u_1(x)}{W} \int_x^b u_2(y) f(y) \, dy.$$

Now,

$$u'(x) = \frac{u_2(x)u_1(x)}{W}f(x) + \frac{u_2'(x)}{W}\int_a^x u_1(y)f(y)\,dy - \frac{u_1(x)u_2(x)}{W}f(x)$$
$$+ \frac{u_1'(x)}{W}\int_x^b u_2(y)f(y)\,dy$$
$$= \frac{u_2'(x)}{W}\int_a^x u_1(y)f(y)\,dy + \frac{u_1'(x)}{W}\int_x^b u_2(y)f(y)\,dy,$$

and then, since $W = p[u_1'u_2 - u_2'u_1]$,

$$u''(x) = \frac{u_2'(x)u_1(x)}{W}f(x) + \frac{u_2''(x)}{W}\int_a^x u_1(y)f(y)\,dy - \frac{u_1'(x)u_2(x)}{W}f(x)$$
$$+ \frac{u_1''(x)}{W}\int_x^b u_2(y)f(y)\,dy$$
$$= -\frac{f(x)}{p(x)} + \frac{u_2''(x)}{W}\int_a^x u_1(y)f(y)\,dy + \frac{u_1''(x)}{W}\int_x^b u_2(y)f(y)\,dy.$$

These expressions show that $u \in C^2([a,b])$; since $G(a,y) = G(b,y) = 0$ for every $y \in [a,b]$ it follows that $u(a) = u(b) = 0$, so $u \in \mathcal{D}$.

Since $L[u] = -pu'' - p'u' + qu$ and L is linear with $L[u_1] = L[u_2] = 0$ it follows that $L[u] = f(x)$ as claimed. □

The existence of the functions u_1 and u_2 needed for this result can be guaranteed using standard results in the theory of ordinary differential equations (see e.g. Hartman, 1973). We can rewrite (17.3) as the coupled system

$$u' = v$$
$$v' = -\frac{p'}{p}v + \frac{q}{p}u,$$

and solve this as an initial value problem with $u_1(a) = 1$, $v_1(a) = 0$ for u_1 and $u_2(b) = 0$, $v_2(b) = 1$ for u_2.

As a simple illustrative example, consider the case $L[u] = -u''$ on the interval $[0,1]$. The general solution of $-y'' = 0$ is $y(x) = Ax + B$, so in the setting of Theorem 17.5 we can take $u_1(x) = x$ and $u_2(x) = x - 1$, for which $W = u_1'u_2 - u_2'u_1 = -1$. The Green's function for the equation $-u'' = f$ is therefore

$$G(x,y) = \begin{cases} x(1-y) & 0 \le x < y \\ (1-x)y & y \le x \le 1. \end{cases}$$

17.3 Eigenvalues of the Sturm–Liouville Problem

The result of Theorem 17.5 shows that we can define a linear operator $T: C([a, b]) \to \mathcal{D}$ by setting

$$[Tf](x) = \int_a^b G(x, y) f(y) \, dy.$$

However, in order to apply the theory from the previous chapter we need to have an operator that is defined on an appropriate Hilbert space, in this case $L^2(a, b)$. We therefore want to consider an operator of the same form, but defined on this larger space.

We will need the fact that the space $\mathcal{D}$ is dense in $L^2(a, b)$.

Lemma 17.6 *The space*

$$\mathcal{D} := \{f \in C^2([a, b]): \ f(a) = f(b) = 0\}$$

is dense in $L^2(a, b)$ for any $1 \le p < \infty$.

Proof We know from Lemma 7.7 that $\mathcal{P}([a, b])$ is dense in $L^2(a, b)$; since $\mathcal{P}([a, b]) \subset C^2([a, b])$ it follows that $C^2([a, b])$ is dense in $L^2(a, b)$. To show that $\mathcal{D}$ is dense we define a family of 'cutoff functions' $\psi_\delta \in C^2([a, b])$ by setting

$$\psi_\delta(x) = \begin{cases} \phi((x - a)/\delta) & a \le x < a + \delta \\ 1 & a + \delta \le x \le b - \delta \\ \phi((b - x)/\delta) & b - \delta < x \le b, \end{cases}$$

where

$$\phi(x) := 3x - 3x^2 + x^3$$

is such that $\phi(0) = 0$, $\phi(1) = 1$, and $\phi'(1) = \phi''(1) = 0$; see Figure 17.1.

For any $\delta > 0$ and any $g \in C^2([a, b])$ we now have $\psi_\delta(x) g(x) \in \mathcal{D}$, and $\|\psi_\delta g - g\|_{L^2} \to 0$ as $\delta \to 0$. Given any $\varepsilon > 0$ we can first approximate $f \in L^2(a, b)$ by $g \in C^2([a, b])$, and then choose δ small enough that

$$\|\psi_\delta g - f\|_{L^2} \le \|\psi_\delta g - g\|_{L^2} + \|g - f\|_{L^2} < \varepsilon. \qquad \square$$

Theorem 17.7 *The operator $\mathcal{T}: L^2(a, b) \to L^2(a, b)$ defined by setting*

$$\mathcal{T}f(x) := \int_a^b G(x, y) f(y) \, dy,$$

where G is as in Theorem 17.5, is compact and self-adjoint and has trivial kernel, i.e. $\mathrm{Ker}(\mathcal{T}) = \{0\}$. The eigenvalues of $\mathcal{T}$ form a countable sequence tending

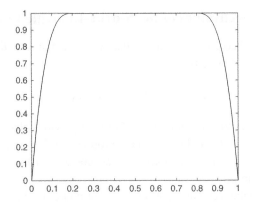

Figure 17.1 The cutoff function ψ_δ (here shown for $a = 0$, $b = 1$, and $\delta = 0.2$) is used to turn a function $g \in C^2([0, 1])$ into an element $\psi_\delta g$ that is zero at $x = 0$ and $x = 1$.

to zero, all the eigenvectors of $\mathfrak{T}$ are elements of $\mathcal{D}$, and these eigenvectors (suitably normalised) form an orthonormal basis for $L^2(a, b)$.

Proof Since $G(x, y)$ is symmetric and bounded, it follows from Example 11.10 and Lemma 13.6 that $\mathfrak{T}$ is a bounded self-adjoint operator from $L^2(a, b)$ into itself; Proposition 15.4 shows that $\mathfrak{T}$ is compact. That the eigenvalues form a countable sequence tending to zero follows from Proposition 16.4.

To show that $\mathfrak{T}f = 0$ implies that $f = 0$, we note first that $\mathcal{D}$ is contained in the range of $\mathfrak{T}$. Indeed, given any $u \in \mathcal{D}$ we can define $g = L[u] \in C([a, b])$, and then by Theorem 17.5 we have $\mathfrak{T}g = u$. If $\mathfrak{T}f = 0$, then for any $g \in \mathcal{D}$ we have $g = \mathfrak{T}\phi$ for some $\phi \in L^2(a, b)$, so we have

$$0 = (\mathfrak{T}f, \phi) = (f, \mathfrak{T}\phi) = (f, g);$$

since $\mathcal{D}$ is dense in $L^2(a, b)$ it follows from Exercise 9.9 that $f = 0$. So $\mathrm{Ker}(\mathfrak{T}) = \{0\}$; in particular, $\mathfrak{T}$ has no non-zero eigenvalues.

As a first step towards showing that the eigenvectors of $\mathfrak{T}$ are elements of $\mathcal{D}$, we show that $\mathfrak{T}f \in C^0([a, b])$ for any $f \in L^2(a, b)$. Since G is continuous on the compact set $[a, b] \times [a, b]$, it is uniformly continuous, so given any $\varepsilon > 0$ there exists $\delta > 0$ such that

$$|G(x, y) - G(x', y)| < \varepsilon \qquad \text{whenever} \quad |x - x'| < \delta, \text{ for all } y \in [a, b].$$

Therefore, if we take $x, x' \in [a, b]$ with $|x - x'| < \delta$ we obtain

$$|\mathcal{T}f(x) - \mathcal{T}f(x')| = \int_a^b [G(x, y) - G(x', y)] f(y) \, dy$$

$$\leq \int_a^b |G(x, y) - G(x', y)| \, |f(y)| \, dy$$

$$\leq \varepsilon \int_a^b |f(y)| \, dy \leq \varepsilon \left(\int_a^b 1 \, dy \right)^{1/2} \left(\int_a^b |f(y)|^2 \, dy \right)^{1/2}$$

$$\leq \varepsilon (b - a)^{1/2} \|f\|_{L^2},$$

so $\mathcal{T}f \in C([a, b])$.

Now observe that if $f \in L^2(a, b)$ with $\mathcal{T}f = \mu f$, then $\mu \neq 0$, so we can write

$$f = \frac{1}{\mu} \mathcal{T}f. \tag{17.5}$$

We have just shown that if $f \in L^2$, then $\mathcal{T}f \in C([a, b])$, so it follows immediately from (17.5) that $f \in C([a, b])$. But $\mathcal{T}$ is the same as T whenever $f \in C([a, b])$, and we know from Theorem 17.5 that when $f \in C([a, b])$ we have $Tf \in \mathcal{D}$; using (17.5) again we conclude that $f \in \mathcal{D}$ as required. (This is an example of the well-established method of 'bootstrapping' to prove regularity of solutions of differential equations.)

Since $\mathrm{Ker}(\mathcal{T}) = \{0\}$, it follows from Corollary 16.7 that the eigenvectors of $\mathcal{T}$ form an orthonormal basis for H. □

We now combine all the results in this chapter.

Theorem 17.8 *The eigenfunctions of the Sturm–Liouville problem form an orthonormal basis for $L^2(a, b)$. Furthermore,*

(i) every eigenvalue is strictly positive;
(ii) each eigenvalue is simple (i.e. has multiplicity one); and
(iii) the eigenvalues can be ordered to form a countable sequence that tends to infinity.

Proof We have already proved (i) in Corollary 17.2 and (ii) in Corollary 17.4. Property (iii) and the fact that the eigenfunctions form a basis derive from the observation that if $u \in \mathcal{D}$ with $L[u] = \lambda u$, then $u = T(\lambda u)$ and vice versa. Since zero is not an eigenvalue of either L or $\mathcal{T}$, it follows that

$$L[u] = \lambda u \qquad \Leftrightarrow \qquad \mathcal{T}u = \frac{1}{\lambda} u, \tag{17.6}$$

i.e. the eigenfunctions of L are precisely the eigenvectors of $\mathcal{T}$, with the eigen-values following the reciprocal relationship in (17.6). Given this, Theorem 17.7 concludes the proof.   □

This gives us another proof that we can expand functions in $L^2(0, 1)$ using a Fourier sine series (see Corollary 6.6). Indeed, if we consider

$$-\frac{d^2u}{dx^2} = \lambda u \qquad u(0) = 0, \quad u(1) = 0, \tag{17.7}$$

which is (17.1) with $p = 1$, $q = 0$, it follows that the eigenfunctions of this equation will form a basis for $L^2(0, 1)$.

To find the eigenvalues and eigenfunctions for this problem, observe that the solutions of $-u'' = \lambda u$ are potentially of three kinds, depending on λ. If $\lambda > 0$, then $u(x) = Ae^{\sqrt{\lambda}x} + Be^{-\sqrt{\lambda}x}$. To satisfy the boundary conditions in (17.7) we would need

$$A + B = Ae^{\sqrt{\lambda}} + Be^{-\sqrt{\lambda}} = 0,$$

which implies that $A = B = 0$ so no $\lambda < 0$ can be an eigenvalue. Similarly if $\lambda = 0$, then the general solution of $-u'' = 0$ is $u(x) = Ax + B$, and the boundary conditions require $B = A + B = 0$ which again yields $A = B = 0$ and $\lambda = 0$ is not an eigenvalue. If $\lambda < 0$, then the general solution is

$$u(x) = A \cos \sqrt{\lambda}x + B \sin \sqrt{\lambda}x,$$

and the boundary conditions require $A = 0$ and $B \sin \sqrt{\lambda} = 0$; to ensure that $B \neq 0$ we have to take λ such that $\sin \sqrt{\lambda} = 0$, i.e. $\lambda = (k\pi)^2$.

The eigenvalues are therefore $\lambda_k = (k\pi)^2$, with corresponding eigenfunctions $\sin k\pi x$. It follows that the set of normalised eigenfunctions

$$\left\{ \frac{1}{\sqrt{2}} \sin k\pi x \right\}_{k=1}^{\infty}$$

form an orthonormal basis for $L^2(0, 1)$, so any $f \in L^2(0, 1)$ can be expanded in the form

$$f(x) = \sum_{k=1}^{\infty} \alpha_k \sin k\pi x,$$

with the sum converging in $L^2(a, b)$.

PART IV

Banach Spaces

18

Dual Spaces of Banach Spaces

In Chapter 12 we introduced the dual space X^* of a normed linear space X (over a field $\mathbb{K}$), the space $B(X, \mathbb{K})$. Recall that since $\mathbb{K}$ is complete, Theorem 11.11 guarantees that when X is a normed space its dual space is always a Banach space with norm

$$\|\phi\|_{X^*} = \sup_{\|x\| \le 1} |\phi(x)|. \tag{18.1}$$

We remarked in Chapter 12 that knowledge of the action of elements of the dual space X^* can tell us a lot about the space X. We give a simple example.

Example 18.1 Take $X = C([a, b])$, and for any choice of $p \in [a, b]$ consider $\delta_p \colon X \to \mathbb{R}$ defined by setting by

$$\delta_p(\phi) = \phi(p) \qquad \text{for all} \qquad \phi \in X.$$

Then

$$|\delta_p(\phi)| = |\phi(p)| \le \|\phi\|_\infty,$$

so that $\delta_p \in X^*$ with $\|\delta_p\|_{X^*} \le 1$. Choosing a function $\phi \in C([a, b])$ such that $|\phi(p)| = \|\phi\|_\infty$ shows that in fact $\|\delta_p\|_{X^*} = 1$.

Observe here that if we knew the value of $f(\phi)$ for every $f \in X^*$, then we would know ϕ.

In this first chapter of Part IV we return to the more general theory of dual spaces, which will play a major role in our study of Banach spaces. We identify the dual spaces of the ℓ^p sequence spaces, and discuss briefly the corresponding results for the spaces L^p of Lebesgue integrable functions.

We showed in Chapter 12 that in a Hilbert space H, given any $y \in H$ we can define an element $f \in H^*$ by setting

$$f(x) = (x, y), \tag{18.2}$$

and that the resulting map $R\colon H \to H^*$ ('the Riesz map') defined by

$$y \mapsto (\cdot, y)$$

is a bijective isometry, which is linear if $\mathbb{K} = \mathbb{R}$ and conjugate-linear if $\mathbb{K} = \mathbb{C}$. In particular, given any $f \in H^*$ there exists some $y \in H$ such that (18.2) holds and $\|y\|_H = \|f\|_{H^*}$.

When $H = \ell^2(\mathbb{R})$ this shows that $\ell^2 \equiv (\ell^2)^*$ via the map $x \mapsto f_x$, where

$$f_x(y) = (y, x) = \sum_{j=1}^{\infty} x_j y_j.$$

We now use the same map to investigate the dual spaces of ℓ^p, $1 \le p < \infty$, and of c_0. First we prove two very useful inequalities.

18.1 The Young and Hölder Inequalities

We say that two indices $1 \le p, q \le \infty$ are *conjugate* if

$$\frac{1}{p} + \frac{1}{q} = 1,$$

allowing $p = 1, q = \infty$ and $p = \infty, q = 1$. The following simple inequality is fundamental.

Lemma 18.2 (Young's inequality) *Let $a, b > 0$ and let (p, q) be conjugate indices with $1 < p, q < \infty$. Then*

$$ab \le \frac{a^p}{p} + \frac{b^q}{q}.$$

Proof The function $x \mapsto e^x$ is convex (see Exercise 3.4) and so

$$ab = \exp(\log a + \log b) = \exp\left(\frac{1}{p}\log a^p + \frac{1}{q}\log b^q\right)$$
$$\le \frac{1}{p}e^{\log(a^p)} + \frac{1}{q}e^{\log(b^q)} = \frac{a^p}{p} + \frac{b^q}{q}. \qquad \square$$

Using Young's inequality we can prove Hölder's inequality in ℓ^p.

Lemma 18.3 (Hölder's inequality in ℓ^p spaces) *If $x \in \ell^p$ and $y \in \ell^q$ with (p, q) conjugate, $1 \le p, q \le \infty$, then*

$$\sum_{j=1}^{\infty} |x_j y_j| \le \|x\|_{\ell^p} \|y\|_{\ell^q}. \tag{18.3}$$

Proof For $1 < p < \infty$, consider

$$\sum_{j=1}^{n} \frac{|x_j|}{\|x\|_{\ell^p}} \frac{|y_j|}{\|y\|_{\ell^q}} \leq \sum_{j=1}^{n} \frac{1}{p} \frac{|x_j|^p}{\|x\|_{\ell^p}^p} + \frac{1}{q} \frac{|y_j|^q}{\|y\|_{\ell^q}^q} \leq 1.$$

So for each $n \in \mathbb{N}$

$$\sum_{j=1}^{n} |x_j y_j| \leq \|x\|_{\ell^p} \|y\|_{\ell^q}$$

and (18.3) follows on letting $n \to \infty$. For $p = 1, q = \infty$,

$$\sum_{j=1}^{n} |x_j y_j| \leq \max_{j=1,\dots,n} |y_j| \left(\sum_{j=1}^{n} |x_j| \right) \leq \|x\|_{\ell^1} \|y\|_{\ell^\infty}. \qquad \square$$

The Hölder inequality in L^p is proved in a similar way.

Theorem 18.4 (Hölder's inequality in L^p spaces) *Suppose that $f \in L^p(\Omega)$ and $g \in L^q(\Omega)$, with (p, q) conjugate, $1 \leq p, q \leq \infty$; then $fg \in L^1(\Omega)$ with*

$$\|fg\|_{L^1} \leq \|f\|_{L^p} \|g\|_{L^q}. \qquad (18.4)$$

Proof If $1 < p < \infty$, then we use Young's inequality ($ab \leq a^p/p + b^q/q$ from Lemma 18.2) to give

$$
\begin{aligned}
\int_\Omega \frac{|fg|}{\|f\|_{L^p} \|g\|_{L^q}} &= \int_\Omega \frac{|f|}{\|f\|_{L^p}} \frac{|g|}{\|g\|_{L^q}} \\
&\leq \int_\Omega \frac{1}{p} \frac{|f|^p}{\|f\|_{L^p}^p} + \frac{1}{q} \frac{|g|^q}{\|g\|_{L^q}^q} \\
&\leq \frac{1}{p\|f\|_{L^p}^p} \int_\Omega |f|^p + \frac{1}{q\|g\|_{L^q}^q} \int_\Omega |g|^q \\
&= 1,
\end{aligned}
$$

from which (18.4) follows. If $p = 1, q = \infty$, then

$$\int_\Omega |f(x)g(x)| \, dx \leq \int_\Omega |f(x)| \|g\|_{L^\infty} \, dx = \|f\|_{L^1} \|g\|_{L^\infty},$$

and similarly if $p = \infty$ and $q = 1$. $\qquad \square$

Hölder's inequality allows for an alternative proof of Minkowski's inequality (the triangle inequality in ℓ^p and L^p) that we proved using a convexity argument in Lemma 3.9; see Exercise 18.2.

18.2 The Dual Spaces of ℓ^p

We can now identify (up to isometric isomorphism) the dual space of ℓ^p for $1 < p < \infty$.

Theorem 18.5 *For $1 < p < \infty$ we have $(\ell^p) \equiv (\ell^q)^*$, with (p, q) conjugate, via the mapping $x \mapsto L_x$, where*

$$L_x(y) = \sum_{j=1}^{\infty} x_j y_j. \qquad (18.5)$$

We denote this mapping $x \mapsto L_x$ by $T_q \colon \ell^p \to (\ell^q)^*$.

(The case $p = q = 2$ here follows from the Riesz Representation Theorem when H is real.)

Proof Given $x \in \ell^p$ define L_x as in (18.5) above; this is clearly a linear map. Then from Hölder's inequality

$$|L_x(y)| = \left| \sum_{j=1}^{\infty} x_j y_j \right| \leq \sum_{j=1}^{\infty} |x_j y_j| \leq \|x\|_{\ell^p} \|y\|_{\ell^q}, \qquad (18.6)$$

so we do indeed have $L_x \in (\ell^q)^*$ with

$$\|L_x\|_{(\ell^q)^*} \leq \|x\|_{\ell^p}. \qquad (18.7)$$

To show that we have $\|L_x\|_{(\ell^q)^*} = \|x\|_{\ell^p}$, consider the element $y \in \ell^q$ given by

$$y_j = \begin{cases} |x_j|^p / x_j & x_j \neq 0 \\ 0 & x_j = 0; \end{cases}$$

this is in ℓ^q since

$$\|y\|_{\ell^q}^q = \sum_{j=1}^{\infty} |y_j|^q = \sum_{j=1}^{\infty} |x_j|^{q(p-1)} = \sum_{j=1}^{\infty} |x_j|^p = \|x\|_{\ell^p}^p < \infty,$$

where we have used the fact that $q(p - 1) = p$. We can rewrite this equality as

$$\|y\|_{\ell^q} = \|x\|_{\ell^p}^{p/q} = \|x\|_{\ell^p}^{p-1},$$

and so therefore

$$|L_x(y)| = \left| \sum_{j=1}^{\infty} x_j y_j \right| = \sum_{j=1}^{\infty} |x_j|^p = \|x\|_{\ell^p}^p = \|x\|_{\ell^p}^{p-1} \|x\|_{\ell^p} = \|y\|_{\ell^q} \|x\|_{\ell^p}.$$

This shows that $\|L_x\|_{(\ell^q)^*} \geq \|x\|_{\ell^p}$ and hence (combining this with the bound in (18.7)) that $x \mapsto L_x$ (the map T_q) is a linear isometry; it follows that

T_q is injective. So we only need to show that it is surjective to show that it is an isomorphism: we need to show that any $f \in (\ell^q)^*$ can be written as L_x for some $x \in \ell^p$.

If we do in fact have $f = L_x$ for some $x \in \ell^p$, then for each $e^{(j)}$ (recall that $e_i^{(j)} = \delta_{ij}$) we must have

$$f(e^{(j)}) = L_x(e^{(j)}) = x_j,$$

which tells us what x must be. Let x be the sequence with $x_j = f(e^{(j)})$; if we can show that $x \in \ell^p$, then we will have finished the proof. Indeed, for any $y \in \ell^q$ we can write $y = \sum_{j=1}^{\infty} y_j e^{(j)}$ and then, using the continuity of f,

$$f(y) = f\left(\sum_{j=1}^{\infty} y_j e^{(j)}\right) = f\left(\lim_{n\to\infty} \sum_{j=1}^{n} y_j e^{(j)}\right) = \lim_{n\to\infty} f\left(\sum_{j=1}^{n} y_j e^{(j)}\right)$$

$$= \lim_{n\to\infty} \sum_{j=1}^{n} y_j f(e^{(j)}) = \lim_{n\to\infty} \sum_{j=1}^{n} y_j x_j = L_x(y),$$

as required.

So we need only show that x defined by setting $x_j = f(e^{(j)})$ as above is an element of ℓ^p. To do this we consider the sequence $\phi^{(n)}$ of elements of ℓ^q defined by

$$\phi_j^{(n)} = \begin{cases} |x_j|^p / x_j & j \le n \text{ and } x_j \ne 0 \\ 0 & j > n \text{ or } x_j = 0; \end{cases}$$

then

$$f(\phi^{(n)}) = f\left(\sum_{j=1}^{n} \phi_j^{(n)} e^{(j)}\right) = \sum_{j=1}^{n} \phi_j^{(n)} f(e^{(j)}) = \sum_{j=1}^{n} \phi_j^{(n)} x_j = \sum_{j=1}^{n} |x_j|^p,$$

and so

$$\sum_{j=1}^{n} |x_j|^p = |f(\phi^{(n)})| \le \|f\|_{(\ell^q)^*} \|\phi^{(n)}\|_{\ell^q}$$

$$= \|f\|_{(\ell^q)^*} \left(\sum_{j=1}^{n} |x_j|^{q(p-1)}\right)^{1/q}$$

$$= \|f\|_{(\ell^q)^*} \left(\sum_{j=1}^{n} |x_j|^p\right)^{1/q},$$

which shows that

$$\left(\sum_{j=1}^{n} |x_j|^p\right)^{1/p} \leq \|f\|_{(\ell^q)^*} \qquad \text{for every } n \in \mathbb{N}$$

and so $x \in \ell^p$ as required. □

We have a similar result for one of the endpoint values (ℓ^∞) but not for the other (ℓ^1).

Theorem 18.6 *We have $\ell^\infty \equiv (\ell^1)^*$ via the mapping (18.5), which we denote by $T_1\colon \ell^\infty \to (\ell^1)^*$.*

Proof Given $x \in \ell^\infty$, Hölder's inequality as in (18.6) shows that L_x defined as in (18.5) is an element of $(\ell^1)^*$ with

$$\|L_x\|_{(\ell^1)^*} \leq \|x\|_{\ell^\infty}.$$

The equality of norms follows by choosing, for each $\varepsilon > 0$, a $j \in \mathbb{N}$ such that $|x_j| > \|x\|_{\ell^\infty} - \varepsilon$, and then considering $y \in \ell^1$ with

$$y_i = \begin{cases} x_i/|x_i| & x_i \neq 0 \\ 0 & x_i = 0. \end{cases}$$

Then

$$|L_x(y)| = |x_j| \geq (\|x\|_{\ell^\infty} - \varepsilon)\,\|y\|_{\ell^1},$$

since $\|y\|_{\ell^1} = 1$. Since this is valid for any $\varepsilon > 0$, it follows that

$$\|L_x\|_{(\ell^1)^*} = \|x\|_{\ell^\infty}.$$

To show that the map $x \mapsto L_x$ is onto we use the same argument as before and, given $f \in (\ell^1)^*$, consider x defined by setting $x_j = f(e^{(j)})$. It is easy to see that this is an element of ℓ^∞, since $e^{(j)} \in \ell^1$ and

$$|x_j| = |f(e^{(j)})| \leq \|f\|_{(\ell^1)^*}\|e^{(j)}\|_{\ell^1} = \|f\|_{(\ell^1)^*}. \qquad \square$$

The arguments used above do not work for ℓ^∞; primarily because the elements $(e^{(j)})$ do not form a basis (see Example 9.2). We will prove later that this is not just a failure in our method of proof, and that $(\ell^\infty)^* \not\cong \ell^1$. However, the dual of c_0 is isometrically isomorphic to ℓ^1; see Exercise 18.4 for a proof.

Theorem 18.7 $(c_0)^* \equiv \ell^1$.

18.3 Dual Spaces of $L^p(\Omega)$

Using Hölder's inequality it is straightforward to show that any $g \in L^q(\Omega)$ gives rise to an element $\Phi_g \in (L^p(\Omega))^*$ by setting

$$\Phi_g(f) := \int_\Omega fg \, dx, \qquad f \in L^p(\Omega). \tag{18.8}$$

That this mapping is surjective requires more advanced results from measure theory, so we only give a very brief sketch of this part of the following result; the full proof can be found in Appendix B. (See Exercise 18.6 for a proof of this result in the range $1 < p < 2$ that uses the Riesz Representation Theorem and the Monotone Convergence Theorem; see also Exercise 26.5 for an alternative proof that uses uniform convexity.)

Theorem 18.8 *For $1 \le p < \infty$ the space $L^q(\Omega)$ is isometrically isomorphic to $(L^p(\Omega))^*$, where (p, q) are conjugate, via the mapping $g \mapsto \Phi_g$ defined in (18.8).*

Proof If we take $g \in L^q$ and define Φ_g as in (18.8), then this map is linear in f since the integral is linear and, using Hölder's inequality,

$$|\Phi_g(f)| = \left| \int_\Omega fg \right| \le \|fg\|_{L^1} \le \|f\|_{L^p} \|g\|_{L^q}$$

so that $\|\Phi_g\|_{(L^p)^*} \le \|g\|_{L^q}$. Furthermore, if we choose

$$f(x) = \begin{cases} g(x)|g(x)|^{q-2} & g(x) \ne 0 \\ 0 & g(x) = 0 \end{cases}$$

then, since

$$|f(x)|^p \le |g(x)|^{p(q-1)} = |g(x)|^q,$$

it follows that $f \in L^p$ with

$$\|f\|_{L^p}^p = \int |f(x)|^p = \int |g(x)|^q = \|g\|_{L^q}^q$$

and

$$|\Phi_g(f)| = \int_\Omega |g|^q = \|g\|_{L^q}^q.$$

Therefore

$$\|g\|_{L^q}^q = |\Phi_g(f)| \le \|\Phi_g\|_{(L^p)^*} \|f\|_{L^p} = \|\Phi_g\|_{(L^p)^*} \|g\|_{L^q}^{q/p},$$

which, since $q - q/p = q(1 - 1/p) = 1$, yields $\|\Phi_g\|_{(L^p)^*} \geq \|g\|_{L^q}$, from which it follows that

$$\|\Phi_g\|_{(L^p)^*} = \|g\|_{L^q}.$$

That any element of $(L^p)^*$ can be obtained in this way requires some more powerful results from measure theory. For any $A \subseteq \Omega$ let

$$\chi_A(x) = \begin{cases} 1 & x \in A \\ 0 & x \notin A \end{cases}$$

denote the characteristic function of A. Then, for any $\Phi \in (L^q)^*$, the map

$$A \mapsto \Phi(\chi_A),$$

defined on the measurable sets, determines a signed measure on Ω; the Radon–Nikodym Theorem then implies that there is a $g \in L^1(\Omega)$ such that

$$\Phi(\chi_A) = \int_A g \, dx,$$

and one can then verify that $g \in L^q(\Omega)$ and that $\Phi_g(f) = \Phi(f)$ for every $f \in L^p(\Omega)$. For details see Theorem B.16 in Appendix B. □

Note that coupled with (18.1) the result of the previous theorem gives one way (which is sometimes useful) to find the L^q norm of f for any $1 \leq q < \infty$:

$$\|f\|_{L^q(\Omega)} = \sup_{\|g\|_{L^p(\Omega)}=1} \left| \int_\Omega f(x)g(x) \, dx \right|,$$

where p and q are conjugate.

(In a space of real functions we do not need the modulus signs around the integral.)

Exercises

18.1 Prove Young's inequality by minimising the function

$$f(t) = \frac{t^p}{p} + \frac{1}{q} - t$$

and then setting $t = ab^{-q/p}$.

18.2 One of the standard ways of proving the Minkowski inequality for the ℓ^p norm on $\mathbb{K}^n$ (and similarly for the L^p norm on $L^p(\Omega)$) starts by writing

$$\sum_{j=1}^{n} |x_j + y_j|^p = \sum_{j=1}^{n} |x_j + y_j|^{p-1} |x_j + y_j|$$

$$\leq \sum_{j=1}^{n} |x_j + y_j|^{p-1} |x_j| + \sum_{j=1}^{n} |x_j + y_j|^{p-1} |y_j|;$$

use Hölder's inequality to complete the proof.

18.3 Suppose that $|\Omega| = \int_{\Omega} 1 \, dx < \infty$. Use Hölder's inequality to show that for $1 \leq p \leq q \leq \infty$ we have $L^q(\Omega) \subset L^p(\Omega)$ with

$$\|f\|_{L^p} \leq |\Omega|^{(q-p)/pq} \|f\|_{L^q}. \tag{18.9}$$

Show, however, that $L^q(\mathbb{R}^n) \not\subset L^p(\mathbb{R}^n)$ for p, q in the same range.

18.4 By following the proof of Theorem 18.5 show that $(c_0)^* \equiv \ell^1$.

18.5 Show that $C([-1, 1])^*$ is not separable.[1] (Find an uncountable collection of functions in $C([-1, 1])^*$ that are a distance 2 apart in the $C([-1, 1])^*$ norm.) (Lax, 2002)

18.6 For another proof that $(L^p)^* \equiv L^q$ for $1 < p < 2$, whose only measure-theoretic ingredient is the Monotone Convergence Theorem, show that
   (i) $\|f\|_{L^p} \leq |\Omega|^{1/p-1/2} \|f\|_{L^2}$, where $|\Omega| = \int_{\Omega} 1 \, dx$;
   (ii) if $\ell \in (L^p)^*$, then $\ell \in (L^2)^*$.
   The Riesz Representation Theorem therefore guarantees that there exists $g \in L^2$ such that

$$\ell(f) = (f, g) \qquad \text{for every } f \in L^2. \tag{18.10}$$

(iii) By considering the functions

$$f_k(x) = \begin{cases} |g_k(x)|^{q-2} g_k(x) & g_k(x) \neq 0 \\ 0 & g_k(x) = 0, \end{cases}$$

   where

$$g_k(x) = \begin{cases} -k & g(x) < -k \\ g(x) & |g(x)| \leq k \\ k & g(x) > k, \end{cases}$$

use the Monotone Convergence Theorem (Theorem B.7) to show that $g \in L^q$. (Lax, 2002)

[1] For identification of the space $C([a, b])^*$ see e.g. Theorem 5.6.5 in Brown and Page (1970) or Chapter 13 in Meise and Vogt (1997).

19

The Hahn–Banach Theorem

Often we want to define linear functionals on a Banach space that have particular properties. One way of doing this is to use the Hahn–Banach Theorem, which guarantees that a linear functional defined on a subspace U of a normed space X can be extended to a linear functional defined on the whole of X without increasing its norm. In other words, given a bounded linear map $\phi \colon U \to \mathbb{K}$, we can find another bounded linear map $f \colon X \to \mathbb{K}$ such that

$$f(u) = \phi(u) \text{ for every } u \in U \qquad \text{and} \qquad \|f\|_{X^*} = \|\phi\|_{U^*}.$$

Note that since U is a subspace of X it naturally inherits the norm from X, and $\|\phi\|_{U^*}$ is understood in this way, i.e.

$$\|\phi\|_{U^*} = \sup\{|\phi(u)| : u \in U, \ \|u\|_X = 1\},$$

cf. the definition of the norm in X^* in (18.1).

We will prove the Hahn–Banach Theorem in a more general form than this, first in Section 19.1 for real vector spaces, and then – after some preparation – for complex spaces in Section 19.2. For the very simple proof in a Hilbert space see Exercise 19.1.

19.1 The Hahn–Banach Theorem: Real Case

We now prove the Hahn–Banach Theorem in an arbitrary vector space, using Zorn's Lemma.[1]

---

[1] It is possible to prove a version of the Hahn–Banach Theorem in a *separable* normed space without using Zorn's Lemma; see Exercise 19.7. Note, however, that this would exclude the cases of ℓ^∞ and L^∞ (among others).

We first prove the theorem in a real vector space and then use this version (after some careful preparation) to deduce a similar result in a complex vector space.

Definition 19.1 If V is a vector space, then a function $p \colon V \to \mathbb{R}$ is

- *sublinear* if for $x, y \in V$

$$p(x + y) \le p(x) + p(y) \qquad \text{and} \qquad p(\lambda x) = \lambda p(x), \ \lambda \in \mathbb{R}, \ \lambda \ge 0;$$

- a *seminorm* if for $x, y \in V$

$$p(x + y) \le p(x) + p(y) \qquad \text{and} \qquad p(\lambda x) = |\lambda| p(x), \ \lambda \in \mathbb{K}.$$

Note that if p is a seminorm, then $p(x) \ge 0$ for every $x \in X$ (see Exercise 19.4 for this and other properties of seminorms). Note also that if $\| \cdot \|$ is a norm on X, then $M \| \cdot \|$ defines a seminorm on X for any $M \ge 0$.

Theorem 19.2 (Real Hahn–Banach Theorem) *Let X be a real vector space and U a subspace of X. Suppose that $\phi \colon U \to \mathbb{R}$ is linear and satisfies*

$$\phi(x) \le p(x) \qquad \text{for all} \quad x \in U$$

for some sublinear map $p \colon X \to \mathbb{R}$. Then there exists a linear map $f \colon X \to \mathbb{R}$ such that $f(x) = \phi(x)$ for all $x \in U$ and

$$f(x) \le p(x) \qquad \text{for all} \quad x \in X.$$

Furthermore, if p is a seminorm, then

$$|f(x)| \le p(x) \qquad \text{for all} \quad x \in X.$$

In particular, if X is a normed vector space, then any $\phi \in U^*$ has an extension $f \in X^*$ with $\|f\|_{X^*} = \|\phi\|_{U^*}$.

The applications in the following chapter mostly use the final version of the result as stated above (the extension of bounded linear functionals in normed vector spaces), but we will also use the more general version that allows bounds by a sublinear map in Section 20.5 and Chapter 21.

We already used Zorn's Lemma in Chapter 1 to show that every vector space has a basis. We will not, therefore, recall all the terminology here, but do restate the lemma itself.

Lemma 19.3 (Zorn's Lemma) *Let $\mathcal{P}$ be a non-empty partially ordered set. If every chain in $\mathcal{P}$ has an upper bound, then $\mathcal{P}$ has at least one maximal element.*

Proof of the real Hahn–Banach Theorem. We will consider all possible linear extensions g of ϕ satisfying the bound $g(x) \le p(x)$, and apply Zorn's Lemma to deduce that there is a 'maximal extension'. We then argue by contradiction to show that this maximal extension must be defined on the whole of X.

More precisely, we consider the collection $\mathcal{P}$ of all pairs (G, g), where G is a subspace of X that contains U and $g \colon G \to \mathbb{R}$ is a linear functional such that $g = \phi$ on U with

$$g(x) \le p(x) \qquad \text{for every } x \in G.$$

Clearly $\mathcal{P}$ is non-empty, since $(U, \phi) \in \mathcal{P}$.

We define an order on $\mathcal{P}$ by declaring that $(G, g) \preceq (H, h)$ if h is an extension of g, i.e. $H \supseteq G$ and $h = g$ on G.

To apply Zorn's Lemma we have to show that every chain has an upper bound. Indeed, if $\mathcal{C} = \{(G_\alpha, g_\alpha) : \alpha \in \mathbb{A}\}$ is a chain, then an upper bound for $\mathcal{C}$ is the pair (G_∞, g_∞), where

$$G_\infty = \bigcup_{\alpha \in \mathbb{A}} G_\alpha$$

and

$$g_\infty(x) := g_\alpha(x) \qquad \text{whenever} \qquad x \in G_\alpha. \tag{19.1}$$

Note that g_∞ is well defined, since any two elements in $\mathcal{C}$ are ordered: indeed, if $x \in G_\beta \cap G_\alpha$, then either $(G_\beta, g_\beta) \preceq (G_\alpha, g_\alpha)$ (or vice versa), and we know that $g_\alpha = g_\beta$ on G_β, since g_α extends g_β, so there is no ambiguity in the definition in (19.1)

Similarly, g_∞ satisfies $g_\infty(x) \le p(x)$ for every $x \in G_\infty$, since this bound holds for g_α whenever $x \in G_\alpha$. Finally, to check that g_∞ is linear, observe that if $x, y \in G_\infty$, then we have $x \in G_\alpha$ and $y \in G_\beta$, and then either $G_\alpha \supseteq G_\beta$ or $G_\beta \supseteq G_\alpha$. Supposing the former then $x, y \in G_\alpha$ and so

$$g_\infty(x + \lambda y) = g_\alpha(x + \lambda y) = g_\alpha(x) + \lambda g_\alpha(y) = g_\infty(x) + \lambda g_\infty(y);$$

we argue similarly if $G_\beta \supseteq G_\alpha$.

Since any chain has an upper bound, Zorn's Lemma now guarantees that $\mathcal{P}$ has a maximal element (Y, f). We want to show that $Y = X$.

Suppose, for a contradiction, that $Y \ne X$, in which case there exists an element $z \in X \setminus Y$; we want to show that in this case we can extend f to the linear span of $Y \cup \{z\}$, which is a space strictly larger than Y. This will contradict the maximality of (Y, f), and so it will follow that $Y = X$ and f is the extension of ϕ to the whole of X required by the theorem.

Any such linear extension F of f must satisfy

$$F(u + \alpha z) = F(u) + \alpha F(z) = f(u) + \alpha F(z) \quad \text{for every } u \in Y, \ \alpha \in \mathbb{R};$$

the only freedom we have is to choose $F(z) = c$ for some $c \in \mathbb{R}$; the issue is how to choose c so that F is still bounded above by p, i.e. so that

$$f(u) + \alpha c \le p(u + \alpha z) \tag{19.2}$$

for every choice of $\alpha \in \mathbb{R}$ and every $u \in Y$. We know (by assumption) that (19.2) holds for $\alpha = 0$, so we have to guarantee that we can find c such that (i) for $\alpha > 0$ we have (dividing by α)

$$c \le p\left(\frac{u}{\alpha} + z\right) - \frac{f(u)}{\alpha} \quad \text{for every } u \in Y$$

and (ii) for every $\alpha < 0$ we have (dividing by $-\alpha$)

$$c \ge \frac{f(u)}{-\alpha} - p\left(\frac{u}{-\alpha} - z\right) \quad \text{for every } u \in Y.$$

Since f is linear, this is the same as requiring

$$f\left(\frac{u}{\alpha}\right) - p\left(\frac{u}{\alpha} - z\right) \le c \le p\left(\frac{u}{\alpha} + z\right) - f\left(\frac{u}{\alpha}\right)$$

for every $u \in Y$ and every $\alpha > 0$, and since Y is a linear subspace this is just the same as

$$f(v) - p(v - z) \le c \le p(v + z) - f(v) \quad \text{for every } v \in Y. \tag{19.3}$$

To show that we can find such a c we use the triangle inequality for p: take $v_1, v_2 \in Y$, and then

$$\begin{aligned} f(v_1) + f(v_2) &= f(v_1 + v_2) \\ &\le p(v_1 + v_2) = p(v_1 - z + v_2 + z) \\ &\le p(v_1 - z) + p(v_2 + z); \end{aligned}$$

therefore

$$f(v_1) - p(v_1 - z) \le p(v_2 + z) - f(v_2) \quad v_1, v_2 \in Y.$$

It follows that for any fixed $v_1 \in Y$ we have

$$f(v_1) - p(v_1 - z) \le \inf_{v \in Y} p(v + z) - f(v)$$

and hence we also have

$$\sup_{v \in Y} f(v) - p(v - z) \le \inf_{v \in Y} p(v + z) - f(v).$$

Therefore we can find a $c \in \mathbb{R}$ to ensure that (19.3) holds.

However, this shows that we can extend (Y, f) to a linear functional F defined on Y', the span of Y and z, that still satisfies $F(x) \leq p(x)$ for all $x \in Y'$. This contradicts the maximality of (Y, f), so in fact $Y = X$.

This proves the result for sublinear functionals.

If p is in fact a seminorm, then $p(-x) = p(x)$ and we have

$$f(x) \leq p(x) \qquad \text{and} \qquad -f(x) = f(-x) \leq p(-x) = p(x),$$

and so our extension satisfies $|f(x)| \leq p(x)$.

To prove the result for bounded linear functionals on normed spaces, i.e. when $p(x) = \|\phi\|_{U^*}\|x\|$, first we observe that this does indeed define a seminorm, so that there exists an extension with

$$|f(x)| \leq \|\phi\|_{U^*}\|x\|,$$

i.e. such that $\|f\|_{X^*} \leq \|\phi\|_{U^*}$. Since $U \subseteq X$, we have $\|f\|_{X^*} \geq \|\phi\|_{U^*}$ if f extends ϕ, and so $\|f\|_{X^*} = \|\phi\|_{U^*}$ as claimed. $\qquad \square$

19.2 The Hahn–Banach Theorem: Complex Case

To extend the Hahn–Banach Theorem to the complex case we will have to restrict to the case that p is a seminorm (rather than just a sublinear functional).

In order to use the real version of the theorem, we first observe that any complex vector space V can be viewed as a real vector space by only allowing scalar multiplication by real numbers. This does not effect the elements of the space itself, but has a significant effect on what it means for a map to be 'linear', i.e. the values of α, β allowed in the expression

$$\phi(\alpha x + \beta y) = \alpha\phi(x) + \beta\phi(y) \qquad x, y \in V \qquad (19.4)$$

become restricted to real numbers. We therefore make the distinction in this section between 'real-linear' maps (allowing only $\alpha, \beta \in \mathbb{R}$ in (19.4)) and 'complex-linear' maps (for which we can take $\alpha, \beta \in \mathbb{C}$ in (19.4)).

While the 'real version' of a normed space V has the same elements, its dual space contains more maps, since they are only required to real-linear. We therefore use the notation $V_{\mathbb{R}}^*$ for the collection of all bounded real-linear functionals on V in order to distinguish it from V^* (which contains only complex-linear functionals when V is complex).

Lemma 19.4 *Let V be a complex vector space. Given any complex-linear functional $f : V \to \mathbb{C}$ there exists a unique real-linear $\psi : V \to \mathbb{R}$ such that*

$$f(v) = \psi(v) - i\psi(iv) \qquad \text{for all} \qquad v \in V. \qquad (19.5)$$

Furthermore,

(i) *if* $p\colon V \to \mathbb{R}$ *is a seminorm and* $|f(v)| \le p(v)$, *then* $|\psi(v)| \le p(v)$;

(ii) *if* V *is a normed space and* $f \in V^*$, *then* $\psi \in V_{\mathbb{R}}^*$ *with*

$$\|\psi\|_{V_{\mathbb{R}}^*} = \|f\|_{V^*}.$$

Conversely, if $\psi\colon V \to \mathbb{R}$ *is real-linear and satisfies* $|\psi(v)| \le p(v)$, *then* $f\colon V \to \mathbb{C}$ *defined by (19.5) is complex-linear and satisfies* $|f(v)| \le p(v)$ *for all* $v \in V$. *Moreover, if* $\psi \in V_{\mathbb{R}}^*$, *then* $f \in V^*$ *with* $\|f\|_{V^*} = \|\psi\|_{V_{\mathbb{R}}^*}$.

Proof If $v \in V$, then we can write

$$f(v) = \psi(v) + i\phi(v),$$

where $\psi, \phi\colon V \to \mathbb{R}$ are real-linear. Since

$$\psi(iv) + i\phi(iv) = f(iv) = if(v) = i\psi(v) - \phi(v),$$

it follows that $\psi(iv) = -\phi(v)$, which yields (19.5).

If $|f(v)| \le p(v)$, then this is inherited by ψ, since

$$|f(v)|^2 = |\psi(v)|^2 + |\psi(iv)|^2 \quad \Rightarrow \quad |\psi(v)| \le |f(v)| \le p(v);$$

for bounded linear functionals on a normed space the same argument gives $\|\psi\|_{V_{\mathbb{R}}^*} \le \|f\|_{V^*}$, but the equality of these norms requires a neat trick.

To show that $\|\psi\|_{V_{\mathbb{R}}^*} \ge \|f\|_{V^*}$, observe that for any x we can write

$$|f(x)| = e^{i\theta} f(x)$$

for some $\theta \in \mathbb{R}$. So

$$|f(x)| = e^{i\theta} f(x) = f(e^{i\theta} x) = \psi(e^{i\theta} x) - i\psi(ie^{i\theta} x).$$

Since $|f(x)|$ is real, we must have, for all $x \in V$,

$$|f(x)| = \psi(e^{i\theta} x) \le |\psi(e^{i\theta} x)| \le \|\psi\|_{V_{\mathbb{R}}^*} \|e^{i\theta} x\|_V = \|\psi\|_{V_{\mathbb{R}}^*} \|x\|_V,$$

and so $\|f\|_{V^*} \le \|\psi\|_{V_{\mathbb{R}}^*}$.

To prove the converse results (from ψ to f), we first show that, given a real-linear map $\psi\colon V \to \mathbb{R}$, the map $f\colon V \to \mathbb{C}$ defined in (19.5) is complex-linear. It is clear that

$$f(u + v) = f(u) + f(v),$$

since ψ has this property. We need only show that $f(\lambda u) = \lambda f(u)$ for $\lambda \in \mathbb{C}$, and for this it suffices to check that $f(iu) = if(u)$, since f is clearly real-linear because ψ is. We have

$$f(iu) = \psi(iu) - i\psi(-u)$$
$$= i[\psi(u) - i\psi(iu)] = if(u).$$

To show that $|f(x)| \le p(x)$ we use the same trick that we used above to show that $\|f\|_{V^*} \le \|\psi\|_{V^*_{\mathbb{R}}}$. Suppose that

$$|f(x)| = e^{i\theta} f(x);$$

then

$$|f(x)| = f(e^{i\theta}x) = \psi(e^{i\theta}x) - i\psi(ie^{i\theta}x),$$

and since $|f(x)|$ is real we have

$$|f(x)| = \psi(e^{i\theta}x) \le |\psi(e^{i\theta}x)| \le p(e^{i\theta}x) = |e^{i\theta}|p(x) = p(x).$$

This also implies the result for the dual norms.   □

As remarked above, for the complex version of the Hahn–Banach Theorem we require our bounding functional p to be at least a seminorm (Definition 19.1). Since we are now dealing with a complex space X, throughout the statement, 'linear' means 'complex-linear'.

Theorem 19.5 (Complex Hahn–Banach Theorem) *Let X be a complex vector space, U a subspace of X, and p a seminorm on X. Suppose that $\phi : U \to \mathbb{C}$ is linear and satisfies*

$$|\phi(x)| \le p(x) \qquad \text{for all} \quad x \in U.$$

Then there exists a linear map $f : X \to \mathbb{C}$ such that $f(x) = \phi(x)$ for all $x \in U$ and

$$|f(x)| \le p(x) \qquad \text{for all} \quad x \in X.$$

In particular, if X is a normed space, then any $\phi \in U^*$ can be extended to some $f \in X^*$ with $\|f\|_{X^*} = \|\phi\|_{U^*}$.

Proof By Lemma 19.4 there exists a real-linear $\psi : U \to \mathbb{R}$ such that

$$\phi(v) = \psi(v) - i\psi(iv), \tag{19.6}$$

and

$$|\psi(w)| \le p(w)$$

for all $w \in U$.

We can now use the real Hahn–Banach Theorem to extend ψ from U to X to give a real-linear map $\Psi : X \to \mathbb{R}$ that satisfies

$$|\Psi(x)| \le p(x) \qquad \text{for every} \quad x \in X.$$

Finally, if we define

$$f(u) = \Psi(u) - i\Psi(iu),$$

then $f \colon X \to \mathbb{C}$ is complex-linear, extends ϕ, and satisfies

$$|f(w)| \le p(w),$$

using the second half of Lemma 19.4.

If $\phi \in U^*$, then we follow a very similar argument, first writing ϕ as in (19.6), where now $\psi \in U_{\mathbb{R}}^*$ with $\|\psi\|_{U_{\mathbb{R}}^*} = \|\phi\|_{U^*}$. We then extend ψ to an element $\Psi \in X_{\mathbb{R}}^*$ with $\|\Psi\|_{X_{\mathbb{R}}^*} = \|\psi\|_{U_{\mathbb{R}}^*}$ and use the second part of Lemma 19.4 to guarantee that the complex-linear functional f on X defined by setting $f(u) := \Psi(u) - i\Psi(iu)$ satisfies

$$\|f\|_{X^*} = \|\Psi\|_{X_{\mathbb{R}}^*} = \|\psi\|_{U^*} = \|\phi\|_{U^*}. \qquad \square$$

Exercises

19.1 Let H be a Hilbert space and U a closed linear subspace of H. Use the Riesz Representation Theorem to show that any $\phi \in U^*$ has an extension to an element $f \in H^*$ such that $f(x) = \phi(x)$ for every $x \in U$ and $\|f\|_{H^*} = \|\phi\|_{U^*}$.

19.2 Show that the extension obtained in the previous exercise is unique.

19.3 Let X be a normed space and U a subspace of X that is not closed. If $\hat{\phi} \colon U \to \mathbb{K}$ is a linear map such that

$$|\hat{\phi}(x)| \le M\|x\| \qquad \text{for every} \quad x \in U$$

show that $\hat{\phi}$ has a unique extension ϕ to $\overline{U}$ (the closure of U in X) that is linear and satisfies

$$|\phi(x)| \le M\|x\|$$

for every $x \in \overline{U}$. (For any $x \in \overline{U}$ there exists a sequence $(x_n) \in U$ such that $x_n \to x$. Define

$$\phi(x) := \lim_{n \to \infty} \hat{\phi}(x_n).$$

Show that this is well defined and has the required properties.)

19.4 Show that a seminorm p on a vector space X satisfies

  (i) $p(0) = 0$;

  (ii) $|p(x) - p(y)| \le p(x - y)$;

  (iii) $p(x) \ge 0$; and

  (iv) $\{x : p(x) = 0\}$ is a subspace of X.

(Rudin, 1991)

19.5 Suppose that U is a subspace of a normed space X and that $\phi \in U^*$. Show that the set of all 'Hahn–Banach extensions' f of ϕ (using the notation from Theorem 19.2) is convex. (Costara and Popa, 2003)

19.6 Suppose that X is a real separable normed space, and W a closed linear subspace of X. Show that there exists a sequence of unit vectors $(z_j) \in X$ such that

$$z_{j+1} \notin W_j := \operatorname{Span} W \cup \{z_1, \ldots, z_j\},$$

and if we define

$$W_\infty = \operatorname{Span} W \cup \{z_j\}_{j=1}^\infty,$$

then $\overline{W_\infty} = X$.

19.7 Suppose that X is a real separable normed space, and that W is a closed linear subspace of X. Use the results of Exercises 19.3 and 19.6, along with the 'extension to one more dimension' part of the proof of the Hahn–Banach Theorem given in Section 19.1 to show that any $\phi \in W^*$ has an extension to an $f \in X^*$ with $\|f\|_{X^*} = \|\phi\|_{W^*}$. (For a separable space this gives a proof of the Hahn–Banach Theorem that does not require Zorn's Lemma.) (Rynne and Youngson, 2008)

20

Some Applications of the Hahn–Banach Theorem

We now explore some consequences of the Hahn–Banach Theorem in normed spaces. In Sections 20.1–20.4 we will only need the simplest version of the theorem, which we state here in a compact form.

Theorem 20.1 (Hahn–Banach) *Let X be a normed space and U a subspace of X. If $\phi \in U^*$, then ϕ has an extension to an element $f \in X^*$, i.e. $f(x) = \phi(x)$ for every $x \in U$, such that $\|f\|_{X^*} = \|\phi\|_{U^*}$.*

In the final section of the chapter we give an application that requires the extension of a linear map bounded by a sublinear functional.

20.1 Existence of a Support Functional

As a first application of the Hahn–Banach Theorem, we prove the existence of a particularly useful class of linear functionals (the 'support functionals') that return the norm of a particular element $x \in X$. This allows us to show that linear functionals can distinguish between elements of X, and that under-standing all the linear functionals on X is in some way enough to understand X itself.

Lemma 20.2 (Support functional) *If X is a normed space, then given any $x \in X$ there exists an $f \in X^*$ such that $\|f\|_{X^*} = 1$ and $f(x) = \|x\|$.*

We term this f the 'support functional at x'.

Proof Define ϕ on the linear space $U = \mathrm{Span}(\{x\})$ by setting

$$\phi(\alpha x) = \alpha \|x\| \qquad \text{for all } \alpha \in \mathbb{K}.$$

Then $\phi(x) = \|x\|$ and $|\phi(z)| \leq \|z\|$ for all $z \in U$, which shows that $\|\phi\|_{U^*} = 1$. Use the Hahn–Banach Theorem to extend ϕ to an $f \in X^*$ such that $\|f\|_{X^*} = 1$, and then note that $f(x) = \phi(x) = \|x\|$. $\qquad\square$

The following simple corollary shows that X^* is rich enough to distinguish between elements of X.

Corollary 20.3 (X^* separates points) *If $x, y \in X$ with $x \neq y$, then there exists $f \in X^*$ such that $f(x) \neq f(y)$. Consequently, if $x, y \in X$ and $f(x) = f(y)$ for every $f \in X^*$, then $x = y$.*

Proof If $x \neq y$, then by the previous lemma, there exists an f with $\|f\|_{X^*} = 1$ such that $f(x) - f(y) = f(x - y) = \|x - y\| \neq 0$. $\qquad\square$

This result shows that, rather than being particular to $C([a, b])^*$, the observation we made in Example 18.1 that understanding the action of elements of X^* on X is enough to 'understand' the whole of X is true in a general context.

20.2 The Distance Functional

The next result is a key ingredient in many subsequent proofs. Given a closed linear subspace Y and a point $x \notin Y$ it provides a linear functional that vanishes on Y and encodes the distance of x from Y. We call this functional a 'distance functional'.

Proposition 20.4 (Distance functional) *Let X be a normed space and Y be a proper closed subspace of X. Take $x \in X \setminus Y$ and set*

$$d = \mathrm{dist}(x, Y) := \inf\{\|x - y\| : y \in Y\} > 0. \qquad (20.1)$$

Then there is an $f \in X^*$ such that $\|f\|_{X^*} = 1$, $f(y) = 0$ for every $y \in Y$, and $f(x) = d$.

Proof First note that $d > 0$ since Y is closed (see the proof of Lemma 5.4). Let $U = \mathrm{Span}\{Y \cup \{x\}\}$ and define $\phi \colon U \to \mathbb{K}$ by setting

$$\phi(y + \lambda x) := \lambda d, \qquad y \in Y, \ \lambda \in \mathbb{K}.$$

To see that ϕ is bounded on U, observe that

$$|\phi(y + \lambda x)| = |\lambda| d \leq |\lambda| \, \|x - (-y/\lambda)\| = \|\lambda x + y\|$$

since $(-y/\lambda) \in Y$ and d is the distance between x and Y; it follows that $\|\phi\|_{U^*} \leq 1$.

To see that $\|\phi\|_{U^*} \geq 1$, take $y_n \in Y$ such that

$$\|x - y_n\| \leq d \left(1 + \frac{1}{n}\right)$$

(the existence of such a sequence follows from the definition of d in (20.1)). Then

$$\phi(-y_n + x) = d \geq \frac{n}{n+1} \|x - y_n\|$$

and so $\|\phi\|_{U^*} \geq n/(n+1)$ for every n, i.e. $\|\phi\|_{U^*} \geq 1$.

We have therefore shown that $\|\phi\|_{U^*} = 1$. We now extend ϕ to an element $f \in X^*$ using the Hahn–Banach Theorem; the resulting f satisfies $\|f\|_{X^*} = 1$, $f(x) = d$, and $f(y) = 0$ for every $y \in Y$, as required. $\qquad\square$

20.3 Separability of X^* Implies Separability of X

We now use the existence of a distance functional to prove the more substantial result that separability of X^* implies separability of X. Be warned that the converse is not true in general: we have already seen that $(\ell^1)^* = \ell^\infty$, and we know that ℓ^1 is separable but that ℓ^∞ is not (Lemma 3.24).

Lemma 20.5 *If X^* is separable, then X is separable.*

Proof Since X^* is separable,

$$S_{X^*} = \{f \in X^* : \|f\|_{X^*} = 1\}$$

(the unit sphere in X^*) is separable (by (i) $\Rightarrow$ (ii) in Lemma 3.23). Let $\{f_n\}$ be a countable dense subset of S_{X^*}. Since $\|f_n\|_{X^*} = 1$, for each n, there exists an $x_n \in X$ with $\|x_n\| = 1$ such that $|f_n(x_n)| \geq 1/2$, by the definition of the norm in X^*.

We now show that

$$M := \mathrm{clin}(\{x_n\})$$

is all of X, which will imply (by (iii) $\Rightarrow$ (i) of Lemma 3.23) that X is separable.

Suppose for a contradiction that $M \neq X$. Then M is a proper closed subspace of X, and so Proposition 20.4 (existence of a distance functional) provides an $f \in X^*$ with $\|f\|_{X^*} = 1$ (i.e. $f \in S_{X^*}$) and $f(x) = 0$ for every $x \in M$. But then $f(x_n) = 0$ for every n and so

$$\frac{1}{2} \leq |f_n(x_n)| = |f_n(x_n) - f(x_n)| \leq \|f_n - f\|_{X^*} \|x_n\| = \|f_n - f\|_{X^*}$$

for every n, which contradicts the fact that $\{f_n\}$ is dense in S_{X^*}. $\qquad\square$

Note that one immediate consequence of this result is that ℓ^1 is *not* the dual of ℓ^∞.

Corollary 20.6 $(\ell^\infty)^* \not\simeq \ell^1$ *and* $(L^\infty)^* \not\simeq L^1$.

Proof We know that ℓ^1 is separable, so if we had $(\ell^\infty)^* \simeq \ell^1$, then $(\ell^\infty)^*$ would be separable, since separability is preserved under isomorphisms (Exercise 3.13). Lemma 20.5 would then imply that ℓ^∞ was itself separable, but we know that this is not true (Lemma 3.24). The same arguments work with L^1 and $(L^\infty)^*$.   □

The three results we have just proved will be used repeatedly in the rest of Part IV. We now give two more applications which demonstrate some of the power of the Hahn–Banach Theorem, although we will not explore their consequences further.

20.4 Adjoints of Linear Maps between Banach Spaces

When H and K are Hilbert spaces we defined in Theorem 13.1 the adjoint $T^* \in B(K, H)$ of a linear map $T \in B(H, K)$, and we showed there that $\|T^*\|_{B(K,H)} = \|T\|_{B(H,K)}$. We can do something similar for linear maps between Banach spaces.

Suppose that X and Y are Banach spaces and $T \in B(X, Y)$. Given any $g \in Y^* = B(Y, \mathbb{K})$, the map $g \circ T$ is a bounded linear functional on X (see (11.5)) with

$$\|g \circ T\|_{X^*} \le \|g\|_{Y^*} \|T\|_{B(X,Y)}. \tag{20.2}$$

We define the adjoint $T^\times$ of T to be the map $T^\times : Y^* \to X^*$ given by

$$T^\times g = g \circ T.$$

This is a linear map on Y^*, since for every $x \in X$

$$T^\times(\alpha g_1 + \beta g_2)(x) = \alpha g_1(T(x)) + \beta g_2(T(x)) = \alpha T^\times g_1(x) + \beta T^\times g_2(x)$$

and

$$T^\times(\alpha g) = \alpha g \circ T = \alpha T^\times g.$$

The inequality in (20.2) shows immediately that

$$\|T^\times\|_{B(Y^*, X^*)} \le \|T\|_{B(X,Y)}.$$

To show that the norms are in fact equal, for every non-zero $x \in X$ we can find $g \in Y^*$ such that $\|g\|_{Y^*} = 1$ and $g(Tx) = \|Tx\|_Y$, using Lemma 20.2. Note that $g(Tx) = (T^\times g)(x)$, and so

$$\|Tx\|_Y = g(Tx) = (T^\times g)(x) \le \|T^\times\|_{B(Y^*, X^*)} \|g\|_{Y^*} \|x\|_X.$$

Since $\|g\|_{Y^*} = 1$, it follows that for every $x \in X$ we have

$$\|Tx\|_Y \le \|T^\times\|_{B(Y^*, X^*)} \|x\|_X,$$

which implies that $\|T^\times\|_{B(Y^*, X^*)} \ge \|T\|_{B(X, Y)}$ and hence that

$$\|T^\times\|_{B(Y^*, X^*)} = \|T\|_{B(X, Y)}. \tag{20.3}$$

(The above shows that the mapping $T \mapsto T^\times$ is a linear isometry from $B(X, Y)$ onto a subspace of $B(Y^*, X^*)$; in general this map need not be onto.)

The following result will be useful later when we discuss reflexive spaces in Chapter 26.

Lemma 20.7 *If $T : X \to Y$ is an isomorphism, then $T^\times : Y^* \to X^*$ is an isomorphism. If T is an isometric isomorphism, then so is $T^\times$.*

Proof First we show that $T^\times$ is an isomorphism by showing that

$$(T^\times)^{-1} = (T^{-1})^\times.$$

Since $T^{-1} \in B(Y, X)$, it follows that $(T^{-1})^\times \in B(X^*, Y^*)$. Now observe that for any $f \in X^*$ and $x \in X$

$$[T^\times((T^{-1})^\times f)](x) = ((T^{-1})^\times f)(Tx) = f \circ T^{-1}(Tx) = f(x),$$

i.e. $T^\times \circ (T^{-1})^\times = \mathrm{id}_{X^*}$; an almost identical argument can be used to show that $(T^{-1})^\times \circ T^\times = \mathrm{id}_{Y^*}$. Thus $(T^\times)^{-1} = (T^{-1})^\times$ and so in particular $T^\times$ is a bijection.

To show that $T^\times$ is an isometry we use Exercise 11.8, which shows that a bijection $T \in B(X, Y)$ is an isometry if $\|T\|_{B(X, Y)} = \|T^{-1}\|_{B(Y, X)} = 1$. We use (20.3) and the fact that T is an isometry to deduce that

$$\|T^\times\| = \|T\| = 1 \qquad \text{and} \qquad \|(T^\times)^{-1}\| = \|T^{-1}\| = 1;$$

it follows that $T^\times$ is also an isometry. $\qquad\qquad\qquad\qquad\qquad\qquad\qquad\square$

For the relationship between the Banach adjoint and the Hilbert adjoint see Exercise 20.11.

20.5 Generalised Banach Limits

Let $X = \ell^\infty(\mathbb{R})$ be the space of all bounded real sequences, and let $\mathfrak{c}(\mathbb{R})$ be the subspace of X consisting of all convergent real sequences.

Let $\ell \in \mathfrak{c}(\mathbb{R})^*$ be defined by setting $\ell(x) = \lim_{n\to\infty} x_n$, i.e. $\ell(x)$ gives the limit of the sequence (x_n). We want to show that we can define an extension of ℓ to the whole of X that retains some of the most important properties of the usual limit. (Since ℓ is a bounded linear functional on $\mathfrak{c}(\mathbb{R})^*$, which is a subspace of $\ell^\infty(\mathbb{R})$, we could just extend ℓ to a linear functional on $\ell^\infty(\mathbb{R})$ using the Hahn–Banach Theorem to give some sort of 'generalised limit'. However, the more involved construction we now give ensures that our generalised limit also satisfies (b) and (c), below.)

A Banach (generalised) limit on X is any $\mathcal{L} \in X^*$ such that for $x \in X$

(a) $\mathcal{L}(x) \geq 0$ if $x_n \geq 0$ for all n;
(b) $\mathcal{L}(x) = \mathcal{L}(\mathsf{s}_l x)$, where s_l is the left shift from Example 11.7; and
(c) $\mathcal{L}((1, 1, 1, \ldots)) = 1$.

Lemma 20.8 *If $\mathcal{L}$ is a Banach limit on X, then*

$$\liminf_{n\to\infty} x_n \leq \mathcal{L}(x) \leq \limsup_{n\to\infty} x_n$$

for every $x \in X$. In particular, if $x \in \mathfrak{c}(\mathbb{R})$, then

$$\mathcal{L}(x) = \ell(x).$$

Proof Since

$$\liminf_{n\to\infty} x_n = \lim_{k\to\infty} \inf_{n\geq k} x_n \qquad \text{and} \qquad \limsup_{n\to\infty} x_n = \lim_{k\to\infty} \sup_{n\geq k} x_n$$

it follows from property (b) that it suffices to show that

$$\inf_n x_n \leq \mathcal{L}(x) \leq \sup_n x_n.$$

Take $\varepsilon > 0$; then there exists n_0 such that

$$\sup_n x_n - \varepsilon < x_{n_0} \leq \sup_n x_n,$$

and so

$$x_{n_0} - x_n + \varepsilon > 0 \qquad \text{for all } n.$$

Using properties (a) and (c) we obtain

$$0 \leq \mathcal{L}(\{x_{n_0} + \varepsilon - x_n\}) = x_{n_0} + \varepsilon - \mathcal{L}(x),$$

which implies that

$$\mathcal{L}(x) - \varepsilon \le x_{n_0} \le \sup_n x_n.$$

Since this holds for all $\varepsilon > 0$, it follows that $\mathcal{L}(x) \le \sup_n x_n$. A similar argument yields $\inf_n x_n \le \mathcal{L}(x)$.

That $\mathcal{L}(x) = \ell(x)$ whenever $x \in \mathfrak{c}(\mathbb{R})$ now follows since for $x \in \mathfrak{c}(\mathbb{R})$ we have $\liminf_{n\to\infty} x_n = \limsup_{n\to\infty} x_n = \lim_{n\to\infty} x_n = \ell(x)$. □

We now use the Hahn–Banach Theorem with an appropriate sublinear functional to show that generalised Banach limits exist.

Proposition 20.9 *Banach limits exist.*

Proof Consider the functional $p\colon X \to \mathbb{R}$ defined by setting

$$p(x) := \limsup_{n\to\infty} \frac{x_1 + \cdots + x_n}{n}.$$

Then we have

$$p(x+y) = \limsup_{n\to\infty} \frac{(x_1 + y_1) + \cdots + (x_n + y_n)}{n}$$
$$\le \limsup_{n\to\infty} \frac{x_1 + \cdots + x_n}{n} + \limsup_{n\to\infty} \frac{y_1 + \cdots + y_n}{n}$$
$$= p(x) + p(y),$$

and clearly $p(\lambda x) = \lambda p(x)$, so p is a sublinear functional on X. We also have

$$-p(-x) = \liminf_{n\to\infty} \frac{x_1 + \cdots + x_n}{n}.$$

Now note that if $x \in \mathfrak{c}(\mathbb{R})$, then

$$\lim_{n\to\infty} \frac{x_1 + \cdots + x_n}{n} = \ell(x)$$

(see Exercise 20.13), and so in particular $p(x) = \ell(x)$ for every $x \in \mathfrak{c}(\mathbb{R})$.

We can now use the Hahn–Banach Theorem to extend ℓ to some $\mathcal{L} \in X^*$ such that

$$-p(-x) \le \mathcal{L}(x) \le p(x) \qquad x \in X.$$

That $\mathcal{L}$ satisfies properties (a) and (c) required for a generalised Banach limit follows immediately from this inequality; for property (b) note that

$$\mathcal{L}(x - s_l x) \le p(x - s_l x) = \limsup_{n\to\infty} \frac{x_{n+1} - x_1}{n} = 0,$$

since $x \in \ell^\infty$, and similarly for the lower bound. □

Goffman and Pedrick (1983) discuss conditions under which the Banach limit is unique, calling such sequences 'almost convergent' (Section 2.10).

Exercises

20.1 If H is a Hilbert space, given $x \in H$, find an explicit form for the functional $f \in H^*$ such that $\|f\|_{H^*} = 1$ and $f(x) = \|x\|$ (as in Lemma 20.2).

20.2 Let X be a normed space, $\{e_j\}_{j=1}^n \in X$ a linearly independent set, and $\{a_j\}_{j=1}^n \in \mathbb{K}$. Show that there exists $f \in X^*$ such that

$$f(e_j) = a_j \qquad j = 1, \ldots, n.$$

20.3 If X is a Banach space show that $\|x\| \leq M$ if and only if $|f(x)| \leq M$ for all $f \in X^*$ with $\|f\|_{X^*} = 1$, and hence show that

$$\|x\| = \sup\{|f(x)| : f \in X^*, \|f\|_{X^*} = 1\}.$$

(Sometimes this provides a useful way to bound $\|x\|$ 'by duality'.)

20.4 Find an explicit form for the distance functional of Proposition 20.4 when X is a Hilbert space.

20.5 Show that if X^* is strictly convex (see Exercise 10.3), then for each $x \in X$ with $x \neq 0$ the set

$$\{f \in X^* : \|f\|_{X^*} = 1 \text{ and } f(x) = \|x\|_X\}$$

consists of a single linear functional.

20.6 If X is a Banach space and $T \in B(X)$ then the *numerical range* of T, $V(T)$, is defined by setting

$$V(T) := \{f(Tx) : f \in X^* : \|f\|_{X^*} = \|x\|_X = f(x) = 1\}.$$

Show that this reduces to $V(T) = \{(Tx, x) : \|x\| = 1\}$ in a Hilbert space (see Exercise 16.3).

20.7 Deduce the existence of a support functional as a corollary of Proposition 20.4.

20.8 Show that if $f \in X^*$ and $f \neq 0$ then

$$\operatorname{dist}(x, \operatorname{Ker}(f)) = \frac{|f(x)|}{\|f\|_{X^*}}.$$

(Pryce, 1973)

20.9 Let X be a separable Banach space. Show that X is isometrically isomorphic to a subspace of ℓ^∞. [Hint: let (x_n) be a dense sequence in the unit sphere of X, let (ϕ_n) be support functionals at (x_n), and show that $T : X \to \ell^\infty$ defined by setting

$$Tx := (\phi_1(x), \phi_2(x), \ldots)$$

is a linear isometry.] (Heinonen, 2003)

20.10 Show that a point $z \in X$ belongs to $\operatorname{clin}(E)$ if and only if $f(z) = 0$ for every $f \in X^*$ that vanishes on E, i.e. $f(x) = 0$ for every $x \in E$ implies that $f(z) = 0$. (Taking $E = Y$ with Y a linear subspace shows that $x \in \overline{Y}$ if and only if $f(x) = 0$ whenever $f \in X^*$ with $f(y) = 0$ for every $y \in Y$.)

20.11 Suppose that H and K are two Hilbert spaces and $T \in B(H, K)$. Show that

$$T^* = R_H^{-1} \circ T^\times \circ R_K,$$

where T^* is the Hilbert adjoint, $T^\times$ the Banach adjoint, and R_H and R_K the Riesz maps from $H \to H^*$ and from $K \to K^*$, respectively.

20.12 Use the Arzelà–Ascoli Theorem to show that if T is compact then $T^\times$ is compact. [Hint: if B is a bounded subset B of Y^* then we can regard B as a subset of $C(K)$, where K is the compact metric space $\overline{T(\mathbb{B}_X)}$.]

20.13 Show that if $x \in c(\mathbb{R})$ with $\lim_{n \to \infty} x_n = \alpha$ then

$$\lim_{n \to \infty} \frac{x_1 + \cdots + x_n}{n} = \alpha.$$

21

Convex Subsets of Banach Spaces

In this chapter we investigate some properties of convex subsets of Banach spaces. We begin by introducing the Minkowski functional, which is a sublinear functional on x that can be defined for any open convex subset that contains the origin. We then use this, along with the Hahn–Banach Theorem, to show that any two disjoint convex sets can be separated by a hyperplane (a translation of the kernel of a linear functional), and finally, we prove the Krein–Milman Theorem, which shows that a compact convex subset of a Banach space is determined by its extreme points.

21.1 The Minkowski Functional

For the proof of the 'separation theorem' for convex sets we will need the following lemma, which allows us to find a sublinear functional that somehow 'encodes' any convex set.

Lemma 21.1 *If C is an open convex subset of a Banach space X with $0 \in C$, then we define the* Minkowski functional *of C by setting*

$$p(x) := \inf\{\lambda > 0 : \lambda^{-1} x \in C\} \quad \textit{for each } x \in X;$$

see Figure 21.1. Then p is a sublinear functional on X and there exists a constant $c > 0$ such that

$$0 \le p(x) \le c\|x\| \quad \textit{for every } x \in X. \tag{21.1}$$

Furthermore,

$$C = \{x : p(x) < 1\}. \tag{21.2}$$

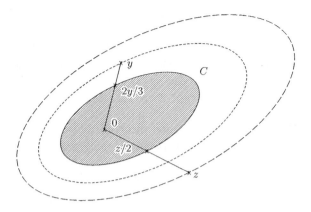

Figure 21.1 The Minkowski functional: p is constant on scaled versions of ∂C. On the dotted line $p = 3/2$; on the dashed line $p = 2$; and $p < 1$ in C.

Proof To see that p is sublinear, first observe that it follows easily from the definition that $p(\lambda x) = \lambda p(x)$ for $\lambda > 0$. For the triangle inequality, take $\alpha > p(x)$ and $\beta > p(y)$; then $\alpha^{-1}x, \beta^{-1}y \in C$, and since C is convex

$$\frac{\alpha}{\alpha + \beta}\alpha^{-1}x + \frac{\beta}{\alpha + \beta}\beta^{-1}y = \frac{x + y}{\alpha + \beta} \in C.$$

It follows that $p(x + y) \leq \alpha + \beta$, and since this holds for any $\alpha > p(x)$, $\beta > p(y)$ we obtain

$$p(x + y) \leq p(x) + p(y),$$

as required.

Since C is open and $0 \in C$, C contains an open ball $B(0, \delta)$ for some $\delta > 0$, and so

$$\|z\| < \delta \quad \Rightarrow \quad z \in C \quad \Rightarrow \quad |p(z)| \leq 1,$$

and then (21.1) follows, since for any non-zero $x \in X$ we can consider

$$z = \frac{\delta}{2}\frac{x}{\|x\|};$$

since $\|z\| < \delta$ it follows that $|p(z)| \leq 1$, i.e.

$$\left|p\left(\frac{\delta}{2}\frac{x}{\|x\|}\right)\right| = \frac{\delta}{2\|x\|}|p(x)| \leq 1,$$

which yields (21.1) with $c = 2/\delta$.

Finally, we prove (21.2). If $x \in C$, then, since C is open, we have $\lambda^{-1}x \in C$ for some $\lambda < 1$, and so $p(x) \leq \lambda < 1$, while if $p(x) < 1$, then $\lambda^{-1}x \in C$ for some $\lambda < 1$, and since $0 \in C$ and C is convex, it follows that

$$x = \lambda(\lambda^{-1}x) + (1 - \lambda)0 \in C. \qquad \square$$

21.2 Separating Convex Sets

We now use the Minkowski functional (applied to an appropriate set) as the sublinear functional in the Hahn–Banach Theorem, to show that any two convex sets in a Banach space X can be 'separated' by some $f \in X^*$ (see Figure 21.2). We start with the case of a real Banach space.

Theorem 21.2 (Functional separation theorem) *Suppose that X is a real Banach space and $A, B \subset X$ are non-empty, disjoint, convex sets.*

(i) *If A is open, then there exist $f \in X^*$ and $\gamma \in \mathbb{R}$ such that*

$$f(a) < \gamma \leq f(b), \qquad a \in A, \; b \in B.$$

(ii) *If A is compact and B is closed, then there exist $f \in X^*$, $\gamma \in \mathbb{R}$, and $\delta > 0$ such that*

$$f(a) \leq \gamma - \delta < \gamma + \delta \leq f(b), \qquad a \in A, \; b \in B.$$

A simple but important example of case (ii) is when $A = \{a\}$ is a point and B is closed. (Compare this result with the Hilbert-space prototype of Corollary 10.2.)

Proof (i) Choose $a_0 \in A$ and $b_0 \in B$, and let $w_0 = b_0 - a_0$. Now consider

$$C := w_0 + A - B,$$

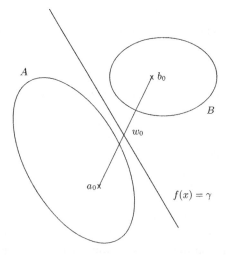

Figure 21.2 Illustration of the functional separation theorem (case (ii)).

i.e.

$$C = \{w_0 + a - b : a \in A, \ b \in B\}.$$

Then it is easy to check (see Exercise 3.2) that C is an open convex set that contains 0; we let $p(\cdot)$ be the Minkowski functional for C defined in Lemma 21.1. Since $A \cap B = \varnothing$, $w_0 \notin C$, and so $p(w_0) \geq 1$.

Let $U = \text{Span}(w_0)$, and define a linear functional ϕ on U by setting

$$\phi(\alpha w_0) = \alpha, \qquad \alpha \in \mathbb{R}.$$

If $\alpha \geq 0$, then

$$\phi(\alpha w_0) = \alpha \leq \alpha p(w_0) = p(\alpha w_0),$$

while if $\alpha < 0$, then

$$\phi(\alpha w_0) < 0 \leq p(\alpha w_0),$$

and so $\phi(w) \leq p(w)$ for every $w \in U$.

We can therefore use the Hahn–Banach Theorem (Theorem 19.2) to find a linear extension $f : X \to \mathbb{R}$ of ϕ such that

$$f(x) \leq p(x) \qquad \text{for every } x \in X.$$

Since we have (21.1), this f satisfies

$$f(x) \leq p(x) \leq c\|x\|.$$

Since f is linear and p is sublinear, it follows that

$$-f(x) = f(-x) \leq p(-x) \leq c\| - x\| = c\|x\|,$$

and so $|f(x)| \leq c\|x\|$, i.e. f is actually an element of X^*.

By definition for any $a \in A$ and $b \in B$ we have $w_0 + a - b \in C$, and so, since $f(w_0) = \phi(w_0) = 1$,

$$1 + f(a) - f(b) = f(w_0 + a - b) \leq p(w_0 + a - b) < 1.$$

This shows that $f(a) < f(b)$, and so if we define $\gamma = \inf_{b \in B} f(b)$ we obtain

$$f(a) \leq \gamma \leq f(b) \qquad a \in A, \ b \in B. \tag{21.3}$$

To guarantee that the left-hand inequality is in fact strict, suppose not, i.e. that there exists an $a \in A$ such that $f(a) = \gamma$. Since A is open, we must have $a + \delta w_0 \in A$ for some $\delta > 0$, and then we would have

$$f(a + \delta w_0) = f(a) + \delta \phi(w_0) = \gamma + \delta > \gamma,$$

which contradicts (21.3).

Note that if A and B are both open, then the same argument shows that

$$f(a) < \gamma < f(b), \qquad a \in A, \; b \in B.$$

To prove (ii) we set

$$\varepsilon = \frac{1}{4} \inf\{\|a - b\| : a \in A, \; b \in B\} > 0,$$

which is strictly positive since A is compact and B is closed. Now consider the two open convex disjoint sets

$$A_\varepsilon := A + B(0, \varepsilon) \qquad \text{and} \qquad B_\varepsilon := B + B(0, \varepsilon),$$

where here $B(0, \varepsilon)$ is the open ball of radius ε.

We can now apply part (i) to the open sets A_ε and B_ε. Set $w_0 = a_0 - b_0$ for some $a_0 \in A_\varepsilon$ and $b_0 \in B_\varepsilon$, and then follow the same argument as in part (i) to find $f \in X^*$ and $\gamma \in \mathbb{R}$ such that

$$f(a) < \gamma < f(b), \qquad a \in A_\varepsilon, \; b \in B_\varepsilon.$$

If we let $\delta = \varepsilon/2\|w_0\|$, then for any $a \in A$ we have $a + \delta w_0 \in A_\varepsilon$, and so

$$f(a) = f(a + \delta w_0) - \delta\phi(w_0) \leq \gamma - \delta,$$

and similarly, $\gamma + \delta \leq f(b)$ for any $b \in B$. □

We have a very similar result in a complex Banach space, except that we replace the linear functional f by its real part throughout.

Theorem 21.3 (Functional separation theorem – complex case) *Suppose that X is a complex Banach space. Then Theorem 21.2 still holds, except that in the inequalities f should be replaced by $\mathrm{Re}\, f$ throughout.*

Proof We consider real-linear maps on the space X, as in the proof of the complex version of the Hahn–Banach Theorem. So, for example, in case (i) (when A is open) we can use Theorem 21.2 to find $\phi \in X_{\mathbb{R}}^*$ and $\gamma \in \mathbb{R}$ such that

$$\phi(a) < \gamma \leq \phi(b), \qquad a \in A, \; b \in B.$$

Now we use the argument from the proof of the complex version of the Hahn–Banach Theorem to show that the linear functional

$$f(x) := \phi(x) - i\phi(ix)$$

is in X^*. This clearly has the required properties, since $\mathrm{Re}\, f = \phi$. □

21.3 Linear Functionals and Hyperplanes

We now show how linear functionals are related to hyperplanes (subspaces of codimension one). The discussion at the end of Chapter 12 has already given some indication of the nature of this relationship.

Definition 21.4 A *hyperplane* U in a vector space X is a codimension-one subspace of X, i.e. a maximal proper subspace: $U \neq X$ and if Z is a subspace with $U \subseteq Z \subseteq X$, then $Z = U$ or $Z = X$.

This definition allows us to give a more geometric interpretation of the above separation theorems (which should be unsurprising given Figure 21.2).

Lemma 21.5 *The following are equivalent:*

 (i) *U is a hyperplane in X;*
 (ii) *U is a subspace of X with $U \neq X$ but for any $x \in X \setminus U$, the span of $(U, \{x\})$ is X; and*
(iii) *$U = \mathrm{Ker}(\phi)$ for some non-zero linear functional $\phi \colon X \to \mathbb{K}$.*

Note that in (iii) the linear functional ϕ does not have to be bounded; we will show in the next result that it is bounded if and only if U is closed.

Proof (i) $\Leftrightarrow$ (ii) If U is a hyperplane and $x \notin U$, then $\mathrm{span}(U, \{x\})$ is a subspace that strictly contains U, so must be X. Conversely, if $U \subseteq Z \subseteq X$, either $Z = U$ or there exists $z \in Z \setminus U$, and then $Z \supset \mathrm{Span}(U, \{z\}) = X$.

(ii) $\Rightarrow$ (iii) Let U be a hyperplane and choose $x \notin U$. Define

$$\phi \colon \mathrm{Span}(U \cup \{x\}) = X \to \mathbb{K}$$

by setting

$$\phi(y + \lambda x) := \lambda, \qquad y \in U, \lambda \in \mathbb{K}.$$

This is well defined since if $y + \lambda x = y' + \lambda' x$, then

$$y - y' = (\lambda' - \lambda)x \quad \Rightarrow \quad (\lambda' - \lambda)x \in U \quad \Rightarrow \quad \lambda = \lambda'.$$

As defined ϕ is linear, non-zero (since $\phi(x) = 1$) and $\phi(y) = 0$ if and only if $y \in U$, so $U = \mathrm{Ker}(\phi)$.

(iii) $\Rightarrow$ (ii) Suppose that $U = \mathrm{Ker}(\phi)$ and take any $x \notin U$; it follows that $\phi(x) \neq 0$. Now given any $z \in X \setminus U$ let

$$y := z - \frac{\phi(z)}{\phi(x)}x,$$

so that $\phi(y) = 0$, i.e. $y \in U$, and $z = y + (\phi(z)/\phi(x))x$. It follows that the span of $U \cup \{x\}$ is X, as claimed. $\qquad\square$

Lemma 21.6 *If $U = \mathrm{Ker}(\phi)$ is a hyperplane in X, then U is closed if and only if ϕ is bounded; otherwise U is dense in X.*

Proof First, note that since $U \neq X$, it cannot be both closed and dense. So it is enough to show that bounded implies closed, and unbounded implies dense.

If ϕ is bounded, then $U = \mathrm{Ker}(\phi)$ is closed by Lemma 11.12.

If ϕ is unbounded, then we can find $(x_n) \in X$ such that $\|x_n\| = 1$ but $\phi(x_n) \geq n$. Now given $x \in X$, consider the sequence

$$y_n = x - \frac{\phi(x)}{\phi(x_n)}x_n.$$

Then $\phi(y_n) = 0$, so $y_n \in U$, and

$$\|x - y_n\| = \left\| \frac{\phi(x)}{\phi(x_n)}x_n \right\| = \frac{|\phi(x)|\|x_n\|}{|\phi(x_n)|} \leq \frac{|\phi(x)|}{n},$$

and so $y_n \to x$ and $n \to \infty$ and U is dense. $\qquad\square$

The *translate by $y \in X$* of a hyperplane U is the set

$$U + y = \{u + y : u \in U\}.$$

Since any hyperplane U is equal to $\mathrm{Ker}(\phi)$ for some $\phi \in X^*$, any translate of U is also given by

$$U + y = \{x \in X : \phi(x) = \phi(y)\}.$$

The following is now just a restatement of Theorem 21.2, if we understand that two sets A and B are separated by $\phi(x) = \gamma$ if $\phi(a) < \gamma < \phi(b)$ for every $a \in A, b \in B$.

Corollary 21.7 *Suppose that A, B are non-empty convex subsets of X with A closed and B compact. Then there exists a closed hyperplane that can be translated so that it separates A and B.*

21.4 Characterisation of Closed Convex Sets

We can use the separation Theorem 21.2 to give a characterisation of any closed convex subset of X as the envelope of its 'supporting hyperplanes'; see Figure 21.3. This result will be useful later when we show that such sets are

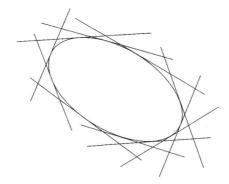

Figure 21.3 A convex set in a real Banach space is the envelope of a collection of translated hyperplanes (Corollary 21.8). Each of the lines represents a set of the form $\{x : f(x) = \inf_{y \in C} f(y)\}$ (in the real case).

also 'weakly closed' (Theorem 27.7), a fact that is very useful in the calculus of variations (maximisation/minimisation problems).

Corollary 21.8 *Suppose that C is a closed convex subset of a Banach space X. Then*

$$C = \{x \in X : \operatorname{Re} f(x) \geq \inf_{y \in C} \operatorname{Re} f(y) \text{ for every } f \in X^*\}.$$

Proof That C is contained in the right-hand side is immediate.

Suppose that $x_0 \notin C$. Then, since $\{x_0\}$ is compact and convex, we can use Theorem 21.2 to find $f \in X^*$ such that

$$\operatorname{Re} f(x_0) \leq \gamma - \delta < \gamma + \delta \leq \operatorname{Re} f(y), \qquad \text{for every } y \in C.$$

In particular, $\operatorname{Re} f(x_0) < \inf_{y \in C} \operatorname{Re} f(y)$. $\qquad \square$

21.5 The Convex Hull

Suppose that U is a subset of a vector space X. Then the *convex hull* of U is the collection of all 'convex linear combinations' of points in U, i.e.

$$\operatorname{conv}(U) := \left\{ \sum_{j=1}^{n} \lambda_j u_j : n \in \mathbb{N}, \ u_j \in U, \ \lambda_j > 0, \ \sum_{j=1}^{n} \lambda_j = 1 \right\}. \quad (21.4)$$

Lemma 21.9 *The convex hull of U is the smallest convex set that contains U.*

Proof First we show that conv(U) contains U and is convex. If $\{u_j\}_{j=1}^n$ and $\{v_k\}_{k=1}^m$ are in U and $\sum_{j=1}^n \lambda_j = 1$ and $\sum_{k=1}^m \mu_k = 1$, then for any $t \in (0, 1)$

$$t\left\{\sum_{j=1}^n \lambda_j u_j\right\} + (1 - t)\left\{\sum_{k=1}^m \mu_j v_j\right\}$$

is another convex linear combination of elements of $\{u_j\}$ (note that the definition in (21.4) does not require the u_j to be distinct, although it could and would yield the same set).

Now suppose that C is convex and contains U. We show by induction that for any $n \in \mathbb{N}$, $\sum_{j=1}^n \lambda_j u_j \in C$ whenever $u_j \in U$, $\lambda_j > 0$, and $\sum_{j=1}^n \lambda_j = 1$. This is true when $n = 2$ since U is convex.

Now take $\{u_j\}_{j=1}^n \in U$ and $\lambda_j > 0$ such that $\sum_{j=1}^n \lambda_j = 1$. By induction we know that $\sum_{j=1}^{n-1} \mu_j u_j \in C$ for any $\mu_j > 0$, $\sum_{j=1}^{n-1} \mu_j = 1$.

Now choose $(1 - t) = \lambda_n$ and μ_j such that $t\mu_j = \lambda_j$; note that

$$\sum_{j=1}^{n-1} \mu_j = t^{-1} \sum_{j=1}^{n-1} \lambda_j = t^{-1}[1 - \lambda_n] = 1.$$

So conv$(U) \subseteq C$. $\square$

In a normed space we can take closures, which leads to the following definition.

Definition 21.10 In a normed space X the *closed convex hull of U* is $\overline{\text{conv}(U)}$.

The closed convex hull of U is the smallest closed convex subset of X that contains U. Exercise 21.4 shows that if U is compact, then its closed convex hull is also compact.

21.6 The Krein–Milman Theorem

We end this chapter with the Krein–Milman Theorem: any non-empty compact convex subset of a Banach space is the closed convex hull of its extreme points.

Definition 21.11 If K is a convex set, then a point $a \in K$ is an *extreme point* (of K) if whenever $a = \lambda x + (1 - \lambda)y$ for some $\lambda \in (0, 1)$, $x, y \in K$, then we must have $x = y = a$.

Extreme points are particular cases of extreme sets.

Definition 21.12 Suppose that K is a convex subset of a Banach space X. A subset $M \subseteq K$ is an *extreme set* (in K) if M is non-empty, closed, and whenever $x, y \in K$ with $\lambda x + (1 - \lambda)y \in M$ for some $\lambda \in (0, 1)$, then $x, y \in M$.

Note that a is an extreme point in K if and only if $\{a\}$ is an extreme set in K, and that – trivially – K is extreme in itself. Note also that if K is compact, then any extreme set in K is also compact, since it is a closed subset of a compact set (Lemma 2.25).

We will use the following lemma twice in our proof of the Krein–Milman Theorem.

Lemma 21.13 *Let K be a non-empty compact convex subset of a Banach space X, let $M \subseteq K$ be an extreme set, and take any $f \in X^*$. Then*

$$M^f := \{x \in M : f(x) = \max_{y \in M} f(y)\}$$

is another extreme set.

Proof First note that M^f is non-empty; since M is compact the continuous map $f : M \to \mathbb{R}$ attains its maximum.

If $x, y \in K$ and $\lambda x + (1 - \lambda)y \in M^f$, $\lambda \in (0, 1)$, then $\lambda x + (1 - \lambda)y \in M$: since M is extreme it follows that $x, y \in M$. Now we have

$$f(\lambda x + (1 - \lambda)y) = \lambda f(x) + (1 - \lambda)f(y) = \max_{z \in M} f(z);$$

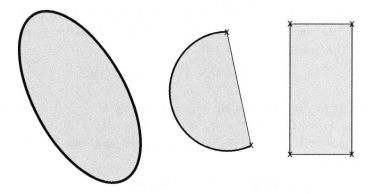

Figure 21.4 The extreme points in three convex sets, identified by bold curves and by crosses: each set is the closed convex hull of its extreme points. Every extreme point is an extreme set; each of the straight edges in the second and third shapes is an extreme set. Note that every extreme set contains an extreme point.

it follows that $f(x) = f(y) = \max_{z \in M} f(z)$, and so $x, y \in M^f$.   □

The following result is the key ingredient in the proof of the Krein–Milman Theorem.

Proposition 21.14 *Every extreme set in a non-empty compact convex set K contains an extreme point.*

Proof Let M be an extreme set in K, and consider the collection $\mathcal{E}$ of all extreme sets in K that are contained in M. Order $\mathcal{E}$ so that $A \preceq B$ if $B \subseteq A$ (so that the 'maximal set' is the smallest).

We use Zorn's Lemma to show that $\mathcal{E}$ has a maximal element.

Given any chain $\mathcal{C}$ we set $B = \cap_{A \in \mathcal{C}} A$. The set B is non-empty: if $B = \varnothing$, then $\{K \setminus A : A \in \mathcal{C}\}$ is an open cover of K, and so has a finite subcover $\{K \setminus A_j : j = 1, \ldots, k\}$. It follows that $\cap_{j=1}^k A_j = \varnothing$; but this is not possible: since $A_j \in \mathcal{C}$ and $\mathcal{C}$ is a chain we have

$$\bigcap_{j=1}^k A_j = A_i$$

for some $i \in \{1, \ldots, k\}$; see Exercise 1.7.

Clearly $A \preceq B$ for every $A \in \mathcal{C}$ (since we order by inclusion). Moreover, B is an extreme set, since if $\lambda x + (1 - \lambda)y \in B$, then for any $A \in \mathcal{C}$ we have $\lambda x + (1 - \lambda)y \in A$, and then $x, y \in A$ (since A is an extreme set), from which it follows that $x, y \in B$.

It follows that B is an upper bound for $\mathcal{C}$, and so – since any chain in $\mathcal{E}$ has an upper bound – Zorn's Lemma guarantees that $\mathcal{E}$ has a maximal element M_*.

Now suppose that M_* is not a single point, so there exist $a, b \in M_*$ with $a \neq b$. Then, since X^* separates points in X (Lemma 20.3), there must exist some $f \in X^*$ such that $f(a) < f(b)$. Consider

$$M_*^f := \{x \in M_* : f(x) = \max_{y \in M_*} f(y)\};$$

Lemma 21.13 guarantees that $M_*^f \in \mathcal{E}$, and certainly $M_* \preceq M_*^f$. Since M_* is maximal, we must have $M_*^f = M_*$; but $a \notin M_*^f$ so $M_*^f \neq M_*$. This contradiction shows that M_* must consist of a single point, and hence M contains an extreme point.   □

We can now prove the Krein–Milman Theorem,

Theorem 21.15 (Krein–Milman) *Suppose that K is a non-empty compact convex subset of a Banach space X. Then K is the closed convex hull of its extreme points.*

Proof Let K' be the closed convex hull of the extreme points of K; then every extreme point is contained in K' and $K' \subseteq K$.

So suppose that there is a point $b \in K$ such that $b \notin K'$. Using the separation form of the Hahn–Banach Theorem (Theorem 21.2 (ii) with $A = K'$ and $B = \{b\}$), there is an $f \in X^*$ such that

$$f(x) < f(b) \qquad \text{for every} \quad x \in K'.$$

Then

$$K^f := \{x \in K : f(x) = \max_{y \in K} f(y)\}$$

is an extreme set of K (by Lemma 21.13) such that $K^f \cap K' = \varnothing$. Then K^f must contain an extreme point of K (Proposition 21.14), but these are all contained in K'. Hence $K^f \cap K' \neq \varnothing$, a contradiction. $\qquad\square$

Exercises

21.1 If $(X, \|\cdot\|)$ is a normed space show that the Minkowski functional of B_X is $\|\cdot\|$. (Lax, 2002)

21.2 Show that if $(K_\alpha)_{\alpha \in \mathbb{A}}$ are a family of convex subsets of a vector space X, then $\bigcap_{\alpha \in \mathbb{A}} K_\alpha$ is also convex.

21.3 Show that if K is a convex subset of a normed space X, then $\overline{K}$ is also convex.

21.4 Show that if U is a compact subset of a Banach space, then its closed convex hull is compact. (Show that conv(U) is totally bounded; see Exercise 6.10.) (Pryce, 1973)

21.5 Suppose that U is a subset of $\mathbb{R}^n$. Show that any point in the convex hull of U can be written as a linear combination of at most $n + 1$ points in U. (Suppose that $x \in \text{conv}(U)$ is a convex combination of more than $n + 1$ points in U, and show that the number of points required can be reduced.)

21.6 If K is a non-empty closed subset of a Banach space X define

$$d(x) = \text{dist}(x, K) = \inf_{k \in K} \|x - k\|.$$

Show that K is convex if and only if $d : X \to \mathbb{R}$ is a convex function. (Giles, 2002)

22

The Principle of Uniform Boundedness

We now use the Baire Category Theorem – a fundamental result about complete spaces – to prove some results about linear maps between Banach spaces. We first state and prove this theorem, and then we use it to prove the Principle of Uniform of Boundedness (in this chapter) and the Open Mapping Theorem and its corollaries (in the next chapter).

22.1 The Baire Category Theorem

The Baire Category Theorem encodes an important property of complete spaces.[1] It comes in two equivalent formulations. The first says that a countable intersection of 'large' sets is still large.

Theorem 22.1 (Baire Category Theorem: residual form) *If $\{G_i\}_{i=1}^{\infty}$ is a countable family of open dense subsets of a complete normed space $(X, \| \cdot \|)$, then*

$$G = \bigcap_{i=1}^{\infty} G_i$$

is dense in X.

Any set that contains a countable intersection of open dense sets (such as the set G in the theorem above) is called *residual*. Note that the intersection of a countable collection of residual sets is still residual.

Proof Take $x \in X$ and $\varepsilon > 0$; we need to show that $B(x, \varepsilon) \cap G$ is non-empty.

---

[1] The theorem also holds in any complete metric space.

First note that for any $n \in \mathbb{N}$, since each G_n is dense, given any $z \in X$ and $r > 0$ we have

$$B(z, r) \cap G_n \neq \varnothing,$$

so there exists some $y \in G_n$ such that $y \in B(z, r) \cap G_n$. Since each G_n is also open, it follows that $B(z, r) \cap G_n$ is open too, and so there exists $r' > 0$ such that $B(y, 2r') \subset B(z, r) \cap G_n$. It follows that

$$\overline{B(y, r')} \subset B(y, 2r') \subset B(z, r) \cap G_n.$$

We now use this observation repeatedly. First choose $x_1 \in G_1$ and $r_1 < 1/2$ such that

$$\overline{B(x_1, r_1)} \subset B(x, \varepsilon) \cap G_1; \tag{22.1}$$

then take $x_2 \in G_2$ and $r_2 < 1/4$ such that

$$\overline{B(x_2, r_2)} \subset B(x_1, r_1) \cap G_2;$$

and inductively find $x_n \in G_n$ and $r_n < 2^{-n}$ such that

$$\overline{B(x_n, r_n)} \subset B(x_{n-1}, r_{n-1}) \cap G_n. \tag{22.2}$$

This yields a sequence of nested closed sets,

$$\overline{B(x_1, r_1)} \supset \overline{B(x_2, r_2)} \supset \overline{B(x_3, r_3)} \supset \cdots. \tag{22.3}$$

The points (x_j) form a Cauchy sequence, since (22.3) shows that for $i, j \geq n$ we have

$$x_i, x_j \in \overline{B(x_n, r_n)} \quad \Rightarrow \quad d(x_i, x_j) < 2^{-(n-1)};$$

since (X, d) is complete the sequence (x_j) must converge to some $x_0 \in X$.

Furthermore, since for each n the point x_i is contained in the closed set $\overline{B(x_n, r_n)}$ for all $i \geq n$ it follows, since $\overline{B(x_n, r_n)}$ is closed, that $x_0 \in \overline{B(x_n, r_n)}$. By (22.1) we have $x_0 \in B(x, \varepsilon)$, and from (22.2) we must have $x_0 \in G_n$ for every $n \in \mathbb{N}$. Therefore $x_0 \in B(x, \varepsilon) \cap G$, so this set is not empty and G is dense in X as claimed. $\qquad\square$

An alternative formulation of the Baire Category Theorem says that a complete metric space cannot be from the countable union of 'small' sets. To be more precise, we say that a subset W of (X, d) is *nowhere dense* if $(\overline{W})^\circ = \varnothing$, i.e. if the closure of W contains no open sets. Observe that if W is nowhere dense, then $X \setminus \overline{W}$ is open and dense: that this set is open is clear (since its complement is closed); if it were not dense there would be a point $x \in X$ such that $B(x, r) \cap X \setminus \overline{W} = \varnothing$ for all r sufficiently small, which would imply that $\overline{W} \supset B(x, r)$, a contradiction.

Theorem 22.2 (Baire Category Theorem: meagre form) *Let* $\{F_j\}_{j=1}^{\infty}$ *be a countable collection of nowhere dense subsets of a complete normed space* $(X, \|\cdot\|)$. *Then*

$$\bigcup_{j=1}^{\infty} F_j \neq X.$$

In particular, if $\{F_j\}_{j=1}^{\infty}$ *are closed and* $\bigcup_{j=1}^{\infty} F_j = X$, *then at least one of the* F_j *contains a non-empty open set.*

A countable union of nowhere dense subsets is called[2] *meagre*.

Proof The sets $X \setminus \overline{F_j}$ form a countable collection of open dense sets. It follows using Theorem 22.1 that

$$\bigcap_{j=1}^{\infty} X \setminus \overline{F_j} = X \setminus \left\{ \bigcup_{j=1}^{\infty} \overline{F_j} \right\}$$

is dense, and in particular non-empty.   □

One classical application of the Baire Category Theorem is to prove the existence of continuous functions that are nowhere differentiable; see e.g. Exercise 13 in Chapter 9 of Costara and Popa (2003) or Section 1.10 in Goffman and Pedrick (1983).

22.2 The Principle of Uniform Boundedness

We now use the Baire Category Theorem in the form of Theorem 22.2 to prove the Banach–Steinhaus Principle of Uniform Boundedness.

Theorem 22.3 (Principle of Uniform Boundedness) *Let* X *be a Banach space and* Y *a normed space. Let* $\mathcal{S} \subset B(X, Y)$ *be a collection of bounded linear operators such that*

$$\sup_{T \in \mathcal{S}} \|Tx\|_Y < \infty \qquad \text{for each} \quad x \in X.$$

Then

$$\sup_{T \in \mathcal{S}} \|T\|_{B(X,Y)} < \infty.$$

---

[2] A less memorable terminology is that such sets are 'of the first category'. A set that is not of the first category is 'of the second category'.

Some care is needed in applying this theorem: each element of $\mathcal{S}$ has to be a bounded linear map from X to Y (this is easy to forget).

Proof Consider the sets

$$F_j = \{x \in X : \|Tx\|_Y \leq j \text{ for all } T \in \mathcal{S}\}.$$

Each F_j is closed, since for each $T \in S$ the set

$$\{x \in X : \|Tx\|_Y \leq j\}$$

is closed because T is continuous, and then

$$F_j = \bigcap_{T \in \mathcal{S}} \{x \in X : \|Tx\|_Y \leq j\}$$

is the intersection of closed sets and therefore closed (Lemma 2.6).

By assumption

$$X = \bigcup_{j=1}^{\infty} F_j,$$

and so Theorem 22.2 implies that at least one of the F_j must contain a non-empty open set; so there must exist $y \in X$ and $r > 0$ such that $B_X(y, r) \subset F_n$ for some n.

Then for any x with $\|x\|_X < r$ we have $y + x \in B_X(y, r) \subset F_n$, and so for every $T \in \mathcal{S}$ we must have

$$\|Tx\|_Y = \|T(y + x) + T(-y)\|_Y \leq n + \|Ty\|_Y \leq 2n,$$

since $y \in F_n$. So for any x with $\|x\|_X = r/2$ we have

$$\|Tx\|_Y \leq 2n \qquad \text{for every } T \in \mathcal{S}.$$

Since T is linear, we can write any $y \in X$ as $y = (2\|y\|_X/r)(ry/2\|y\|_X)$, and then

$$\|Ty\|_Y = \frac{2\|y\|_X}{r} \left\| T \frac{ry}{2\|y\|_X} \right\|_Y \leq \frac{4n}{r} \|y\|_X,$$

and the conclusion follows. $\qquad\qquad\qquad\qquad\qquad\qquad\qquad\qquad \square$

Corollary 22.4 *Suppose that X is a Banach space, Y a normed space, and that $T_n \in B(X, Y)$. Suppose that*

$$Tx := \lim_{n \to \infty} T_n x$$

exists for every $x \in X$. Then $T \in B(X, Y)$.

Proof The operator T is linear, since if $x, y \in X$ and $\alpha, \beta \in \mathbb{K}$, then

$$T(\alpha x + \beta y) = \lim_{n \to \infty} T_n(\alpha x + \beta y) = \lim_{n \to \infty} \alpha T_n x + \beta T_n y$$
$$= \alpha \lim_{n \to \infty} T_n x + \beta \lim_{n \to \infty} T_n y$$
$$= \alpha T x + \beta T y.$$

To show that T is bounded, observe that since

$$\lim_{n \to \infty} \|T_n x\|_Y$$

exists it follows that for every $x \in X$ the sequence $(T_n x)_{n=1}^{\infty}$ is bounded. The Principle of Uniform Boundedness now shows that $\|T_n\|_{B(X,Y)} \leq M$ for every $n \in \mathbb{N}$. It follows that

$$\|Tx\|_Y = \lim_{n \to \infty} \|T_n x\|_Y \leq M \|x\|_X$$

and so T is bounded. $\qquad\qquad\qquad\qquad\qquad\qquad\qquad\qquad\qquad\qquad\qquad\square$

An almost immediate corollary of the Principle of Uniform Boundedness, the 'Condensation of Singularities', is often useful in applications; we will use it in the next section to show that there are continuous functions whose Fourier series do not converge at every point (i.e. that diverge at at least one point).

Corollary 22.5 (Condensation of Singularities) *Suppose that X is a Banach space, Y a normed space, and $\mathcal{S} \subset B(X, Y)$ with*

$$\sup_{T \in \mathcal{S}} \|T\|_{B(X,Y)} = \infty.$$

Then there exists $x \in X$ such that

$$\sup_{T \in \mathcal{S}} \|T_n x\|_Y = \infty. \tag{22.4}$$

Proof If (22.4) does not hold for any $x \in X$, then the Principle of Uniform Boundedness would imply that $\sup_{T \in \mathcal{S}} \|T\|_{B(X,Y)} < \infty$. $\qquad\qquad\square$

Exercise 22.2 shows that under the conditions of this corollary the set of $x \in X$ for which (22.4) holds is a residual subset of X.

22.3 Fourier Series of Continuous Functions

As an application of the Principle of Uniform Boundedness we now prove that there is a 2π-periodic continuous function $f : [-\pi, \pi] \to \mathbb{R}$ such that the

Fourier series of f at 0 does not converge, i.e. the partial sums are unbounded. This does not contradict the result of Corollary 6.10 that any such function can be uniformly approximated by an expression of the form

$$\sum_{k=-n}^{n} c_k e^{ikx};$$

rather, it shows that these approximations are not simply the partial sums of a single infinite 'series expansion' of f. Nor does it contradict the result of Lemma 9.16, where we showed that for any $f \in L^2(-\pi, \pi)$ the expansion

$$\sum_{k=-\infty}^{\infty} c_k e^{ikx}$$

converges to f in L^2, since we will be showing rather that this expansion does not converge uniformly on $[-\pi, \pi]$.

Using the expression (9.7) for the coefficients c_k from Lemma 9.16 the nth partial sum is

$$f_n(x) = \frac{1}{2\pi} \sum_{k=-n}^{n} \left(\int_{-\pi}^{\pi} f(t) e^{ikt} \, dt \right) e^{-ikx}.$$

At $x = 0$ this gives

$$f(0) = \frac{1}{2\pi} \sum_{k=-n}^{n} \left(\int_{-\pi}^{\pi} f(t) e^{ikt} \, dt \right)$$

$$= \frac{1}{2\pi} \int_{-\pi}^{\pi} f(t) \left(\sum_{k=-n}^{n} e^{ikt} \right) dt$$

$$= \frac{1}{2\pi} \int_{-\pi}^{\pi} f(t) K_n(t) \, dt,$$

where we define

$$K_n(t) := \sum_{k=-n}^{n} e^{ikt}.$$

Note that $K_n(t)$ is real and continuous with $|K_n(t)| \le 2n + 1$.

Let

$$X = \{ f \in C([-\pi, \pi]) : \; f(-\pi) = f(\pi) \};$$

this is a Banach space when equipped with the supremum norm. We consider the collection of maps $S_n \colon X \to \mathbb{R}$ given by $f \mapsto f_n(0)$. Note that we have $S_n \in B(X, \mathbb{R})$, since

$$|S_n(f)| = \left| \frac{1}{2\pi} \int_{-\pi}^{\pi} f(t)\, K_n(t)\, dt \right|$$

$$\leq \left(\frac{1}{2\pi} \int_{-\pi}^{\pi} |K_n(t)|\, dt \right) \|f\|_\infty$$

$$\leq (2n+1)\|f\|_\infty,$$

since $|K_n(t)| \leq (2n+1)$.

Using the 'Condensation of Singularities' (Corollary 22.5) it is enough to show that $\sup_n \|S_n\| = \infty$: if this is true, then there must exist an $f \in X$ such that $|S_n f| = |f_n(0)|$ is unbounded, and therefore $f_n(0)$ cannot converge to a limit as $n \to \infty$. In Example 11.9 we showed that

$$\|S_n\| = I_n := \frac{1}{2\pi} \int_{-\pi}^{\pi} |K_n(t)|\, dt;$$

we now prove that $I_n \to \infty$ as $n \to \infty$.

We first simplify the expression for K_n: if $t \neq 0$, then

$$K_n(t) := \sum_{k=-n}^{n} e^{ikt} = e^{-int}(1 + \cdots + e^{2int})$$

$$= e^{-int}\, \frac{e^{i(2n+1)t} - 1}{e^{it} - 1}$$

$$= \frac{e^{i(n+\frac{1}{2})t} - e^{-i(n+\frac{1}{2})t}}{e^{\frac{1}{2}it} - e^{-\frac{1}{2}it}} = \frac{\sin(n+\frac{1}{2})t}{\sin \frac{1}{2}t};$$

and clearly $K_n(0) = 2n + 1$.

It follows that

$$I_n = \frac{1}{2\pi} \int_{-\pi}^{\pi} \left| \frac{\sin(n+\frac{1}{2})t}{\sin \frac{1}{2}t} \right| dt.$$

To estimate I_n from below, observe that $|\sin(t/2)| \leq |t/2|$, and so

$$\left| \frac{\sin(n+\frac{1}{2})t}{\sin \frac{1}{2}t} \right| \geq \left| \frac{\sin(n+\frac{1}{2})t}{\frac{1}{2}t} \right|.$$

The right-hand side can be bounded below, for each $k = 1, \ldots, n$, by

$$\left| \frac{\sin(n+\frac{1}{2})t}{\frac{1}{2}t} \right| \geq \left| \frac{\sin(n+\frac{1}{2})t}{\frac{1}{2}k\pi/(n+\frac{1}{2})} \right| = \frac{2n+1}{k\pi} \left| \sin\left(n+\frac{1}{2}\right)t \right|,$$

$$\text{for all } t \in \left[\frac{(k-1)\pi}{n+\frac{1}{2}}, \frac{k\pi}{n+\frac{1}{2}} \right].$$

So

$$2\pi I_n > \int_0^{n\pi/(n+\frac{1}{2})} \left| \frac{\sin(n+\frac{1}{2})t}{\sin\frac{1}{2}t} \right| dt$$

$$\geq \sum_{k=1}^n \frac{2n+1}{k\pi} \int_{(k-1)\pi/(n+\frac{1}{2})}^{k\pi/(n+\frac{1}{2})} \left| \sin(n+\tfrac{1}{2})t \right| dt$$

$$= \sum_{k=1}^n \frac{2n+1}{k\pi} \int_0^{\pi/(n+\frac{1}{2})} \left| \sin(n+\tfrac{1}{2})t \right| dt$$

$$= \sum_{k=1}^n \frac{2}{k\pi} \int_0^\pi \sin t \, dt = \frac{4}{\pi} \sum_{k=1}^n \frac{1}{k},$$

which is unbounded as $n \to \infty$.

Exercises

22.1 Let X be an infinite-dimensional Banach space and let $(x_i)_{i=1}^\infty$ be a sequence in X. Let $Y_n = \mathrm{Span}(x_1, \ldots, x_n)$. Using the Baire Category Theorem show that the linear span of (x_i) is not the whole of X. (No infinite-dimensional Banach space can have a countable Hamel basis.)

22.2 Show that if $\mathcal{S} \subset B(X, Y)$ is such that $\sup_{T \in \mathcal{S}} \|T\| = \infty$, then the set of $x \in X$ for which $\sup_{T \in \mathcal{S}} \|Tx\| = \infty$ is residual. (Consider the collection of sets $\{x \in X : \sup_{T \in \mathcal{S}} \|Tx\| \leq n\}$.)

22.3 Consider the space X of all real polynomials

$$p(x) = \sum_{j=0}^\infty a_j x^j, \tag{22.5}$$

where $a_j = 0$ for all $j \geq N$ for some N, equipped with the norm

$$\|p\| = \max_j |a_j|.$$

Consider the sequence of linear functionals $T_n \colon X \to \mathbb{R}$ defined by setting

$$T_n p = \sum_{j=0}^{n-1} a_j \quad \text{when} \quad p(x) = \sum_{j=0}^\infty a_j x^j.$$

Show that $\sup_n |T_n p| < \infty$ for every $p \in X$, but $\sup_n \|T_n\| = \infty$. Deduce that X is not complete with the norm $\|\cdot\|$.

22.4 Suppose that X, Y, and Z are normed spaces, and that one of X and Y is a Banach space. Suppose that $b \colon X \times Y \to Z$ is bilinear and continuous. Use the Principle of Uniform Boundedness to show that there exists an $M > 0$ such that

$$|b(x, y)| \le M \|x\|_X \|y\|_Y \qquad \text{for every } x \in X, \ y \in Y.$$

(Rudin, 1991)

22.5 Show that ℓ^p is meagre in ℓ^q if $1 \le p < q \le \infty$. (Show that for every n the set $\{x \in \ell^p : \|x\|_{\ell^p} \le n\}$ is closed but has empty interior in ℓ^q.) (Rudin, 1991)

22.6 Suppose that x is a sequence (in $\mathbb{K}$) with the property that

$$\sum_{j=1}^{\infty} x_j y_j$$

converges for every $y \in \ell^p$, $1 < p < \infty$. Show that $x \in \ell^q$, where q is conjugate to p. (Meise and Vogt, 1997)

23

The Open Mapping, Inverse Mapping, and Closed Graph Theorems

In this chapter we prove another three fundamental theorems about linear operators between Banach spaces. First we prove the Open Mapping Theorem, using the Baire Category Theorem once again. As a consequence we prove the Inverse Mapping Theorem, which guarantees that any bounded linear map between Banach spaces that has an inverse is 'invertible' (i.e. its inverse is bounded) and then the Closed Graph Theorem, which provides what can be a relatively simple way of proving that a linear map between Banach spaces is bounded.

23.1 The Open Mapping and Inverse Mapping Theorems

We start by proving the Open Mapping Theorem.

Theorem 23.1 (Open Mapping Theorem) *If X and Y are Banach spaces and $T: X \to Y$ is a bounded surjective linear map, then T maps open sets in X to open sets in Y.*

Recall that we use $\mathbb{B}_X$ for the closed unit ball in X, and $B_Y(x, r)$ for the open ball in Y of radius r around x.

Proof It suffices to show that $T(\mathbb{B}_X)$ includes an open ball around 0 in Y, say $B_Y(0, r)$ for some $r > 0$: if U is an open set in X and $y \in TU$, then $y = Tx$ for some $x \in U$, and so there exists a $\delta > 0$ such that $x + \delta \mathbb{B}_X \subset U$. Then

$$T(U) \supset T(x + \delta \mathbb{B}_X) = Tx + \delta T(\mathbb{B}_X) \supset y + \delta B_Y(0, r) = B_Y(y, \delta r).$$

First we show that $\overline{T(\mathbb{B}_X)}$ contains a non-empty open ball around 0.

249

Notice that the closed sets

$$\overline{T(n\mathbb{B}_X)} = n\overline{T(\mathbb{B}_X)}$$

cover Y (since $T(X) = Y$ as T is surjective). It follows by using the Baire Category Theorem (in the form of Theorem 22.2) that at least one of them contains a non-empty open ball.

All these sets are scaled copies of $\overline{T(\mathbb{B}_X)}$, so $\overline{T(\mathbb{B}_X)}$ contains $B_Y(z, r)$ for some $z \in Y$ and some $r > 0$. Since $\overline{T(\mathbb{B}_X)}$ is symmetric, it follows that $B_Y(-z, r) \subset \overline{T(\mathbb{B}_X)}$ too. Since $\overline{T(\mathbb{B}_X)}$ is convex, we can deduce that $B_Y(0, r) \subset \overline{T(\mathbb{B}_X)}$: any $v \in B_Y(0, r)$ can be written as $\frac{1}{2}(v + z) + \frac{1}{2}(v - z)$.

Now we show that $T(2\mathbb{B}_X)$ must include $B_Y(0, r)$. Note that given any $y \in B_Y(0, \alpha r)$, since $\overline{T(\alpha\mathbb{B}_X)} \ni y$, for any $\varepsilon > 0$ there exists $x \in \alpha\mathbb{B}_X$ such that

$$\|y - Tx\|_Y < \varepsilon.$$

We use this argument repeatedly. Given $u \in B_Y(0, r)$, find $x_1 \in \mathbb{B}_X$ with $\|u - Tx_1\|_Y < r/2$. Then, since

$$u - Tx_1 \in B(0, r/2),$$

we can find $x_2 \in \frac{1}{2}\mathbb{B}_X$ such that

$$\|(u - Tx_1) - Tx_2\| < r/4.$$

Now find $x_3 \in \frac{1}{4}\mathbb{B}_X$ such that

$$\|(u - Tx_1 - Tx_2) - Tx_3\| < r/8,$$

and so on, yielding a sequence (x_n) with $x_n \in 2^{-n}\mathbb{B}_X$ such that

$$\left\| u - T\left(\sum_{j=1}^{n} x_j\right) \right\| < r2^{-n}. \tag{23.1}$$

Now, since X is a Banach space and $\sum_{j=1}^{n} \|x_j\| \leq 2 < \infty$ it follows using Lemma 4.13 that $\sum_{j=1}^{\infty} x_j$ converges to some $x \in 2\mathbb{B}_X$; since T is bounded it is continuous, so taking limits in (23.1) yields $Tx = u$.

Since $T(2\mathbb{B}_X) \supset B_Y(0, r)$, it follows that $T(\mathbb{B}_X) \supset B_Y(0, r/2)$. $\qquad\square$

The Inverse Mapping Theorem (also known as the Banach Isomorphism Theorem) is an almost immediate corollary. If $T \in B(X, Y)$ with X and Y both Banach spaces this result removes the seeming anomaly that a map can have an inverse but not be 'invertible'.

Theorem 23.2 (Inverse Mapping Theorem) *If X and Y are Banach spaces and $T \in B(X, Y)$ is bijective, then $T^{-1} \in B(Y, X)$, so T is invertible in the sense of Definition 11.13.*

Proof The map T has an inverse since it is bijective, and this inverse is necessarily linear, as discussed in Section 11.5. The Open Mapping Theorem shows that whenever U is open in X, $(T^{-1})^{-1}(U) = T(U)$ is open in Y, so T^{-1} is continuous (Lemma 2.13) and hence bounded (Lemma 11.3). For a less topological argument, note that the Open Mapping Theorem shows that $T(\mathbb{B}_X)$ includes $\theta \mathbb{B}_Y$ for some $\theta > 0$, so

$$T^{-1}(\theta \mathbb{B}_Y) \subseteq \mathbb{B}_X \qquad \Rightarrow \qquad T^{-1}(\mathbb{B}_Y) \subseteq \frac{1}{\theta} \mathbb{B}_X,$$

and so $\|T^{-1}\|_{B(Y,X)} \leq \theta^{-1}$. $\qquad\qquad\qquad\square$

This result has an immediate application in spectral theory. Recall (see Chapter 14) that when X is an infinite-dimensional normed space, the resolvent set of a bounded linear operator $T \colon X \to X$ consists of all those $\lambda \in \mathbb{C}$ for which $T - \lambda I$ is 'invertible', i.e. has a bounded inverse. The Inverse Mapping Theorem tells us that if X is a Banach space and $T \in B(X)$, then the boundedness of the inverse of $T - \lambda I$ is automatic if the inverse exists, so in fact

$$\rho(T) := \{\lambda \in \mathbb{C} : T - \lambda I : X \to X \text{ is a bijection}\} \qquad (23.2)$$

and

$$\sigma(T) = \{\lambda \in \mathbb{C} : T - \lambda I : X \to X \text{ is not a bijection}\}.$$

We will use this result in the next chapter to investigate the spectrum of compact operators on Banach spaces.

The following corollary of the Inverse Mapping Theorem, concerning equivalence of norms on Banach spaces, follows by considering the identity map $I_X \colon (X, \| \cdot \|_1) \to (X, \| \cdot \|_2)$.

Corollary 23.3 *If X is a Banach space that is complete with respect to two different norms $\| \cdot \|_1$ and $\| \cdot \|_2$ and $\|x\|_2 \leq C \|x\|_1$, then the two norms are equivalent.*

A quick example before a much longer one: you cannot put the ℓ^1 norm on ℓ^2 and make a complete space: we know that $\|x\|_{\ell^2} \leq \|x\|_{\ell^1}$. But $(1, 1/2, 1/3, \ldots) \in \ell^2$ and not in ℓ^1, so the norms are not equivalent.

23.2 Schauder Bases in Separable Banach Spaces

We now use Corollary 23.3 to investigate bases in separable Banach spaces. Suppose that[1] X is a separable Banach space with a countable Schauder basis $\{e_n\}_{n=1}^{\infty}$ so that (see Definition 9.1) any element $x \in X$ can be written uniquely in the form

$$x = \sum_{j=1}^{\infty} a_j e_j,$$

for some coefficients $a_j \in \mathbb{K}$, where the sum converges in X. Suppose that we consider the 'truncated' expansions

$$P_n x = \sum_{j=1}^{n} a_j e_j.$$

Can we find a constant C such that $\|P_n x\| \leq C\|x\|$ for every n and every $x \in X$? If $\{e_n\}_{n=1}^{\infty}$ is an orthonormal basis in a Hilbert space, then this follows immediately from Bessel's Inequality with $C = 1$; see Lemma 9.11. (The minimal C such that $\|P_n x\| \leq C\|x\|$ for every n and x is known as the basis constant.)

If we knew that P_n was in $B(X)$ for every n (i.e. that for each n we had $\|P_n x\| \leq C_n \|x\|$ for every $x \in X$) this would follow from the Principle of Uniform Boundedness; but this is not so easy to show directly. Instead we use Corollary 23.3.

Proposition 23.4 *Suppose that $\{e_j\}_{j=1}^{\infty}$ is a countable Schauder basis for a Banach space X, with $\|e_j\| = 1$ for every j. Then, setting*

$$|||x||| := \sup_n \left\| \sum_{j=1}^{n} a_j e_j \right\|_X \qquad \text{when} \quad x = \sum_{j=1}^{\infty} a_j e_j$$

defines a norm $|||\cdot|||$ on X, and X is complete with respect to this norm.

Proof To check that $|||\cdot|||$ is a norm, the only issue is the triangle inequality; but note that by the triangle inequality for $\|\cdot\|$

$$\left\| \sum_{i=1}^{n} (x_i + y_i) e_i \right\| \leq \left\| \sum_{i=1}^{n} x_i e_i \right\| + \left\| \sum_{i=1}^{n} y_i e_i \right\|,$$

---

[1] Contrary to intuition, there are separable Banach spaces that do not have a Schauder basis; a counterexample was constructed by Enflo (1973).

so

$$\sup_n \left\| \sum_{i=1}^n (x_i + y_i)e_i \right\| \leq \sup_n \left\| \sum_{i=1}^n x_i e_i \right\| + \sup_n \left\| \sum_{i=1}^n y_i e_i \right\|,$$

i.e.

$$\|\|x + y\|\| \leq \|\|x\|\| + \|\|y\|\|.$$

To show that $(X, \|\|\cdot\|\|)$ is complete is more involved. A key observation, however, is that if $x = \sum_{j=1}^{\infty} a_j e_j$, then for every j

$$|a_j| = \|a_j e_j\| = \left\| \sum_{j=1}^m a_j e_j - \sum_{j=1}^{m-1} a_j e_j \right\|$$

$$\leq \left\| \sum_{j=1}^m a_j e_j \right\| + \left\| \sum_{j=1}^{m-1} a_j e_j \right\| \leq 2\|\|x\|\|. \qquad (23.3)$$

Now suppose that $x^{(n)}$ is a Cauchy sequence in $(X, \|\|\cdot\|\|)$, with

$$x^{(n)} = \sum_{j=1}^{\infty} a_j^{(n)} e_j :$$

given any $\varepsilon > 0$ there exists an $N = N(\varepsilon)$ such that

$$\left\| x^{(n)} - x^{(m)} \right\| = \sup_k \left\| \sum_{j=1}^k [a_j^{(n)} - a_j^{(m)}] e_j \right\| < \varepsilon$$

for all $n, m \geq N(\varepsilon)$.

It follows that for each j the sequence $(a_j^{(n)})$ is Cauchy, so $a_j^{(n)} \to \alpha_j$ for some $\alpha_j \in \mathbb{K}$ as $n \to \infty$. We now show that $\sum_{j=1}^{\infty} \alpha_j e_j$ is convergent to some element $x \in X$ and that $\|\|x^{(n)} - x\|\| \to 0$ as $n \to \infty$.

Take $n \geq N(\varepsilon)$ and any $k \geq 1$. Then

$$\left\| \sum_{j=1}^k (a_j^{(n)} - \alpha_j)e_j \right\| = \left\| \sum_{j=1}^k (a_j^{(n)} - \lim_{m \to \infty} a_j^{(m)})e_j \right\|$$

$$= \lim_{m \to \infty} \left\| \sum_{j=1}^k (a_j^{(n)} - a_j^{(m)})e_j \right\| < \varepsilon. \qquad (23.4)$$

We will use this to show that the partial sums $\left(\sum_{j=1}^k \alpha_j e_j \right)_k$ form a Cauchy sequence in $(X, \|\cdot\|)$ and hence, since $(X, \|\cdot\|)$ is complete, they converge to some $x = \sum_{j=1}^{\infty} \alpha_j e_j$.

Now set $n := N(\varepsilon)$; we know that $\sum_{j=1}^{\infty} a_j^{(n)} e_j$ converges in X, so there exists $M(\varepsilon)$ such that if $r > s \geq M(\varepsilon)$ we have

$$\left\| \sum_{i=s}^{r} a_i^{(n)} e_i \right\| < \varepsilon. \tag{23.5}$$

Therefore

$$\left\| \sum_{i=s}^{r} \alpha_i e_i \right\| = \left\| \sum_{i=s}^{r} (a_i^{(n)} - \alpha_i) e_i \right\| + \left\| \sum_{i=s}^{r} a_i^{(n)} e_i \right\|$$

$$\leq \left\| \sum_{i=1}^{r} (a_i^{(n)} - \alpha_i) e_i - \sum_{i=1}^{s-1} (a_i^{(n)} - \alpha_i) e_i \right\| + \varepsilon$$

$$\leq 3\varepsilon,$$

using (23.4) and (23.5), which shows that $(\sum_{j=1}^{k} \alpha_j e_j)_k$ is a Cauchy sequence.

It now follows from (23.4) that $x^{(n)}$ converges to x in the norm $\|\|\cdot\|\|$, as required. ☐

Corollary 23.5 *If $\{e_j\}_{j=1}^{\infty}$ satisfies the conditions of Proposition 23.4, then there exists a constant $C > 0$ such that*

$$\|P_n x\| \leq C \|x\| \qquad \text{for every } n \in \mathbb{N}, \tag{23.6}$$

where

$$P_n x := \sum_{j=1}^{n} a_j e_j \qquad \text{for} \qquad x = \sum_{j=1}^{\infty} a_j e_j.$$

In particular, the map $x \mapsto a_j$ is an element of X^* for each j (and its norm can be bounded independently of n).

Proof Since both $(X, \|\cdot\|)$ and $(X, \|\|\cdot\|\|)$ are complete and

$$\|x\| = \lim_{n \to \infty} \left\| \sum_{j=1}^{n} a_j e_j \right\|_X \leq \sup_n \left\| \sum_{j=1}^{n} a_j e_j \right\|_X = \|\|x\|\|,$$

it follows from Corollary 23.3 that $\|\|x\|\| \leq C \|x\|$ for some $C > 0$. The linearity of the map $x \mapsto a_j$ follows from the uniqueness of the expansion $x = \sum_{j=1}^{\infty} a_j e_j$; this map is bounded since $|a_j| \leq 2\|\|x\|\| \leq 2C\|x\|$ (see (23.3)). ☐

Corollary 23.3 seems to suggest that you cannot put two different norms on a vector space X to make it a Banach space, but this is not the case. For simplicity consider a Banach space with a Schauder basis $\{e_n\}_{n=1}^\infty$, and define a linear map $T \colon X \to X$ by setting $T e_n = n e_n$ for each n. Exercise 5.2 guarantees that $\| \cdot \|_T$ defined by setting

$$\|x\|_T := \|Tx\|$$

is also a norm on X, with which X is again complete. However, $\| \cdot \|$ and $\| \cdot \|_T$ cannot be equivalent: $\|x_n\|_T = n\|x_n\|$, so there is no constant C such that $\|x\|_T \le C\|x\|$ for every $x \in X$. (In a general infinite-dimensional Banach space one can use Riesz's Lemma to find a countable linearly independent set, and then use a similar construction.)

23.3 The Closed Graph Theorem

The Closed Graph Theorem gives a way to check whether a linear map $T \colon X \to Y$ is bounded when both X and Y are Banach spaces by considering its 'graph' in the product space $X \times Y$.

Theorem 23.6 (Closed Graph Theorem) *Suppose that $T \colon X \to Y$ is a linear map between Banach spaces and that the graph of T,*

$$G := \{(x, Tx) \in X \times Y : x \in X\},$$

is a closed subset of $X \times Y$ (with norm $\|(x, y)\|_{X \times Y} = \|x\|_X + \|y\|_Y$). Then T is bounded.

If the graph G is closed, then this means that if $x_n \to x$ and $Tx_n \to y$, then $Tx = y$. Continuity is stronger, since it does not require $Tx_n \to y$ (so whenever $T \in B(X, Y)$ the set G is automatically closed).

Proof Since $X \times Y$ is a Banach space (Lemma 4.6) and G is a subspace of $X \times Y$ (since T is linear), it follows from the assumption that G is closed that G is a Banach space when equipped with the norm of $X \times Y$ (Lemma 4.3).

Now consider the projection map $\Pi_X \colon G \to X$, defined by

$$\Pi_X(x, y) = x,$$

which is both linear and bounded. This map is clearly surjective and it is one-to-one, since

$$\Pi_X(x, Tx) = \Pi_X(y, Ty) \quad \Rightarrow \quad x = y \quad \Rightarrow \quad (x, Tx) = (y, Ty).$$

By the Inverse Mapping Theorem (Theorem 23.2) the map Π_X^{-1} is bounded. It follows that

$$\|x\|_X + \|Tx\|_Y = \|(x, Tx)\|_{X \times Y} = \|\Pi_X^{-1} x\|_{X \times Y} \leq M \|x\|_X,$$

and so $\|Tx\|_Y \leq M \|x\|_X$ as required. $\qquad\qquad\qquad\square$

Exercises

23.1 Show that if (α_n) is a sequence of strictly positive real numbers such that $\sum_{n=1}^{\infty} \alpha_n < \infty$, then there is a sequence (y_n) with $y_n \to \infty$ as $n \to \infty$ such that $\sum_{n=1}^{\infty} \alpha_n y_n < \infty$. [Hint: consider the map $T : \ell^\infty \to \ell^1$ defined by setting

$$(Tx)_j = \alpha_j x_j;$$

this is clearly one-to-one. Assume that no (y_n) as above exists and deduce a contradiction by applying the Inverse Mapping Theorem to T.] (Pryce, 1973)

23.2 Show that a countable set $\{e_j\}_{j=1}^\infty$ of norm one elements of a Banach space X is a Schauder basis for X if and only if (i) the linear span of $\{e_j\}_{j=1}^\infty$ is dense in X and (ii) there is a constant K such that

$$\left\| \sum_{i=1}^{n} a_i e_i \right\| \leq K \left\| \sum_{j=1}^{m} a_j e_j \right\|$$

for all $\{a_j\} \subset \mathbb{K}$ and every $n < m$. (One way follows from Corollary 23.5. For the other direction, let Y be the linear span of the $\{e_j\}$ and for each n define a projection $\Pi_n : Y \to \mathrm{Span}(e_1, \ldots, e_n)$ by setting

$$\Pi_n \left(\sum_{j=1}^{m} a_j e_j \right) = \sum_{j=1}^{\min(m,n)} a_j e_j.$$

Extend each Π_n to a map $P_n : X \to \mathrm{Span}(e_1, \ldots, e_n)$ using Exercise 19.3 and then show that $P_n x \to x$ for every $x \in X$.) (Carothers, 2005)

23.3 Use the Closed Graph Theorem to show that if H is a Hilbert space and $T : H \to H$ is a linear operator that satisfies

$$(Tx, y) = (x, Ty) \qquad \text{for every} \qquad x, y \in H$$

then T is bounded. (This is the Hellinger–Toeplitz Theorem.)

23.4 Let $X = C^1([0, 1])$ with the supremum norm (so this is not a Banach space) and $Y = C([0, 1])$ with the supremum norm (which is a Banach space). If we define $T: X \to Y$ by $Tf = f'$ show that the graph of T is closed but that T is not bounded. (This does not contradict the Closed Graph Theorem since X is not a Banach space.)

23.5 Use the Closed Graph Theorem to show that if X is a real Banach space and $T: X \to X^*$ is a linear map such that

$$(Tx)(x) \geq 0 \qquad \text{for every } x \in X,$$

then T is bounded. (Brezis, 2011)

24

Spectral Theory for Compact Operators

In Chapter 16 we investigated the spectrum of compact self-adjoint operators on Hilbert spaces. We showed there that the spectrum of these operators consists entirely of eigenvalues, apart perhaps from zero. We also showed that each eigenvalue has finite multiplicity and that the eigenvalues have no accumulation points except zero. In this chapter we will prove the same results but for a compact operator T on a Banach space X; we drop the self-adjointness and the requirement that the operator acts on a Hilbert space. In the next chapter we will instead drop the compactness (and more besides), and consider self-adjoint unbounded operators on Hilbert spaces.

Recall (see Definition 15.1) that $T \colon X \to X$ is compact if whenever (x_n) is a bounded sequence in X, (Tx_n) has a convergent subsequence. Our primary tool throughout this chapter (in addition to the compactness of T itself) will be Riesz's Lemma (Lemma 5.4): if Y is a proper closed subspace of X, then there exists $x \notin Y$ with $\|x\| = 1$ such that $\|x - y\| \geq 1/2$ for every $y \in Y$.

24.1 Properties of $T - I$ When T Is Compact

In order to investigate the spectrum of T we have to understand properties of the operators $T - \lambda I$ for $\lambda \in \mathbb{C}$. Since we can write $T - \lambda I = \lambda(\frac{1}{\lambda}T - I)$ for any $\lambda \neq 0$, it is enough to understand operators of the form $T - I$ when T is compact. We therefore analyse this case and deduce results for $T - \lambda I$ as a consequence.

Our first result is that all eigenvalues of T have finite multiplicity.

Lemma 24.1 *If* $T \in K(X)$*, then* $\dim \operatorname{Ker}(T - I) < \infty$.

258

Proof Write E for $\mathrm{Ker}(T - I)$, and suppose that $\dim E = \infty$. We show that in this case we can find a sequence $(w_j) \in E_\lambda$ such that $\|w_j\| = 1$ and $\|w_i - w_j\| \geq 1/2$ for $i \neq j$.

If we have a collection $(w_j)_{j=1}^n$ of elements of E such that $\|w_j\| = 1$ and

$$\|w_i - w_j\| \geq 1/2 \qquad 1 \leq i, j \leq n, \ i \neq j,$$

then let Y_n be the finite-dimensional space spanned by $\{w_1, \ldots, w_n\}$. Since this space is finite-dimensional, it is closed (Exercise 5.3), and so we can use Riesz's Lemma (Lemma 5.4) to find $w_{n+1} \in E$ with $\|w_{n+1}\| = 1$ such that $\|w_{n+1} - w_j\| \geq 1/2$ for every $j = 1, \ldots, n$.

This inductive process can be started with any choice of $w_1 \in E$ with $\|w_1\| = 1$. In this way we generate a sequence (w_j) such that $T w_j = w_j$ for every j. However, since $\|w_j\| = 1$ and T is compact the sequence $(T w_j)$ must have a convergent subsequence: so (w_j) must have a convergent subsequence. However,

$$\|w_i - w_j\| \geq 1/2 > 0$$

for every $i \neq j$, so (w_j) cannot have a convergent subsequence.

It follows that $\dim \mathrm{Ker}(T - I)$ is finite, as claimed. □

Recall (see Section 14.1) that the eigenspace corresponding to an eigenvalue λ is

$$E_\lambda := \{x \in X : \ Tx = \lambda x\} = \mathrm{Ker}(T - \lambda I),$$

and that the multiplicity of λ is the dimension of E_λ.

Corollary 24.2 *Suppose that $T \in K(X)$ and $\lambda \neq 0$. Then*

$$\dim \mathrm{Ker}(T - \lambda I) < \infty;$$

in particular, any non-zero eigenvalue of T has finite multiplicity.

We have already seen that in general the range of a bounded operator need not be closed: we gave an example in (11.13), the map $T \colon \ell^2 \to \ell^2$ defined by setting

$$Tx := \left(x_1, \frac{x_2}{2}, \frac{x_3}{3}, \frac{x_4}{4}, \ldots\right).$$

This is also a compact map from ℓ^2 into ℓ^2 (see Exercise 15.3), so in fact the range of a compact map need not be closed either. It is therefore striking that the range of $T - I$ is closed whenever T is compact.

Proposition 24.3 *If $T \in K(X)$, then $\mathrm{Range}(T - I)$ is closed.*

(For use in the proof of Theorem 24.4 we make explicit here the trivial observation that $\text{Range}(I - T)$ is also closed if $T \in K(X)$.)

Proof We use Lemma 24.1, which guarantees that $\text{Ker}(T - I)$ is finite-dimensional.

Take $(y_n) \in \text{Range}(T - I)$ with $y_n = (T - I)x_n$, such that $y_n \to y$ for some $y \in X$. We have to show that $y \in \text{Range}(T - I)$, i.e. that $y = (T - I)x$ for some $x \in X$. Let

$$d_n := \text{dist}(x_n, \text{Ker}(T - I));$$

since $\text{Ker}(T - I)$ is finite-dimensional there exists $z_n \in \text{Ker}(T - I)$ such that $\|x_n - z_n\| = d_n$; see Exercise 5.4.

We want to show that $\|x_n - z_n\|$ is a bounded sequence; if not, then there is a subsequence such that $\|x_{n_j} - z_{n_j}\| \to \infty$ as $j \to \infty$. Note that since $z_n \in \text{Ker}(T - I)$ we have

$$y_n = (T - I)(x_n - z_n) = T(x_n - z_n) - (x_n - z_n). \tag{24.1}$$

Setting $w_j = (x_{n_j} - z_{n_j})/\|x_{n_j} - z_{n_j}\|$ it follows that $\|w_j\| = 1$ and

$$Tw_j - w_j = \frac{y_{n_j}}{\|x_{n_j} - y_{n_j}\|} \to 0 \qquad \text{as } j \to \infty.$$

Since T is compact, we can find a subsequence of (w_j), (w_{j_k}), such that $Tw_{j_k} \to q$ for some $q \in X$. Using the triangle inequality we obtain

$$\|w_{j_k} - q\| \le \|w_{j_k} - Tw_{j_k}\| + \|Tw_{j_k} - q\|,$$

and so we also have $w_{j_k} \to q$. Since T is compact, it is bounded and therefore continuous, so

$$q = \lim_{k \to \infty} Tw_{j_k} = T\left(\lim_{k \to \infty} w_{j_k}\right) = Tq,$$

i.e. $q \in \text{Ker}(T - I)$. So on the one hand $\text{dist}(w_{j_k}, \text{Ker}(T - I)) \to 0$, while on the other

$$\text{dist}(w_{j_k}, \text{Ker}(T - I)) = \frac{\text{dist}(x_{n_{j_k}}, \text{Ker}(T - I))}{\|x_{n_{j_k}} - z_{n_{j_k}}\|} = 1,$$

a contradiction.

Since $(x_n - z_n)$ is a bounded sequence in X, $T(x_n - z_n)$ has a convergent subsequence with $T(x_{n_j} - z_{n_j}) \to p$ for some $p \in X$. Using this observation in (24.1) and recalling that $y_n \to y$ it follows that

$$x_{n_j} - z_{n_j} \to p - y,$$

so setting $x = p - y$ we obtain $y = (T - I)x$ and hence Range$(T - I)$ is closed as claimed. □

We will obtain more information about the spectrum of general compact operators from the following important result.

Theorem 24.4 *If $T \in K(X)$ and* Ker$(T - I) = \{0\}$, *then $T - I$ is invertible.*

Proof Since $T - I$ is a bounded linear operator and the assumption that Ker$(T - I) = \{0\}$ ensures that $T - I$ is injective (Lemma 1.21), $T - I$ will be a bijection from X onto X if we show that it is surjective. The Inverse Mapping Theorem (Theorem 23.2) will then guarantee that $(T - I)^{-1}$ is bounded, and hence that $T - I$ is invertible. So all we need to prove is that $T - I$ maps X onto X.

We begin by observing that

$$(T - I)^n = \sum_{k=0}^{n} (-1)^{n-k} \binom{n}{k} T^k = (-1)^n [S_n - I],$$

where $S_n \in K(X)$ (since T^k is compact for each k and $K(X)$ is a vector space). Now let

$$X_n := \text{Range}((T - I)^n);$$

it follows using Proposition 24.3 that each X_n is closed. So we have

$$X \supseteq X_1 \supseteq X_2 \supseteq X_3 \supseteq \cdots , \tag{24.2}$$

with X_{n+1} a closed linear subspace of X_n. To prove the theorem we will show that $X_1 = \text{Range}(T - I) = X$.

If $X_{n+1} \neq X_n$ for every n, then, using Riesz's Lemma (Lemma 5.4), it follows that for each n there exists $x_n \in X_n$ with $\|x_n\| = 1$ such that

$$\text{dist}(x_n, X_{n+1}) \geq \frac{1}{2}.$$

Now observe that if $n > m$, then

$$\|Tx_m - Tx_n\| = \|x_m - [(T - I)x_n + x_n - (T - I)x_m]\| \geq \frac{1}{2}, \tag{24.3}$$

since $(T - I)x_n + x_n - (T - I)x_m \in X_{m+1}$ (the first term is in X_{n+1}, the second in X_n, the third in X_{m+1}, and these spaces are nested as in (24.2)).

The inequality in (24.3) shows that (Tx_j) cannot have a convergent subsequence, which contradicts the compactness of T since the sequence (x_j) is bounded. It follows that $X_{n+1} = X_n$ for some $n \in \mathbb{N}$, and then $X_j = X_n$ for all $j \geq n$.

Therefore, given any $x \in X$, we have

$$(T - I)^n x = (T - I)^{2n} y$$

for some $y \in X$. Since $\mathrm{Ker}(T - I) = \{0\}$, it follows that $\mathrm{Ker}(T - I)^n = \{0\}$, and so $(T - I)^n$ is one-to-one, which shows that

$$x = (T - I)^n y,$$

i.e. $X_n = X$. That $X_1 = X$ now follows immediately from (24.2). $\qquad\square$

24.2 Properties of Eigenvalues

After all the above preparation, we can now quickly deduce that the spectrum of a compact operator consists entirely of eigenvalues, perhaps with the exception of zero, their only possible accumulation point.

Corollary 24.5 *If $T \in B(X)$ is compact and $\lambda \in \sigma(T)$ with $\lambda \neq 0$, then λ is an eigenvalue of T.*

Proof If λ is not an eigenvalue of T, then $\mathrm{Ker}(T - \lambda I) = \{0\}$. For $\lambda \neq 0$

$$\mathrm{Ker}\left(\frac{T}{\lambda} - I\right) = \mathrm{Ker}(T - \lambda I) = \{0\};$$

since T/λ is a compact operator we can now apply Theorem 24.4 to guarantee that $(T/\lambda) - I$ is invertible. Since this implies that $T - \lambda I$ is invertible, it follows that $\lambda \notin \sigma(T)$. $\qquad\square$

Proposition 24.6 *If $T \in K(X)$ and $(\lambda_j)_{j=1}^\infty$ is a sequence of distinct non-zero eigenvalues of T, then $\lambda_j \to 0$ as $j \to \infty$.*

Proof Choose eigenvectors (e_j) for (λ_j) with $\|e_j\| = 1$, and define a sequence of closed subspaces of X by setting $E_n := \mathrm{Span}(e_1, \ldots, e_n)$. Since the eigenvectors (e_j) are linearly independent (Lemma 14.2), we have $\dim(E_n) = n$ and the spaces E_n are strictly increasing (i.e. E_n is a proper subspace of E_{n+1}). Using Riesz's Lemma (Lemma 5.4) we can find $x_n \in E_n$ such that $\|x_n\| = 1$ and $\mathrm{dist}(x_n, E_{n-1}) \geq 1/2$.

Now take $n > m \geq 2$; then we have

$$\left\| \frac{Tx_n}{\lambda_n} - \frac{Tx_m}{\lambda_m} \right\| = \left\| x_n + \frac{(T - \lambda_n I)x_n}{\lambda_n} - x_m - \frac{(T - \lambda_m I)x_m}{\lambda_m} \right\|,$$

and since

$$E_{m-1} \subset E_m \subseteq E_{n-1} \subset E_n \quad \text{and} \quad (T - \lambda_n I)E_n = E_{n-1}$$

it follows that

$$-\frac{(T - \lambda_n I)x_n}{\lambda_n} + x_m + \frac{(T - \lambda_m I)x_m}{\lambda_m} \in E_{n-1},$$

and so

$$\left\| \frac{T x_n}{\lambda_n} - \frac{T x_m}{\lambda_m} \right\| \geq \text{dist}(x_n, E_{n-1}) \geq \frac{1}{2}.$$

This shows that $(T x_n / \lambda_n)$ cannot have a convergent subsequence. Since T is compact, this implies that (x_n / λ_n) can have no bounded subsequence (if it did, then there would be a further subsequence $x_{n_{j_k}}$ for which $T x_{n_{j_k}} / \lambda_{n_{j_k}}$ did converge). Since $\|x_n / \lambda_n\| = 1/\lambda_n$, it follows that λ_n^{-1} has no bounded subsequence, so $\lambda_n \to 0$ as $n \to \infty$. $\qquad\square$

We summarise the results for compact operators from Chapters 14 and 15 and this chapter in the following theorem.

Theorem 24.7 *Suppose that X is an infinite-dimensional Banach space and $T \in K(X)$. Then*

(i) *$0 \in \sigma(T)$;*
(ii) *$\sigma(T) = \sigma_p(T) \cup \{0\}$;*
(iii) *all non-zero eigenvalues of T have finite multiplicity;*
(iv) *eigenvectors corresponding to distinct eigenvalues are linearly independent; and*
(v) *the only possible accumulation point of the set $\sigma_p(T)$ is zero.*

Proof (i) is Theorem 15.8; (ii) follows from Corollary 24.5 and (i); (iii) is Corollary 24.2; (iv) is Lemma 14.2; and (v) is Proposition 24.6. $\qquad\square$

25

Unbounded Operators on Hilbert Spaces

We will now look at unbounded self-adjoint operators defined on dense sub-spaces of Hilbert spaces. These arise very naturally in applications, in particu-lar, in differential equations (see e.g. Evans, 1998; Renardy and Rogers, 1993) and quantum mechanics (see e.g. Kreyszig, 1978; Zeidler, 1995). Because we will want to prove some results about the spectrum we assume throughout that H is a complex Hilbert space; in particular, we write ℓ^2 for $\ell^2(\mathbb{C})$ (and similarly for other sequence spaces).

As an illustration throughout we will use a simple example, closely related to the operation of taking two derivatives. If we take $f \in L^2(0, \pi)$, then we can expand it as a Fourier cosine series,

$$f(x) = \sum_{k=0}^{\infty} c_k \cos kx,$$

where the sum converges in L^2 (cf. Exercise 9.11). If we take two derivatives of both sides (assuming that we can differentiate term-by-term on the right-hand side), then we obtain

$$-f''(x) = \sum_{k=0}^{\infty} k^2 c_k \cos kx.$$

Thus the map $f \mapsto -f''$ induces a map on the coefficients,

$$(c_1, c_2, c_3, \ldots) \mapsto (c_1, 2^2 c_2, 3^2 c_3, \ldots).$$

We will therefore consider the operator T_2 that acts on sequences $x \in \ell^2$ with $(T_2 x)_n = n^2 x_n$, i.e.

$$T_2 x := (x_1, 2^2 x_2, 3^2 x_3, \ldots, n^2 x_n, \ldots). \tag{25.1}$$

For $T_2 x$ to be an element of ℓ^2 we certainly need to place some restrictions on x; for example, if $x \in \ell^2$ there is no reason why $T_2 x$ should also be in ℓ^2. We have some freedom to define an appropriate *domain (of definition)* of T_2, which we write as $D(T_2)$, that ensures that $Tx \in \ell^2$. To start with, our only requirement will be that this domain is a dense subspace of ℓ^2; so, for example, we could choose $D(T_2) = c_{00}$, the space of sequences with only a finite number of non-zero terms (this is dense in ℓ^2; see the very last paragraph of Chapter 3). While $T_2 x \in \ell^2$ for every $x \in c_{00}$, $T_2 : c_{00} \to \ell^2$ is unbounded, since $T_2 e_n = n^2 e_n$ for each n.

The density of the domain will prove crucial: we will frequently use the fact that if $(x, y) = (x, z)$ for every x in a dense subset A of H, then $y = z$ (this was proved in Exercise 9.9).

25.1 Adjoints of Unbounded Operators

We will develop a general theory for unbounded linear operators defined on a dense subspace $D(T)$ of H, $T : D(T) \to H$. We use the norm and inner product of H on $D(T)$, so T is 'unbounded' in the sense that there is no constant C such that

$$\|Tx\|_H \leq C \|x\|_H \qquad \text{for every } x \in D(T).$$

The space $D(T)$ is called the *domain* of the operator T, and when an operator is unbounded this domain forms part of the definition of the operator. We therefore refer throughout this chapter to an 'operator' as a pair $(D(T), T)$ to emphasise the importance of the domain.

We want to define an adjoint for unbounded operators. We saw (see (13.2)) that a bounded operator on a Hilbert space is self-adjoint if and only if it is symmetric, but we require something more for unbounded operators.

Lemma 25.1 *If $T : D(T) \to H$ is an unbounded operator, then there exists an adjoint operator $T^* : D(T^*) \to H$ such that*

$$(Tx, y) = (x, T^* y) \qquad \text{for every } x \in D(T), \ y \in D(T^*). \tag{25.2}$$

Note that as well as defining the adjoint we also have to define an appropriate domain $D(T^*)$ on which the adjoint can act. Once we fix the definition of this domain we will show that the adjoint operator is uniquely determined by (25.2).

Proof We define $D(T^*)$ to be all those $y \in H$ for which the map

$$x \mapsto (Tx, y)$$

defines a bounded linear functional $\phi \colon D(T) \to \mathbb{C}$.

Since $D(T)$ is dense in H, we can use the result of Exercise 19.3 to extend ϕ in a unique way to an element $f \in H^*$; now we can use the Riesz Representation Theorem to guarantee that there is a $z \in H$ such that $f(x) = (x, z)$. We set $T^*y := z$, and then by definition

$$(Tx, y) = (x, T^*y) \qquad x \in D(T), \ y \in D(T^*).$$

It is easy to check that T^* is linear and that $D(T^*)$ is a subspace of H.

Note that if (25.2) holds for two operators T^* and T' defined on $D(T^*)$ we have $(x, T^*y) = (x, T'y)$ for every $x \in D(T)$ $y \in D(T^*)$. Since $D(T)$ is dense in H, it follows that $T^*y = T'y$ for every $y \in D(T^*)$, and so $T^* = T'$. $\square$

Note that while we could choose the domain $D(T)$ for the original operator T, once this is done the domain of T^* is determined by the above procedure.

Definition 25.2 An operator $(D(T), T)$ is *self-adjoint* if

$$(D(T^*), T^*) = (D(T), T).$$

This definition means that we need the domains of T and T^* to coincide $(D(T^*) = D(T))$ and for T to be symmetric in the sense that

$$(Tx, y) = (x, Ty) \qquad \text{for every } x, y \in D(T). \tag{25.3}$$

We showed in Exercise 23.3 that if (25.3) holds for every $x, y \in H$, then T must be bounded, so this is the strongest symmetry property we can expect for unbounded operators. The proof used in Exercise 16.1 can easily be adapted to show that $(D(T), T)$ is symmetric if and only if (Tx, x) is real for every $x \in D(T)$.

Our example operator $T_2 \colon c_{00} \to \ell^2$ is symmetric, since

$$(T_2 x, y) = \sum_{j=1}^{\infty} j^2 x_j \overline{y_j} = (x, T_2 y), \qquad x, y \in c_{00},$$

so (c_{00}, T_2) is symmetric. However, it is not self-adjoint. We have

$$(T_2 x, y) = \sum_{j=1}^{\infty} j^2 x_j \overline{y_j}.$$

Thus the map $x \mapsto (T_2 x, y)$ defines a bounded linear functional on $c_{00} \subset \ell^2$ precisely when $(j^2 y_j) = T_2 y \in \ell^2$. So we have

$$D(T_2^*) = \{y : \sum_{j=1}^{\infty} j^4 |y_j|^2 < \infty\} = \{y : T y \in \ell^2\} \neq c_{00}$$

and in this case

$$T_2^* y = (y_1, 2^2 y_2, 3^2 y_3, \ldots) = T_2 y.$$

Essentially the same calculation shows that T_2 *is* self-adjoint if we take its domain to be

$$\mathfrak{h}^2 := \{x \in \ell^2 : (j^2 x_j) \in \ell^2\}. \tag{25.4}$$

25.2 Closed Operators and the Closure of Symmetric Operators

A linear operator is continuous if and only if it is bounded (Lemma 11.3), but there is a substitute for continuity for unbounded operators.

Definition 25.3 A densely defined linear operator $T: D(T) \to H$ is called *closed* if its graph

$$G = \{(x, Tx) \subset H \times H : x \in D(T)\}$$

is closed, i.e. if $(x_n) \in D(T)$ with $x_n \to x$ and $T x_n \to y$ implies that $y = Tx$.

The Closed Graph Theorem (Theorem 23.6) guarantees that if T is closed and $D(T) = H$, then T is bounded. So being 'closed' is like a weak form of being bounded.

The operator T_2 is not closed if we take its domain to be c_{00}: we can take $x_n = (1, 2^{-3}, 3^{-3}, \ldots, n^{-3}, 0, 0, \ldots) \in c_{00}$, and then

$$T_2 x_n = (1, 2^{-1}, 3^{-1}, \ldots, n^{-1}, 0, 0, \ldots).$$

As $n \to \infty$ we have $T_2 x_n \to (1/j)_j \in \ell^2$ and $x_n \to x := (j^{-3})_j \in \ell^2$, but $x \notin c_{00}$, so (c_{00}, T_2) is not closed. However, if we choose to define T_2 on the larger domain $\mathfrak{h}^2$ from (25.4), then it is closed. To prove this, we define $S: \ell^2 \to \ell^2$ by setting

$$S(x_1, x_2, x_3, \ldots) = (x_1, \frac{x_2}{2^2}, \frac{x_3}{3^2}, \ldots).$$

Then $S: \ell^2 \to \ell^2$ is bounded, and if $x \in \mathfrak{h}_2$, we have $S(T_2 x) = x$.

Now suppose that $x^{(n)} \to x$ and $T_2 x^{(n)} \to y$; then

$$x^{(n)} = S(T_2 x^{(n)}) \to S(y) \qquad \text{as } n \to \infty$$

since S is continuous; so $x = S(y)$. It follows that

$$\sum_{j=1}^{\infty} j^4 |x_j|^2 = \sum_{j=1}^{\infty} j^4 \left| \frac{y_j}{j^2} \right|^2 = \sum_{j=1}^{\infty} |y_j|^2 < \infty,$$

so $x \in \mathfrak{h}_2$ and $T_2 x = T_2 S y = y$.

Here we started with a symmetric operator that was not closed, and obtained a closed operator by enlarging its domain. That this is always possible is shown by the following result. We say that (D_2, T_2) is an extension of (D_1, T_1) if $D_1 \subseteq D_2$ and $T_2|_{D_1} = T_1$.

Theorem 25.4 *If $T \colon D(T) \to H$ is symmetric, then it has an extension $(D(\overline{T}), \overline{T})$ that is closed and symmetric.*

Proof Let $D(\overline{T})$ be the set of all $x \in H$ for which we can find a sequence $(x_n) \in D(T)$ such that

$$x_n \to x \qquad \text{and} \qquad T x_n \to y$$

for some $y \in H$. The space $D(\overline{T})$ is a vector space, and $D(T) \subset D(\overline{T})$ (for any $x \in D(T)$ take the constant sequence $x_n = x$). On $D(\overline{T})$ we define $\overline{T}$ by setting $\overline{T} x = y$. We now show that $\overline{T}$ is well defined on $D(\overline{T})$, and is both symmetric and closed.

(i) $\overline{T}$ is well defined. Suppose that (x'_n) is another sequence in $D(T)$ such that

$$x'_n \to x \qquad \text{and} \qquad T x'_n \to z.$$

Now for any $v \in D(T)$ we have

$$(v, T x_n - T x'_n) = (v, T(x_n - x'_n)) = (Tv, x_n - x'_n)$$

since T is symmetric. Letting $n \to \infty$ we obtain $(v, y - z) = 0$ for every $v \in D(T)$. Since $D(T)$ is dense in H, it follows that $y = z$.

(ii) It is easy to check that $\overline{T}$ is linear; so it is an extension of T. To check that $\overline{T}$ is also symmetric, for every $x, x' \in D(\overline{T})$ there exist $(x_n), (x'_n)$ in $D(T)$ such that

$$x_n \to x \qquad \text{and} \qquad T x_n \to \overline{T} x$$

and

$$x'_n \to x' \qquad \text{and} \qquad T x'_n \to \overline{T} x'.$$

Since T is symmetric, $(Tx_n, x_n') = (x_n, Tx_n')$, and so $(\overline{T}x, x') = (x, \overline{T}x')$ as the inner product is continuous.

(iii) Finally, we show that $\overline{T}$ is closed. We take $x_n \in D(\overline{T})$ such that

$$x_n \to x \quad \text{and} \quad \overline{T}x_n \to y;$$

we need to show that $x \in D(\overline{T})$ and $\overline{T}x = y$. For each n we find $\xi_n \in D(T)$ such that

$$\|x_n - \xi_n\| < \frac{1}{n} \quad \text{and} \quad \|\overline{T}x_n - T\xi_n\| < \frac{1}{n}.$$

It follows that $\xi_n \to x$ and $T\xi_n \to y$, and so $x \in D(\overline{T})$ and $\overline{T}x = y$, so $\overline{T}$ is closed. $\qquad\square$

We have actually found the minimal closed extension of $(D(T), T)$, i.e. if $(D(T'), T')$ is such that $D(T') \supseteq D(T)$ and T' is closed, then $D(T') \supseteq D(\overline{T})$. Our extension $(D(\overline{T}), \overline{T})$ is called the *closure* of T. An operator T whose closure is self-adjoint is called essentially self-adjoint.

While it is not the case that every symmetric operator has a self-adjoint closure, if T is symmetric and bounded below then there always exists an extension of T that is self-adjoint (the 'Friedrichs extension'). The proof is lengthy, and we do not give it here; it can be found in Chapter 5 of Zeidler (1995).

Theorem 25.5 (Friedrichs extension) *If* $T: D(T) \to H$ *is symmetric and* $(Tx, x) \geq \alpha \|x\|^2$ *for some* $\alpha \in \mathbb{R}$ *for every* $x \in D(T)$, *then* T *has a self-adjoint extension* $(D(\hat{T}), \hat{T})$.

25.3 The Spectrum of Closed Unbounded Self-Adjoint Operators

We showed in Chapter 23, using the Inverse Mapping Theorem (Theorem 23.2), that for a bounded operator the resolvent set is given by (23.2),

$$\rho(T) := \{\lambda \in \mathbb{C} : T - \lambda I : H \to H \text{ is a bijection}\}, \qquad (25.5)$$

and then

$$\sigma(T) = \{\lambda \in \mathbb{C} : T - \lambda I : H \to H \text{ is not a bijection}\}.$$

For a closed unbounded operator $(D(T), T)$ we cannot expect $T - \lambda I$ to map H onto H, so we make the following alternative definition.

Definition 25.6 Suppose that $(D(T), T)$ is a closed linear operator. The *resolvent set* $\rho(T)$ is

$$\rho(T) := \{\lambda \in \mathbb{C} : T - \lambda I \text{ is injective with a dense range}$$
$$\text{on which } (T - \lambda I)^{-1} \text{ is bounded}\}. \quad (25.6)$$

The spectrum of T, $\sigma(T)$ is the complement of $\rho(T)$; it can be decomposed into

- the *point spectrum*: $\lambda \in \sigma_p(T)$ if $T - \lambda I$ is not injective;
- the *continuous spectrum*: $\lambda \in \sigma_s(T)$ if $T - \lambda I$ is injective with a dense range but $(T - \lambda I)^{-1}$ is not bounded;
- the *residual spectrum*: $\lambda \in \sigma_r(T)$ if $T - \lambda I$ is injective and its range is not dense.

Although this new definition of the resolvent set looks quite different from that in (25.5), they do in fact coincide for bounded operators.

Lemma 25.7 *If $T \in B(H)$, then the two definitions of the resolvent set agree.*

Proof Assume that $S = T - \lambda I$ is injective with a dense range. We need to show that S maps H onto H (Definition 25.5) if and only if S^{-1} is bounded on Range(S) (as in Definition 25.6).

If S maps H onto H, then, by the Inverse Mapping Theorem, S^{-1} is bounded on $H = $ Range(S).

If S^{-1} is bounded on the dense set Range(S), then, given any $y \in H$, we can find $y_n \in $ Range(S) such that $y_n \to y$ and then set

$$x := \lim_{n \to \infty} S^{-1} y_n.$$

The boundedness of S^{-1} implies that $S^{-1} y_n$ is a Cauchy sequence in H, and so this limit exists. Now since S is bounded we have

$$Sx = S\left(\lim_{n \to \infty} S^{-1} y_n\right) = \lim_{n \to \infty} S(S^{-1} y_n) = \lim_{n \to \infty} y_n = y;$$

it follows that S is onto. $\qquad\square$

The point spectrum consists, as before, of eigenvalues of T, i.e. $\lambda \in \mathbb{C}$ such that $Tx = \lambda x$ for some non-zero $x \in D(T)$. The proof of the following result is identical to the case when T is bounded (see Corollary 16.2).

Lemma 25.8 *If $(T, D(T))$ is symmetric, then all of its eigenvalues are real.*

We will now show that the residual spectrum of a self-adjoint operator is empty.

Lemma 25.9 *Suppose that* $(D(T), T)$ *is self-adjoint. If* $\lambda \in \mathbb{C}$ *is such that* $T - \lambda I$ *is injective, then* Range$(T - \lambda I)$ *is dense in* H. *In particular,* $\sigma_r(T) = \varnothing$.

Proof Suppose that $T - \lambda I$ is injective but that its range is not dense, i.e. $R = \overline{\text{Range}(T - \lambda I)} \neq H$. It follows that there exists a $y \in R^\perp$ with $y \neq 0$ such that

$$0 = ((T - \lambda I)x, y) \qquad \text{for every } x \in D(T).$$

So

$$(Tx, y) = \lambda(x, y),$$

which shows that $y \in D(T^*)$. Therefore we can write

$$
\begin{aligned}
0 &= (Tx, y) - \lambda(x, y) \\
&= (x, T^*y) - (x, \bar{\lambda}y) \\
&= (x, Ty) - (x, \bar{\lambda}y) \\
&= (x, Ty - \bar{\lambda}y).
\end{aligned}
$$

Since $D(T)$ is dense in H, it follows that $Ty = \bar{\lambda}y$, i.e. $\bar{\lambda}$ is an eigenvalue of T. But eigenvalues of self-adjoint operators are real, so $Ty = \lambda y$, and since $y \neq 0$ this contradicts the initial assumption that $T - \lambda I$ is injective. Since $\sigma_r(T)$ consists of those λ for which $T - \lambda I$ is injective but Range$(T - \lambda I)$ is not dense, it follows immediately that $\sigma_r(T) = \varnothing$. $\qquad\square$

While not every element of the spectrum of a self-adjoint operator T need be an eigenvalue, because $\sigma_r(T)$ is empty any $\lambda \in \sigma(T)$ is either an eigenvalue or a member of the continuous spectrum. Using this we can show that every element of $\sigma(T)$ is an *approximate eigenvalue*, in the sense made precise in the following result.

Corollary 25.10 *If* T *is self-adjoint, then for every* $\lambda \in \sigma(T)$ *there exists a sequence* $(x_n) \in D(T)$ *such that* $\|x_n\| = 1$ *and* $Tx_n - \lambda x_n \to 0$.

Proof If $\lambda \in \sigma_p(T)$ this is immediate. Otherwise $(T - \lambda I)^{-1}$ is unbounded, so there exists a sequence y_n with $\|y_n\| = 1$ and $\|\xi_n\| \geq n$, where we set $\xi_n = (T - \lambda I)^{-1} y_n$; note that $\xi_n \in D(T)$. Therefore

$$(T - \lambda I)\frac{\xi_n}{\|\xi_n\|} = \frac{y_n}{\|\xi_n\|} \to 0 \qquad \text{as} \quad n \to \infty,$$

from which the result follows with $x_n := \xi_n / \|\xi_n\|$. $\qquad\square$

As a consequence of this we can show that the spectrum is real and closed.

Theorem 25.11 *If* $T : D(T) \rightarrow H$ *is self-adjoint, then its spectrum is real and closed.*

Proof If $\lambda \in \sigma(T)$, then we can find $(x_n) \in D(T)$ such that $\|x_n\| = 1$ and $Tx_n - \lambda x_n \rightarrow 0$ and hence

$$(Tx_n, x_n) - \lambda \|x_n\|^2 \rightarrow 0; \qquad (25.7)$$

taking the complex conjugate and using the fact that T is self-adjoint

$$(x_n, Tx_n) - \bar{\lambda} \|x_n\|^2 = (Tx_n, x_n) - \bar{\lambda} \|x_n\|^2 \rightarrow 0.$$

Combining this with (25.7) it follows that

$$(\lambda - \bar{\lambda}) = (\lambda - \bar{\lambda}) \|x_n\|^2 \rightarrow 0 \qquad \text{as } n \rightarrow \infty,$$

and hence $\lambda = \bar{\lambda}$.

The spectrum is closed since the resolvent is open: if $\lambda \in \rho(T)$, then $T - \lambda I$ has a bounded inverse so

$$\|(T - \lambda I)^{-1} x\| \leq C \|x\| \qquad \Rightarrow \qquad \|Ty - \lambda y\| \geq \frac{1}{C} \|y\|,$$

where $y = (T - \lambda I)^{-1} x$. If $|\mu - \lambda| < 1/2C$, then

$$\|Ty - \mu y\| \geq \frac{1}{2C} \|y\|.$$

The above inequality shows that $T - \mu I$ is injective, and gives a bound on its inverse. Lemma 25.9 shows that if T is self-adjoint and $T - \mu I$ is injective, then the range of $T - \mu I$ is dense; it follows that $\mu \in \rho(T)$. $\qquad \square$

26

Reflexive Spaces

We have seen that $(\ell^q)^* \equiv \ell^p$ for $1 \leq q < \infty$ and (p, q) conjugate [i.e. $p^{-1} + q^{-1} = 1$]. It follows for $1 < q < \infty$ that if we take the 'second dual' (i.e. the dual of the dual), then $[(\ell^q)^*]^* \equiv (\ell^p)^* \equiv \ell^q$: we get back to where we started by taking the dual twice, so $(\ell^q)^{**} \equiv \ell^q$.

We pursue this idea further in this chapter, introducing the notion of a 'reflexive space'. However, we emphasise that being reflexive is more than just having $X^{**} \equiv X$. James (1951) constructed a Banach space X that satisfies $X^{**} \equiv X$ but is not reflexive.

26.1 The Second Dual

Since X^* is always a Banach space (Theorem 11.11), there is nothing to stop us from considering the dual of X^*, i.e. the set of bounded linear functionals from X^* into $\mathbb{K}$. We write X^{**} for this space (which we would otherwise denote by $(X^*)^*$), so

$$X^{**} := B(X^*; \mathbb{K}). \tag{26.1}$$

There is a canonical way[1] of associating any element $x \in X$ with an element $x^{**} \in X^{**}$, by setting

$$x^{**}(f) := f(x) \qquad \text{for each} \qquad f \in X^*.$$

The following lemma shows that x^{**} is indeed an element of X^{**}, and that the mapping $x \mapsto x^{**}$ is a linear isometry; in this way '$X \subseteq X^{**}$' for any normed space X.

---

[1] The process is canonical, but the x^{**} notation adopted here is not. The notation J for the corresponding map from X into X^{**} is more common, but is not universal.

Lemma 26.1 *For any normed space X we can isometrically map X onto a subspace of X^{**} via the canonical linear mapping $x \mapsto x^{**}$, where x^{**} is the element of X^{**} defined by setting*

$$x^{**}(f) = f(x) \qquad \text{for each} \quad f \in X^*.$$

We denote this mapping by $J : X \to X^{**}$.

Proof We have to show that for any $x \in X$, x^{**} defines a linear functional on X^* (i.e. an element of X^{**}) with the same norm as x. Given $x \in X$ we set

$$x^{**}(f) := f(x) \qquad \text{for every } f \in X^*.$$

Then, since

$$|x^{**}(f)| = |f(x)| \leq \|f\|_{X^*} \|x\|_X,$$

it certainly follows that $x^{**} \in X^{**}$ and that $\|x^{**}\|_{X^{**}} \leq \|x\|_X$.

If we take the 'support functional at x' from Corollary 20.2, i.e. $f \in X^*$ for which $\|f\| = 1$ and $f(x) = \|x\|$, then we have

$$|x^{**}(f)| = |f(x)| = \|x\|_X = \|x\|_X \|f\|_{X^*}$$

(since $\|f\|_{X^*} = 1$) and it follows that $\|x^{**}\|_{X^{**}} \geq \|x\|_X$, which yields the required equality of norms. $\qquad\qquad\qquad\qquad\qquad\qquad\qquad\qquad\qquad\square$

In general J does not map X onto X^{**}, but only onto a subspace of X^{**}.

Lemma 26.2 *If X is a Banach space, then $J(X)$ is a closed subspace of X^{**}.*

Proof If $(F_n) \in J(X)$ with $F_n \to F$ in X^{**}, then (F_n) must be Cauchy in X^{**}. Since there exist $x_n \in X$ such that $F_n = x_n^{**}$ and the map J is a linear isometry, we have

$$\|x_n - x_m\|_X = \|F_n - F_m\|_{X^{**}},$$

so (x_n) is Cauchy in X. It follows that there exists $x \in X$ such that $x_n \to x$ in X, and so

$$\|F_n - x^{**}\|_{X^{**}} = \|x_n - x\|_X \to 0 \qquad \text{as} \qquad n \to \infty.$$

By uniqueness of limits it follows that $F = x^{**}$, so $J(X)$ is closed. $\qquad\square$

When J does map X onto X^{**}, we say that X is reflexive. It follows immediately in this case that J is an isometric isomorphism and so $X \equiv X^{**}$; but note that this is a consequence of X being reflexive, and not the definition of reflexivity.

Definition 26.3 A Banach space X is *reflexive* if $J: X \to X^{**}$ is onto, i.e. if every $F \in X^{**}$ can be written as x^{**} for some $x \in X$.

26.2 Some Examples of Reflexive Spaces

We now show that all Hilbert spaces are reflexive, as are the ℓ^p and L^p spaces when $1 < p < \infty$. We start with Hilbert spaces.

Proposition 26.4 *All Hilbert spaces are reflexive.*

Before we begin the proof, recall that we know from the Riesz Representation Theorem (Theorem 12.4) that the map $R: H \to H^*$ defined by setting

$$R(x)(y) := (y, x) \qquad \text{for every } x, y \in H$$

is a conjugate-linear isometric isomorphism. Since R is surjective, given any $f \in H^*$ we can write

$$f(y) = (y, R^{-1}f). \tag{26.2}$$

(Here $R^{-1}f$ is the element $z \in H$ such that $f(y) = (y, z)$ for all $y \in H$; the existence of such a z is the main element of the Riesz Representation Theorem.)

Proof Given $F \in H^{**}$ we need to find $x \in H$ such that $F = x^{**}$, i.e. such that for every $f \in H^*$ we have

$$F(f) = f(x)$$

(this is because $x^{**}(f)$ is defined to be $f(x)$).

Now, $F \circ R: H \to \mathbb{K}$ is a bounded conjugate-linear map, so the map $\overline{F \circ R}: H \to \mathbb{K}$, defined by setting

$$\overline{F \circ R}(y) := \overline{F \circ R(y)}$$

is a bounded linear map, i.e. an element of H^*. So, using the Riesz Representation Theorem, we can find an element $x \in H$ such that

$$\overline{(F \circ R)(y)} = (y, x) \tag{26.3}$$

for every $y \in H$. Since R is a bijection, for any $f \in H^*$ we can choose $y = R^{-1}f$, and then

$$\overline{F(f)} = \overline{(F \circ R)(R^{-1}f)} = (R^{-1}f, x) = \overline{(x, R^{-1}f)} = \overline{f(x)},$$

where we use (26.3) for the first equality and (26.2) for the final equality; thus
$F(f) = f(x)$ as required.   □

We now use a similar argument to show that ℓ^p is reflexive provided that
$1 < p < \infty$.

Proposition 26.5 *The sequence space* $\ell^p(\mathbb{K})$ *is reflexive if* $1 < p < \infty$.

In the proof we use the notation

$$\langle \boldsymbol{x}, \boldsymbol{y} \rangle := \sum_{j=1}^{\infty} x_j y_j;$$

although this agrees with the L^2 inner product when $\mathbb{K} = \mathbb{R}$, it lacks the complex conjugate that we would use in the complex case. We know from Theorem
18.5 that the map $T_q \colon \ell^p \to (\ell^q)^*$ defined by setting

$$[T_q(\boldsymbol{x})](\boldsymbol{y}) = \langle \boldsymbol{x}, \boldsymbol{y} \rangle \qquad \text{for } \boldsymbol{y} \in \ell^q,$$

is a linear isometric isomorphism. So, given any $f \in (\ell^q)^*$ we can write

$$f(\boldsymbol{y}) = \langle T_q^{-1}(f), \boldsymbol{y} \rangle \qquad \text{for all } \boldsymbol{y} \in \ell^q \qquad\qquad (26.4)$$

(and similarly with p and q switched).

Proof Since $(\ell^p)^* \equiv \ell^q$, we also have $(\ell^p)^{**} \equiv (\ell^q)^* \equiv \ell^p$; to prove reflexivity we have to be careful about the maps involved: given any $F \in (\ell^p)^{**}$ we
need to find $\boldsymbol{x} \in \ell^p$ such that $F(f) = f(\boldsymbol{x})$ for every $f \in (\ell^p)^*$.
   We start by relating $F \in (\ell^p)^{**}$ to an element of $(\ell^q)^*$. To do this, note that
$F \circ T_p \colon \ell^q \to \mathbb{K}$ is both linear and bounded; so $F \circ T_p \in (\ell^q)^*$. Now we use
that fact that $(\ell^q)^* \equiv \ell^p$ via (26.4) to find $\boldsymbol{x} \in \ell^p$ such that

$$(F \circ T_p)(\boldsymbol{y}) = \langle \boldsymbol{x}, \boldsymbol{y} \rangle$$

for all $\boldsymbol{y} \in \ell^q$ (it is probably unhelpful to write $\boldsymbol{x} = T_q^{-1}(F \circ T_p)$). Since
$T_p \colon \ell^q \to (\ell^p)^*$ is a bijection, for any $f \in (\ell^p)^*$ we can choose $\boldsymbol{y} = T_p^{-1} f$,
and then (using (26.4) once again but with p and q swapped)

$$F(f) = F \circ T_p(T_p^{-1}(f)) = \langle \boldsymbol{x}, T_p^{-1}(f) \rangle = \langle T_p^{-1}(f), \boldsymbol{x} \rangle = f(\boldsymbol{x}).$$

This shows that $F = \boldsymbol{x}^{**}$, and so ℓ^p is reflexive.   □

A very similar proof, involving little more than a slight change of notation,
shows that L^p is reflexive for $1 < p < \infty$. We write q for the conjugate

exponent to p, and denote by T_p the isometric isomorphism from L^q onto $(L^p)^*$ given by

$$(T_p g)(f) := \int_\Omega fg \, dx \qquad \text{for every } f \in L^p.$$

Proposition 26.6 *The space $L^p(\Omega)$ is reflexive for $1 < p < \infty$.*

Proof Given any $F \in (L^p)^{**}$ we need to find $g \in L^p$ such that

$$F(\phi) = \phi(g) \qquad \text{for every } \phi \in (L^p)^*.$$

First, note that $F \circ T_p \colon L^q \to \mathbb{K}$ is a bounded linear functional, so defines an element $\psi \in (L^q)^*$; we can therefore write

$$F \circ T_p = T_q g$$

for some $g \in L^p$.

Now if we take any $\phi \in (L^p)^*$ we have $\phi = T_p f$ for $f = T_p^{-1} \phi \in L^q$, and then

$$F(\phi) = F \circ T_p(T_p^{-1}\phi) = (T_q g)f$$
$$= \int_\Omega g(x)f(x)\,dx = (T_p f)g = \phi(g).$$

It follows that $g = T_q^{-1}(F \circ T_p)$ is the required element of ℓ^p. $\qquad\square$

We will see shortly that ℓ^1, ℓ^∞, L^1, and L^∞ are not reflexive. For now we give a quick proof that $C([-1,1])$ is not reflexive; it relies on the fact that $C([-1,1])$ is separable but $C([-1,1])^*$ is not.

Lemma 26.7 *The space $C([-1,1])$ (with the usual supremum norm) is not reflexive.*

Proof Let $X = C([-1,1])$. We know that X is separable from Corollary 6.4. If X was reflexive we would as a consequence have $X^{**} \equiv X$, which would imply that X^{**} was separable (by Exercise 3.13). Lemma 20.5 would then imply that X^* was separable; but this contradicts the result of Exercise 18.5 that X^* is not separable. $\qquad\square$

26.3 X Is Reflexive If and Only If X^* Is Reflexive

The following result is very useful; its proof is a good exercise in using the definition of reflexivity.

Theorem 26.8 *Let X be a Banach space. Then X is reflexive if and only if X^* is reflexive.*

In the second part of the proof we use implicitly the fact that

$$(X^*)^{**} = (X^{**})^*.$$

This equality is clear when we note from (26.1) that

$$(X^*)^{**} = B((X^*)^*; \mathbb{K}) = B(X^{**}; \mathbb{K}) = (X^{**})^*.$$

Before we begin the proof a brief remark on notation might be helpful: we will use x, y for elements of X, f, g for elements of X^*, F, G for elements of X^{**}, and Φ, Ψ for elements of X^{***}.

Proof Suppose first that X is reflexive; we want to show that X^* is reflexive, i.e. that for any $\Phi \in (X^*)^{**}$ we can find an $f \in X^*$ such that $f^{**} = \Phi$, i.e. such that

$$\Phi(F) = F(f) \qquad \text{for every } F \in X^{**}.$$

This actually tells us what f should be. Since any $F \in X^{**}$ can be written as x^{**} for some $x \in X$, we require

$$\Phi(x^{**}) = x^{**}(f) \qquad \text{for every } x \in X.$$

But since, by definition, $x^{**}(f) = f(x)$, this says that we must have

$$f(x) = \Phi(x^{**}) \qquad \text{for every } x \in X,$$

and we now use this as the definition of f. We just have to check that f really is an element of X^*, i.e. is a bounded linear map from X into $\mathbb{K}$. But this follows immediately, as it is the composition of J, a bounded linear map from X into X^{**}, with Φ, which is a bounded linear map from X^{**} into $\mathbb{K}$.

For the converse, suppose that X^* is reflexive but X is not, i.e. there is an element $F \in X^{**}$ such that $F \neq x^{**}$ for any $x \in X$. Then the set

$$J(X) = \{x^{**} : x \in X\}$$

is a proper closed linear subspace of X^{**} (see Lemma 26.2), and hence by Proposition 20.4 (existence of a distance functional) there is some non-zero $\Phi \in (X^{**})^*$ such that $\Phi = 0$ on $J(X)$, i.e.

$$\Phi(x^{**}) = 0 \qquad \text{for all } x \in X.$$

Since $(X^{**})^* = (X^*)^{**}$ and X^* is reflexive, we know that $\Phi = f^{**}$ for some $f \in X^*$, and so if $x \in X$, we have

$$f(x) = x^{**}(f) = f^{**}(x^{**}) = \Phi(x^{**}) = 0.$$

But this means that $f = 0$, which in turn implies that $\Phi = 0$, a contradiction.
□

Since any reflexive space satisfies $X^{**} \equiv X$ (but not vice versa, as commented above), and we know that

$$(c_0)^* \equiv \ell^1 \qquad \text{and} \qquad (\ell^1)^* \equiv (\ell^\infty),$$

it follows that the space c_0 is not reflexive. (We know that $c_0 \not\simeq \ell^\infty$ because c_0 is separable and ℓ^∞ is not.)

We would like to say now that ℓ^1 cannot be reflexive (because it is the dual of c_0), and then that ℓ^∞ is not reflexive (because it is the dual of ℓ^1); but we need to be a little careful, since the dual of c_0 is not ℓ^1 but a space isometrically isomorphic to ℓ^1.

Lemma 26.9 *If X is reflexive and $X \equiv Y$, then Y is reflexive.*

(In fact the hypothesis can be weakened to $X \simeq Y$; see e.g. Megginson, 1998.)

Proof Suppose that $\phi \colon X \to Y$ is a linear isometric isomorphism. Then the Banach adjoint of ϕ, $\phi^\times \colon Y^* \to X^*$ defined by setting

$$\phi^\times(g) = g \circ \phi,$$

is a linear isometric isomorphism (see Lemma 20.7). Applying the same argument again, the map $\phi^{\times\times} \colon X^{**} \to Y^{**}$ by setting

$$\phi^{\times\times}(F) = F \circ \phi^\times$$

as again a linear isometric isomorphism.

Take $G \in Y^{**}$, then $G = \phi^{\times\times}(F) = F \circ \phi^\times$ for some $F \in X^{**}$; so for any $g \in Y^*$ we have

$$G(g) = F \circ \phi^\times(g) = F(g \circ \phi) = (g \circ \phi)(x)$$

for some $x \in X$ (since X is reflexive); but $(g \circ \phi)(x) = g(\phi(x)) = g(y)$, where $y := \phi(x) \in Y$. So Y is reflexive.
□

We can now say with confidence that ℓ^1 is not reflexive (since it is isometrically isomorphic to $(c_0)^*$, which is not reflexive) and hence ℓ^∞ is not reflexive (since it is isometrically isomorphic to $(\ell^1)^*$, which is not reflexive).

The following result, whose argument follows similar lines as that used to prove Theorem 26.8, will be useful later.

Lemma 26.10 *Any closed subspace Y of a reflexive Banach space X is reflexive.*

Proof Take $f \in X^*$ and let f_Y denote the restriction of f to Y, so that $f_Y \in Y^*$. Because of the Hahn–Banach Theorem, any element of Y^* can be obtained as such a restriction.

To show that Y is reflexive we need to show that for any $\Psi \in Y^{**}$ there exists a $y \in Y$ such that

$$\Psi(f_Y) = y^{**}(f_Y) \qquad \text{for every} \quad f \in X^*.$$

First define an element $\hat{\Psi} \colon X^* \to \mathbb{R}$ by setting

$$\hat{\Psi}(f) = \Psi(f_Y),$$

and then

$$|\hat{\Psi}(f)| \le \|\Psi\| \|f_Y\| \le \|\Psi\| \|f\| \qquad \text{for any} \quad f \in X^*,$$

so $\hat{\Psi} \in X^{**}$. Now we can use the fact that X is reflexive to find an $x \in X$ such that

$$\hat{\Psi} = x^{**}.$$

We only need now show that $x \in Y$.

Suppose that $x \notin Y$. Then the distance functional from Proposition 20.4 provides an $f \in X^*$ such that $f(x) \ne 1$ and $f(y) = 0$ for every $y \in Y$, i.e. such that $f_Y = 0$. Then

$$f(x) = x^{**}(f) = \hat{\Psi}(f) = \Psi(f_Y) = 0,$$

a contradiction. $\qquad\qquad\qquad\qquad\qquad\qquad\qquad\qquad\qquad\qquad\qquad\square$

Exercises

26.1 Suppose that X and Y are Banach spaces, and that $T_X \colon X \to Y^*$ and $T_Y \colon Y \to X^*$ are both isometric isomorphisms (so that $X^* \equiv Y$ and $Y^* \equiv X$). Show that if

$$[T_X x](y) = [T_Y y](x) \qquad \text{for all} \qquad x \in X, \ y \in Y,$$

then X is reflexive. (The proof, generalising the argument we used to prove reflexivity of the ℓ^p and L^p spaces, gives some indication why $X^{**} \equiv X$ alone is not sufficient for X to be reflexive.)

26.2 Suppose that U is a subset of a Banach space X. Show that U is bounded if and only if for every $f \in X^*$ the set

$$f(U) = \{f(u) : u \in U\} \qquad \text{is bounded in } \mathbb{R}.$$

(Use the Principle of Uniform Boundedness on an appropriately chosen set of elements of X^{**}.)

26.3 If $T: X \rightarrow Y$ is a linear map between Banach spaces and $\phi \circ T$ is bounded for every $\phi \in Y^*$ show that T is bounded. (Prove the contrapositive.)

26.4 Suppose that X is a reflexive real Banach space and $\xi: [0, T] \rightarrow X$ is such that the real-valued function $f(\xi(\cdot)): [0, T] \rightarrow \mathbb{R}$ is integrable for every $f \in X^*$. Show that if $\int_0^T \|\xi(t)\|_X < \infty$, then there exists a unique $y \in X$ such that

$$f(y) = \int_0^T f(\xi(t))\, dt$$

for every $f \in X^*$. If we define $\int_0^T \xi(t)\, dt := y$ show that

$$\left\| \int_0^T \xi(t)\, dt \right\|_X \leq \int_0^T \|\xi(t)\|_X\, dt.$$

26.5 Assuming that L^p is reflexive, show that $(L^p)^* \equiv L^q$ as follows. We showed in Theorem 18.8 that the map $T: L^q \mapsto (L^p)^*$ by setting

$$[T(g)](f) = \int_\Omega fg\, dx \qquad \text{for each } f \in L^p$$

is a linear isometry. Suppose that T is not onto and obtain a contradiction, using the fact that $T(L^q)$ is a closed subset of $(L^p)^*$ along with a variant of the proof of the second part of Theorem 26.8. (This is not necessarily a circular argument, since L^p is uniformly convex (see Exercise 10.6) and the Milman–Pettis Theorem guarantees that any uniformly convex space is reflexive; see e.g. Theorem 5.2.15 in Megginson, 1998.) (Lax, 2002)

27

Weak and Weak-∗ Convergence

We have seen that in any infinite-dimensional space the closed unit ball is not compact (and that this characterises infinite-dimensional spaces). However, in this chapter we will prove that in any reflexive Banach space the closed unit ball is weakly sequentially compact, which is often sufficient in applications. We first introduce the notion of weak convergence: the key idea is to define a convergence based on the action of linear functionals.

27.1 Weak Convergence

The definition of convergence that we have used up until now ($x_n \to x$ if $\|x_n - x\| \to 0$) we will here call 'strong convergence' to distinguish it from the notion of weak convergence that we now introduce.

Definition 27.1 We say that a sequence $(x_n) \in X$ *converges weakly* to $x \in X$, and write $x_n \rightharpoonup x$, if

$$f(x_n) \to f(x) \qquad \text{for all} \qquad f \in X^*.$$

Note that in a Hilbert space, where every linear functional is of the form $x \mapsto (x, y)$ for some $y \in H$, $x_n \rightharpoonup x$ if

$$(x_n, y) \to (x, y) \qquad \text{for all} \qquad y \in H.$$

This observation allows us to provide an example of a sequence that converges weakly but does not converge strongly. Pick any countable orthonormal sequence $(e_j)_{j=1}^\infty$ in H; then for any $y \in H$ Bessel's inequality (Lemma 9.11)

$$\sum_{j=1}^\infty |(y, e_j)|^2 \le \|y\|^2$$

shows that the sum converges; it follows that $(y, e_j) \to 0$ as $j \to \infty$ for any $y \in H$, and hence that $e_j \rightharpoonup 0$. But the sequence (e_j) does not converge (any two elements are a distance $\sqrt{2}$ apart).

Lemma 27.2 *Weak convergence has the following properties.*

  (i) *Strong convergence implies weak convergence;*
  (ii) *in a finite-dimensional normed space weak convergence and strong convergence are equivalent;*
 (iii) *weak limits are unique;*
 (iv) *weakly convergent sequences are bounded; and*
  (v) *if $x_n \rightharpoonup x$, then*

$$\|x\| \le \liminf_{n \to \infty} \|x_n\|. \tag{27.1}$$

Proof (i) If $x_n \to x$, then for any $f \in X^*$

$$|f(x_n) - f(x)| \le \|f\|_{X^*} \|x_n - x\|_X \to 0 \qquad \text{as} \qquad n \to \infty,$$

so $f(x_n) \to f(x)$, and hence $x_n \rightharpoonup x$.

(ii) Due to part (i) we need only show that if V is a finite-dimensional normed space, then weak convergence in V implies strong convergence in V. If $\{e_1, \dots, e_n\}$ is a basis for V, then for each $i = 1, \dots, n$ the map

$$x = \sum_{j=1}^{n} x_j e_j \mapsto x_i$$

is an element of V^*, so if $x^{(k)} \rightharpoonup x$ it follows that $x_j^{(k)} \to x_j$ for each $j = 1, \dots, n$, and so

$$x^{(k)} = \sum_{j=1}^{n} x_j^{(k)} e_j \to \sum_{j=1}^{n} x_j e_j = x.$$

(iii) Suppose that $x_n \rightharpoonup x$ and $x_n \rightharpoonup y$. Then for any $f \in X^*$,

$$f(x) = \lim_{n \to \infty} f(x_n) = f(y),$$

so, by Lemma 20.3 (X^* separates points in X), $x = y$.

(iv) Since $f(x_n)$ converges, it follows that $f(x_n)$ is a bounded sequence (in $\mathbb{K}$) for every $f \in X^*$. If we consider the sequence $(x_n^{**}) \in X^{**}$, then, since

$$x_n^{**}(f) = f(x_n),$$

it follows that $(x_n^{**}(f))_n$ is bounded in $\mathbb{K}$ for every $f \in X^*$. We can now use the Principle of Uniform Boundedness (Theorem 22.3) to deduce that (x_n^{**}) is bounded in X^{**}. Since $\|x^{**}\|_{X^{**}} = \|x\|_X$ (Lemma 26.1), it follows that (x_n) is bounded in X.

(v) Choose $f \in X^*$ with $\|f\|_{X^*} = 1$ such that $f(x) = \|x\|$ (the support functional at x whose existence is guaranteed in Lemma 20.2). Then

$$\|x\| = f(x) = \lim_{n \to \infty} f(x_n),$$

so

$$\|x\| \le \liminf_{n \to \infty} |f(x_n)| \le \liminf_{n \to \infty} \|f\|_{X^*} \|x_n\|_X;$$

the result follows since $\|f\|_{X^*} = 1$. $\qquad\square$

There are two situations in which we can easily convert weak to strong convergence.

The first is in a Hilbert space: if a sequence (x_n) converges weakly to x and we also know that the norms converge, $\|x_n\| \to \|x\|$, then this implies strong convergence. In fact the same result is true in any uniformly convex Banach space (see Exercise 27.4), but the proof in a Hilbert space is particularly simple.

Lemma 27.3 *Let H be a Hilbert space. If $(x_n) \in H$ with $x_n \rightharpoonup x$ and $\|x_n\| \to \|x\|$, then $x_n \to x$.*

Proof Observe that

$$\|x - x_n\|^2 = (x - x_n, x - x_n) = \|x\|^2 - (x, x_n) - (x_n, x) + \|x_n\|^2.$$

Since $x_n \rightharpoonup x$, we have $(x_n, x) \to (x, x) = \|x\|^2$ and $\|x_n\|^2 \to \|x\|^2$ by assumption; so $\|x - x_n\|^2 \to 0$ as $n \to \infty$. $\qquad\square$

Another way to obtain a strongly convergent sequence starting with a weakly convergent one is to apply a compact operator.

Lemma 27.4 *Suppose that $T: X \to Y$ is a compact linear operator. If $(x_n) \in X$ with $x_n \rightharpoonup x$ in X, then $Tx_n \to Tx$ in Y.*

Proof We first show that $Tx_n \rightharpoonup Tx$ in Y; indeed, if $f \in Y^*$, then $f \circ T$ is an element of X^*, so that $x_n \rightharpoonup x$ implies that

$$f(Tx_n) \to f(Tx).$$

Now, suppose that $Tx_n \not\to Tx$; then there is an $\varepsilon > 0$ and a subsequence $(x_{n_j})_j$ such that

$$\|Tx_{n_j} - Tx\| > \varepsilon \qquad \text{for every } j. \tag{27.2}$$

Since x_{n_j} converges weakly, it is a bounded sequence in X (by part (iv) of Lemma 27.2); since T is compact it follows that (Tx_{n_j}) has a subsequence $(Tx_{n_j'})_j$ that converges to some $z \in Y$. Since strong convergence implies weak convergence (Lemma 27.2 (i)), we also have $Tx_{n_j'} \rightharpoonup z$; but weak limits are unique (part (iii) of Lemma 27.2) and we already know that $Tx_{n_j'} \rightharpoonup Tx$ (since $x_{n_j'}$ is a subsequence of x_n and we know that $Tx_n \rightharpoonup Tx$), so we must have $z = Tx$ and

$$\lim_{j \to \infty} \|Tx_{n_j'} - Tx\| \to 0 \qquad \text{as } j \to \infty.$$

Since $x_{n_j'}$ is a subsequence of x_{n_j} the preceding equation contradicts (27.2), and therefore $Tx_n \to Tx$ as claimed. $\qquad\qquad\square$

27.2 Examples of Weak Convergence in Various Spaces

We now look at some examples of weak convergence in particular spaces. We characterise weak convergence in ℓ^p for $1 < p < \infty$, show that weak and strong convergence in ℓ^1 coincide, and make some observations about weak convergence in $C([a, b])$ and how it relates to other notions of convergence.

27.2.1 Weak Convergence in ℓ^p, $1 < p < \infty$

If we take $1 \le p < \infty$, then we know from Theorem 18.5 that any element of $(\ell^p)^*$ can be represented as $\langle \cdot, y \rangle$ for some $y \in \ell^q$, where p and q are conjugate and we use $\langle \cdot, \cdot \rangle$ to denote the pairing

$$\langle x, y \rangle = \sum_{j=1}^{\infty} x_j y_j$$

(whenever this makes sense). So we have

$$x^{(n)} \rightharpoonup x \text{ in } \ell^p \quad \Leftrightarrow \quad \langle x^{(n)}, y \rangle \to \langle x, y \rangle \quad \text{for every } y \in \ell^q. \qquad (27.3)$$

For $1 < p < \infty$ there is an even nicer characterisation. (The following result is not true in ℓ^1 or in ℓ^∞: see Exercise 27.6.)

Lemma 27.5 *Let* $(x^{(n)})_{n=1}^{\infty}$ *be a sequence in* ℓ^p, *with* $1 < p < \infty$. *Then* $x^{(n)} \rightharpoonup x$ *in* ℓ^p *if and only if*

$$\|x^{(n)}\|_{\ell^p} \text{ is bounded} \qquad \text{and} \qquad x_k^{(n)} \to x_k \text{ for every } k \in \mathbb{N}.$$

Proof ⇒ This follows from taking $y = e^{(k)}$ in (27.3), and using the fact that any weakly convergent sequence is bounded (Lemma 27.2 (iii)).

⇐ Suppose that $\|x^{(n)}\|_{\ell^p} \le M$; we first show that $\|x\|_{\ell^p} \le M$. For any k and any $\varepsilon > 0$ there exists N such that for every $n \ge N$ we have

$$\sum_{j=1}^{k} |x_j|^p \le \left(\sum_{j=1}^{k} |x_j^{(n)}|^p \right) + \varepsilon \le M^p + \varepsilon.$$

Since this holds for every $k \in \mathbb{N}$ and $\varepsilon > 0$ is arbitrary, we have $\|x\|_{\ell^p} \le M$ as claimed.

Take any $y \in \ell^q$; then, since $y = \lim_{k \to \infty} \sum_{j=1}^{k} y_j e^{(j)}$, given any $\varepsilon > 0$ there exists k such that

$$\left\| y - \sum_{j=1}^{k} y_j e^{(j)} \right\|_{\ell^q} < \frac{\varepsilon}{4M};$$

then

$$
\begin{aligned}
|\langle x^{(n)} - x, y \rangle| &= \left| \left\langle x^{(n)} - x, \sum_{j=1}^{k} y_j e^{(j)} \right\rangle + \left\langle x^{(n)} - x, y - \sum_{j=1}^{k} y_j e^{(j)} \right\rangle \right| \\
&\le \left| \left\langle x^{(n)} - x, \sum_{j=1}^{k} y_j e^{(j)} \right\rangle \right| + \|x^{(n)} - x\|_{\ell^p} \left\| y - \sum_{j=1}^{k} y_j e^{(j)} \right\|_{\ell^q} \\
&\le \sum_{j=1}^{k} |y_j| |\langle x^{(n)} - x, e^{(j)} \rangle| + 2M \frac{\varepsilon}{4M} \\
&= \sum_{j=1}^{k} |y_j| |\langle x^{(n)} - x, e^{(j)} \rangle| + \frac{\varepsilon}{2}.
\end{aligned}
$$

Since $\langle x^{(n)}, e^{(j)} \rangle \to \langle x, e^{(j)} \rangle$ for each j, it follows that $x^{(n)} \rightharpoonup x$. □

27.2.2 Weak Convergence in ℓ^1: Schur's Theorem

In ℓ^1 weak convergence is equivalent to strong convergence.

Theorem 27.6 (Schur's Theorem) *If $x^{(n)} \rightharpoonup x$ in ℓ^1, then $x^{(n)} \to x$ in ℓ^1.*

Proof By subtracting x from $x^{(n)}$, it suffices to show that if $x^{(n)} \rightharpoonup 0$, then $x^{(n)} \to 0$.

Suppose that this conclusion does not hold: then $\|x^{(n)}\|_{\ell^1} \not\to 0$, so we can find $\varepsilon > 0$ and a subsequence (which we relabel) such that $x^{(n)} \rightharpoonup 0$ and

$$\|x^{(n)}\|_{\ell^1} = \sum_{k=1}^{\infty} |x_k^{(n)}| \geq \varepsilon \qquad \text{for every } n. \qquad (27.4)$$

Note that we know, taking the linear map in $(\ell^1)^*$ given by $x \mapsto x_j$, that each component $x_j^{(n)} \to 0$ as $n \to \infty$.

Now we inductively choose N_j, M_j in the following way: first choose N_1 such that

$$\sum_{k=N_1+1}^{\infty} |x_k^{(1)}| < \frac{\varepsilon}{6}$$

(we can do this since $x^{(1)} \in \ell^1$) and M_1 such that

$$\sum_{k=1}^{N_1} |x_k^{(M_1)}| < \frac{\varepsilon}{6}$$

(we can do this since for every k we have $x_k^{(n)} \to 0$ as $n \to \infty$). Then, given N_{j-1} and M_{j-1}, choose N_j such that

$$\sum_{k=N_j+1}^{\infty} |x_k^{(M_{j-1})}| < \frac{\varepsilon}{6} \qquad (27.5)$$

(which we can do since $x^{(M_{j-1})} \in \ell^1$) and M_j such that

$$\sum_{k=1}^{N_j} |x_k^{(M_j)}| < \frac{\varepsilon}{6} \qquad (27.6)$$

(which we can do since $x_k^{(n)} \to 0$ for each k).

Combining (27.4) with (27.5) and (27.6) this construction ensures that

$$\sum_{k=N_j+1}^{N_{j+1}} |x_k^{(M_j)}| > \frac{2}{3}\varepsilon. \qquad (27.7)$$

Now let $y \in \ell^\infty$ be given by

$$y_k = \begin{cases} |x_k^{(M_j)}|/x_k^{(M_j)}, & x_k^{(M_j)} \neq 0, \\ 0, & x_k^{(M_j)} = 0, \end{cases} \qquad \text{for} \qquad N_{j-1}+1 \leq k \leq N_j$$

(note that $|y_k| \leq 1$ for every k) and let f be the linear functional on ℓ^1 that corresponds to $\mathbf{y}$, i.e.

$$f(\mathbf{x}) = \sum_{j=1}^{\infty} x_j y_j.$$

So we have

$$f\left(x^{(M_j)}\right) = \sum_{k=1}^{\infty} y_k x_k^{(M_j)}$$

$$= \sum_{k=0}^{N_{j-1}} y_k x_k^{(M_j)} + \sum_{k=N_{j-1}+1}^{N_j} y_k x_k^{(M_j)} + \sum_{k=N_j+1}^{\infty} y_k x_k^{(M_j)},$$

from which it follows that for every j

$$\left| f\left(x^{(M_j)}\right)\right| = \left| \sum_{k=1}^{\infty} y_k x_k^{(M_j)}\right|$$

$$\geq \left| \sum_{k=N_j+1}^{N_{j+1}} y_k x_k^{(M_j)}\right| - \left| \sum_{k=1}^{N_j} y_k x_k^{(M_j)}\right| - \left| \sum_{k=N_{j+1}+1}^{\infty} y_k x_k^{(M_j)}\right|$$

$$\geq \sum_{k=N_j+1}^{N_{j+1}} |x_k^{(M_j)}| - \sum_{k=1}^{N_j} |x_k^{(M_j)}| - \sum_{k=N_{j+1}+1}^{\infty} |x_k^{(M_j)}|$$

$$\geq \frac{2}{3}\varepsilon - \frac{\varepsilon}{6} - \frac{\varepsilon}{6} = \frac{\varepsilon}{3} > 0,$$

using (27.7), (27.6), and (27.5) to bound the three terms in the penultimate line. This contradicts the fact that $f(\mathbf{x}^{(n)}) \to 0$ as $n \to \infty$, so we must have $\mathbf{x}^{(n)} \to 0$. $\qquad\square$

27.2.3 Weak versus Pointwise Convergence in $C([0, 1])$

We now consider weak convergence in the space $X = C([0, 1])$, equipped with the supremum norm.

We have already observed (Example 18.1) that each of the maps $\delta_x : X \to \mathbb{R}$ given by

$$\delta_x(f) = f(x)$$

is an element of X^*. So if $f_n \rightharpoonup f$, then $f_n(x) \to f(x)$ for every $x \in X$, i.e. f_n converges pointwise to f.

However, weak convergence has other consequences: if we take any subinterval $[a, b] \subset [0, 1]$, then the map

$$f \mapsto \int_a^b f(x) \, dx$$

is also an element of X^*. So if $f_n \rightharpoonup f$ we have

$$\int_a^b f_n(x) \, dx \to \int_a^b f(x) \, dx \tag{27.8}$$

for any $0 \le a \le b \le 1$.

Now, observe that there are sequences such that $g_n \to g$ pointwise but for which (27.8) does not hold: for example, the sequence of functions

$$g_n(x) = \begin{cases} n^2 x & 0 \le x < 1/n \\ n(2 - nx) & 1/n \le n \le 2/n \\ 0 & 2/n < x \le 1, \end{cases}$$

(see Figure 7.2) converges pointwise to zero, but their integrals on $[0, 1]$ are constant and equal to 1. It follows that weak convergence in $C([0, 1])$ is a stronger notion of convergence than pointwise convergence.

27.3 Weak Closures

We say that a subset A of X is *weakly closed* if whenever $x_n \in A$ and $x_n \rightharpoonup x$, we have $x \in A$. In general this is a stronger property than being closed: weakly closed implies closed but not vice versa. For example, if H is a separable Hilbert space, then S_H, the unit sphere in H, is closed but not weakly closed, since we can take a countable orthonormal set $(e_j)_{j=1}^\infty$; then $e_j \in S_H$, but $e_j \rightharpoonup 0$ and $0 \notin S_H$.

However, for convex subsets being closed and being weakly closed are equivalent. This result has significant consequences in the Calculus of Variations; see Lemma 27.14 for a very simple example.

Theorem 27.7 (Mazur's Theorem) *Closed convex subsets of a Banach space are also weakly closed.*

Proof Using Corollary 21.8 we know that

$$C = \{x : \operatorname{Re} f(x) \ge \inf_{y \in C} \operatorname{Re} f(y) \text{ for every } f \in X^*\}.$$

If $x_n \in C$ and $x_n \rightharpoonup x$, then, since $x_n \in C$, for any $f \in X^*$ we have $\mathrm{Re}\, f(x_n) \geq \inf_{y \in C} \mathrm{Re}\, f(y)$, and so

$$\mathrm{Re}\, f(x) = \lim_{n \to \infty} \mathrm{Re}\, f(x_n) \geq \inf_{y \in C} \mathrm{Re}\, f(y),$$

i.e. $x \in C$. □

A nice consequence of this result is that if $x_n \rightharpoonup x$, then there is a sequence (y_n), where y_n is a convex combination of $\{x_1, \ldots, x_n\}$, i.e.

$$y_n = \sum_{j=1}^{n} \lambda_j x_j, \qquad 0 \leq \lambda_j \leq 1, \qquad \sum_{j=1}^{n} \lambda_j = 1,$$

such that $y_n \to x$ (i.e. y_n converges strongly to x) (see Exercise 27.3).

27.4 Weak-∗ Convergence

There is another notion of weak convergence, weak-∗ ('weak-star') convergence, which deals with sequences in X^*.

Definition 27.8 If $(f_n)_{n=1}^{\infty} \in X^*$, then f_n *converges weakly-∗ ('weakly star')* to f if

$$f_n(x) \to f(x) \qquad \text{for all} \qquad x \in X;$$

we write $f_n \overset{*}{\rightharpoonup} f$.

Note that weak-∗ convergence is a very natural way to define convergence of sequences in X^*: it is the equivalent of pointwise convergence for continuous functions.

We have a result along similar lines to Lemma 27.2 giving various properties of weak-∗ convergence and weak-∗ limits. In parts (iv) and (v) we compare weak-∗ convergence in X^* (apply f_n to elements in X) to weak convergence in X^* (apply elements in X^{**} to the sequence f_n).

Lemma 27.9 *Weak-∗ convergence has the following properties.*

  (i) *Strong convergence in X^* implies weak-∗ convergence in X^*;*
 (ii) *weak-∗ limits are unique;*
(iii) *weakly-∗ convergent sequences are bounded;*
 (iv) *if $f_n \overset{*}{\rightharpoonup} f$, then*

$$\|f\|_{X^*} \leq \liminf_{n \to \infty} \|f_n\|_{X^*};$$

(v) *weak convergence in X^* implies weak-∗ convergence in X^*;*

(vi) *if X is reflexive, then weak-∗ convergence in X^* implies weak convergence in X^*.*

Proof (i) If $f_n \to f$ in X^*, i.e. $\|f_n - f\|_{X^*} \to 0$, then for any $x \in X$ we have

$$|f_n(x) - f(x)| = |(f_n - f)(x)| \le \|f_n - f\|_{X^*} \|x\|_X \to 0$$

and so $f_n \xrightarrow{*} f$.

(ii) If $f_n \xrightarrow{*} f$ and $f_n \xrightarrow{*} g$, then

$$f(x) = \lim_{n \to \infty} f_n(x) = g(x) \qquad \text{for every } x \in X,$$

so $f = g$.

(iii) We have $f_n \in B(X; \mathbb{K})$ for each n, and if $f_n(x) \to f(x)$ for every $x \in X$, then

$$\sup_n |f_n(x)| < \infty \qquad \text{for every } x \in X,$$

so it follows from the Principle of Uniform Boundedness (Theorem 22.3) that $\sup_n \|f_n\|_{X^*} < \infty$.

(iv) Given any $\varepsilon > 0$ we can find an $x \in X$ with $\|x\| = 1$ such that $f(x) > \|f\|_{X^*} - \varepsilon$; then

$$\|f\|_{X^*} - \varepsilon < f(x) = \lim_{n \to \infty} f_n(x) \le \liminf_{n \to \infty} \|f_n\|_{X^*} \|x\| = \liminf_{n \to \infty} \|f_n\|_{X^*},$$

which yields the result since $\varepsilon > 0$ is arbitrary.

(v) $f_n \rightharpoonup f$ in X^* means that for every $F \in X^{**}$ we have

$$F(f_n) \to F(f).$$

Given any element $x \in X$ we can consider the corresponding $x^{**} \in X^{**}$. Since $f_n \rightharpoonup f$ in X^*, we have

$$f_n(x) = x^{**}(f_n) \to x^{**} f = f(x),$$

and so $f_n \xrightarrow{*} f$.

(vi) When X is reflexive any $F \in X^{**}$ is of the form x^{**} for some $x \in X$. So if $f_n \xrightarrow{*} f$ in X^* we have

$$F(f_n) = x^{**}(f_n) = f_n(x) \to f(x) = x^{**}(f) = F(f),$$

using the weak-∗ convergence of f_n to f to take the limit. So $f_n \rightharpoonup f$ in X^*. $\qquad \square$

27.5 Two Weak-Compactness Theorems

We now prove two key compactness theorems. We begin with a preparatory lemma.[1]

Lemma 27.10 *Suppose that* (f_n) *is a bounded sequence in* X^*, *so that* $\|f_n\|_{X^*} \le M$ *for some* $M > 0$, *and suppose that* $f_n(a)$ *converges as* $n \to \infty$ *for every* $a \in A$, *where* A *is a dense subset of* X. *Then* $\lim_{n\to\infty} f_n(x)$ *exists for every* $x \in X$, *and the map* $f : X \to \mathbb{R}$ *defined by setting*

$$f(x) = \lim_{n\to\infty} f_n(x), \qquad \text{for each} \qquad x \in X$$

is an element of X^* *with* $\|f\|_{X^*} \le M$.

Proof We first prove that if $f_n(a)$ converges for every $a \in A$, then $f_n(x)$ converges for every $x \in X$. Given $\varepsilon > 0$ and $x \in X$, first choose a such that

$$\|x - a\|_X \le \varepsilon/3M.$$

Now, using the fact that $f_n(a)$ converges as $n \to \infty$, choose n_0 sufficiently large that $|f_n(a) - f_m(a)| < \varepsilon/3$ for all $n, m \ge n_0$. Then for all $n, m \ge n_0$ we have

$$|f_n(x) - f_m(x)|$$
$$\le |f_n(x) - f_n(a)| + |f_n(a) - f_m(a)| + |f_m(a) - f_m(x)|$$
$$\le \|f_n\|_{X^*}\|x - a\| + \frac{\varepsilon}{3} + \|f_m\|_{X^*}\|a - x\|$$
$$\le \varepsilon.$$

It follows that $(f_n(x))$ is Cauchy and hence converges.

We now define $f : X \to \mathbb{R}$ by setting

$$f(x) := \lim_{n\to\infty} f_n(x).$$

Then f is linear since

$$f(x + \lambda y) = \lim_{n\to\infty} f_n(x + \lambda y) = \lim_{n\to\infty} f_n(x) + \lambda f_n(y) = f(x) + \lambda f(y)$$

and f is bounded since

$$|f(x)| = \lim_{n\to\infty} |f_n(x)| \le M\|x\|. \qquad \square$$

---

[1] Note that this is similar to Corollary 22.4, which translated to this particular setting says that if $(f_n) \in X^*$ and $f_n(x)$ converges for every $x \in X$, then setting $f(x) := \lim_{n\to\infty} f_n(x)$ defines an element $f \in X^*$. The lemma here weakens one hypothesis by requiring convergence only on a dense subset of X, but the boundedness of the sequence (f_n), which was the key thing to be proved in Corollary 22.4, becomes an assumption here and is used to show that convergence on a dense subset implies convergence for every $x \in X$.

Using this we can prove a weak-∗ compactness result when X is separable.[2] The theorem as stated here is due to Helly (1912).

Theorem 27.11 *Suppose that X is separable. Then any bounded sequence in X^* has a weakly-∗ convergent subsequence.*

Proof Let $\{x_k\}$ be a countable dense subset of X, and (f_j) a sequence in X^* such that $\|f_j\|_{X^*} \le M$. As in the proof of Theorem 15.3 we will use a diagonal argument to find a subsequence of the (f_j) (which we relabel) such that $f_j(x_k)$ converges for every k.

Since $|f_n(x_1)| \le M\|x_1\|$, we can use the Bolzano–Weierstrass Theorem to find a subsequence $f_{n_{1,i}}$ such that $f_{n_{1,i}}(x_1)$ converges. Now, since $|f_{n_{1,i}}(x_2)| \le M\|x_2\|$ we can find a subsequence $f_{n_{2,i}}$ of $f_{n_{1,i}}$ such that $f_{n_{2,i}}(x_2)$ converges; $f_{n_{2,i}}(x_1)$ will still converge since it is a subsequence of $f_{n_{1,i}}(x_1)$ which we have already made converge. We continue in this way to find successive subsequences $f_{n_{m,i}}$ such that

$$f_{n_{m,i}}(x_k) \qquad \text{converges as } i \to \infty \text{ for every } k = 1, \dots, m.$$

By taking the diagonal subsequence $f_m^* := f_{n_{m,m}}$ (as in the proof of the Arzelà–Ascoli Theorem) we can ensure that $f_m^*(x_k)$ converges for every $k \in \mathbb{N}$.

The proof concludes using Lemma 27.10. $\qquad\square$

A consequence of Helly's Theorem is the following extremely powerful weak-compactness result that holds in any reflexive space. It finally offers a way around the failure of the 'Bolzano–Weierstrass property' in infinite-dimensional spaces, i.e. in general bounded sequences have no strongly convergent subsequence.

The proof seems short, but it builds on many of the results and techniques that have been introduced in the course of this book. The Hahn–Banach Theorem plays a crucial (though hidden role), since it is this that allowed us to prove the transfer of separability from X^* to X (Lemma 20.5) and the transfer of reflexivity from X to any closed subset of X (Lemma 26.10). Also note that

---

[2] This result can also be derived as a consequence of the more powerful Banach–Alaoglu Theorem, which guarantees that for any Banach space X (no separability assumption required) the closed unit ball in X^* is compact in the weak-∗ topology. While topological compactness and sequential compactness are not equivalent in general, they coincide in metric spaces, and hence also in 'metrisable' topologies, i.e. those that can arise from a metric. When X is separable, it is possible to find a metric that gives rise to the weak-∗ topology on the closed unit ball in X^*, and one can then deduce the sequential compactness of our Theorem 27.11 from the topological compactness of the Banach–Alaoglu Theorem. This is a very long way round to obtain Theorem 27.11 compared to our more direct proof, but the more general result is important in the further development of the theory of Banach spaces (see e.g. Megginson, 1998). Details are given in Appendix C.

while the statement concerns only weak convergence, the proof relies on the definition of weak-∗ convergence.

Theorem 27.12 *Let X be a reflexive Banach space. Then any bounded sequence in X has a weakly convergent subsequence.*

This is equivalent to the statement that the closed unit ball is 'weakly sequentially compact'. Eberlein (1947) proved that the closed unit ball in X is weakly sequentially compact if and only if X is reflexive; see Exercises 27.9 and 27.10.

Proof Take a bounded sequence $(x_n) \in X$ and let

$$Y := \text{clin}\{x_1, x_2, \ldots\}.$$

Then, using Lemma 3.23, Y is separable. Since $Y \subseteq X$ and X is reflexive, so is Y (Lemma 26.10). Therefore $Y^{**} \equiv Y$, which implies that Y^{**} is separable (Exercise 3.13); Lemma 20.5 implies that Y^* is separable.

Now, x_n^{**} is a bounded sequence in Y^{**}, so using Theorem 27.11 there is a subsequence x_{n_k} such that $x_{n_k}^{**}$ is weakly-∗ convergent in Y^{**} to some limit $\Phi \in Y^{**}$. Since Y is reflexive, $\Phi = x^{**}$ for some $x \in Y \subseteq X$.

Now for any $f \in X^*$ we have $f_Y := f|_Y \in Y^*$, so

$$\lim_{k \to \infty} f(x_{n_k}) = \lim_{k \to \infty} f_Y(x_{n_k}) = \lim_{k \to \infty} x_{n_k}^{**}(f_Y)$$
$$= x^{**}(f_Y) = f_Y(x) = f(x),$$

i.e. $x_{n_k} \rightharpoonup x$. □

This theorem can be used to deduce that certain spaces are not reflexive; see e.g. Exercise 27.2.

Here is an example of the use of weak compactness and 'approximation' to prove the existence of a fixed point.

Lemma 27.13 *Let X be a reflexive Banach space, and $T : X \to X$ a compact linear operator. Suppose that (x_n) is a sequence in X such that there exist c_1, c_2 with $0 < c_1 \leq c_2$ so that $c_1 \leq \|x_n\| \leq c_2$ and*

$$\|Tx_n - x_n\| \to 0 \qquad (27.9)$$

as $n \to \infty$. Then there exists a non-zero $x \in X$ such that $Tx = x$.

Note that a linear map always has $x = 0$ as a fixed point.

Proof Since (x_n) is a bounded sequence in a reflexive Banach space, by Theorem 27.12 it has a weakly convergent subsequence, $x_{n_j} \rightharpoonup x$. Since T is compact, it follows from Lemma 27.4 that $Tx_{n_j} \to Tx$ strongly in X. Since

$$\lim_{j \to \infty} T x_{n_j} - x_{n_j} = 0,$$

it follows that $x_{n_j} \to Tx$. Since strong convergence implies weak convergence, we have $x_{n_j} \rightharpoonup Tx$, and since weak limits are unique and we already have $x_{n_j} \rightharpoonup x$ it follows that $x = Tx$.

To ensure that $x \neq 0$, note that since $T x_{n_j}$ converges strongly to $Tx = x$, it follows from (27.9) that x_{n_j} also converges strongly to x, and so $\|x\| \geq c_1$. □

We end with a simple prototype of the sort of minimisation problem that occurs in the calculus of variations. We will use Mazur's Theorem and sequential weak compactness in a reflexive Banach space X to prove the existence of at least one closest point in any closed, convex subset of X.

Lemma 27.14 *Suppose that X is reflexive and that K is closed convex subset of X. Then for any $x \in X \setminus K$ there exists at least one $k \in K$ such that*

$$\|x - k\| = \mathrm{dist}(x, K) = \inf_{y \in K} \|x - y\|.$$

Proof Let (y_n) be a sequence in K such that $\|x - y_n\| \to \mathrm{dist}(x, K)$. Then (y_n) is a bounded sequence in X, so has a subsequence y_{n_k} that converges weakly to some $k \in X$. Since K is closed and convex, it is also weakly closed (Theorem 27.7), and so $k \in K$. Since $y_{n_k} \rightharpoonup k$, $x - y_{n_k} \rightharpoonup x - k$, and so we have

$$\|x - k\| \leq \liminf_{k \to \infty} \|x - y_{n_k}\| = \mathrm{dist}(x, K)$$

using (27.1). Since $\|x - k\| \geq \mathrm{dist}(x, K)$, we have $\|x - k\| = \mathrm{dist}(x, K)$ as required. □

Exercises

27.1 Suppose that U is a closed linear subspace of a Banach space X, and suppose that $(x_n) \in U$ converges weakly to some $x \in X$. Show that $x \in U$. [Hint: use Exercise 20.10.]

27.2 Use Theorem 27.12 to show that ℓ^1 is not reflexive.

27.3 Suppose that X is a Banach space, and that (x_n) is a sequence in X such that $x_n \rightharpoonup x$. Show that there exist $y_n \in X$ such that $y_n \to x$, where each y_n is a convex combination of $(x_1, \ldots, x_n)$. [Hint: use the (obvious) fact that the closed convex hull of the $\{x_n\}$ is closed and convex.]

27.4   Recall (see Exercise 8.10) that a space X is *uniformly convex* if for every $\varepsilon > 0$ there exists $\delta > 0$ s.t.

$$\|x - y\| > \varepsilon, \ x, y \in \mathbb{B}_X \qquad \Rightarrow \qquad \left\| \frac{x+y}{2} \right\| < 1 - \delta,$$

where $\mathbb{B}_X$ is the closed unit ball in X. Show that if X is uniformly convex, then

$$x_n \rightharpoonup x \quad \text{and} \quad \|x_n\| \to \|x\| \qquad \Rightarrow \qquad x_n \to x.$$

(Set $y_n = x_n/\|x_n\|$, $y = x/\|x\|$, and show that $\|(y_n+y)/2\| \to 1$. Then argue by contradiction using the uniform convexity of X.)

27.5   The space $L^1(0, 2\pi)$ is not uniformly convex. Find a counterexample to the result of Exercise 27.4 in this space.

27.6   Find an example to show that Lemma 27.5 does not hold in ℓ^1 and ℓ^∞.

27.7   Show that if (e_n) is an orthonormal sequence in a Hilbert space H and $T: H \to H$ is compact, then $Te_n \to 0$ as $n \to \infty$.

27.8   Show that a sequence $(\phi_n) \in C([-1, 1]; \mathbb{R})$ satisfies

$$\lim_{n\to\infty} \int_{-1}^{1} f(t)\phi_n(t)\, dt = f(0) \tag{27.10}$$

for all $f \in C([-1, 1])$ if and only if

$$\lim_{n\to\infty} \int_{-1}^{1} \phi_n(t)\, dt = 1, \tag{27.11}$$

for every function $g \in C([-1, 1])$ that is zero in a neighbourhood of $x = 0$,

$$\lim_{n\to\infty} \int_{-1}^{1} g(t)\phi_n(t)\, dt = 0, \tag{27.12}$$

and there exists a constant $M > 0$ such that

$$\int_{-1}^{1} \phi_n(t)\, dt \le M \qquad \text{for every } n. \tag{27.13}$$

(Lax, 2002)

27.9   Suppose that X is a real Banach space. A theorem due to James (1964) states that if X is not reflexive, then there exists $\theta \in (0, 1)$ and sequences $(f_n) \in S_{X^*}$, $(x_n) \in S_X$, such that

$$f_n(x_j) \ge \theta, \ n \le j, \qquad f_n(x_j) = 0, \ n > j.$$

Show that the sets $C_n := \overline{\text{conv}}\{x_n, x_{n+1}, x_{n+2}, \ldots\}$ form a decreasing sequence of non-empty closed bounded convex sets in X that satisfies $\cap_j C_j = \varnothing$. (Show that if $x \in C_k$ for some k, then $f_n(x) \to 0$

as $n \to \infty$, but that if $x \in \cap_j C_j$, then $f_n(x) \geq \theta$ for every n.) (Megginson, 1998)

27.10 Let X be a Banach space. Show that if every bounded sequence in X has a weakly convergent subsequence, then whenever (C_n) is a decreasing sequence ($C_{n+1} \subseteq C_n$) of non-empty closed bounded convex sets in X, $\cap_n C_n \neq \varnothing$. Deduce, using the previous exercise, that X is reflexive if and only if its closed unit ball is weakly sequentially compact. [Hint: use Corollary 21.8.] (Megginson, 1998)

APPENDICES

APPENDICES

Appendix A
Zorn's Lemma

We will show here that Zorn's Lemma is a consequence Axiom of Choice (in fact the two are equivalent). We follow the lecture notes of Bergman (1997).

We begin with a formal statement of the Axiom of Choice.

Axiom of Choice If $(X_\alpha)_{\alpha \in \mathbb{A}}$ is any family of non-empty sets, there exists a function $\varphi \colon \mathbb{A} \to \cup_{\alpha \in \mathbb{A}} X_\alpha$ such that $\varphi(\alpha) \in X_\alpha$ for every $\alpha \in \mathbb{A}$.

The statement of Zorn's Lemma requires some more terminology.

A set P is *partially ordered* with respect to the relation $\preceq$ provided that

(i) $x \preceq x$ for all $x \in P$;

(ii) if $x, y, z \in P$ with $x \preceq y$ and $y \preceq z$, then $x \preceq z$;

(iii) if $x, y \in P$, $x \preceq y$, and $y \preceq x$, then $x = y$.

Two elements $x, y \in \mathcal{P}$ are *comparable* if $x \preceq y$ or $y \preceq x$. A subset $C \subseteq P$ is called a *chain* if every two elements of C are comparable, and $\mathcal{P}$ is *totally ordered* if every two elements of P are comparable.

An element $b \in P$ in an *upper bound* for a subset $T \subseteq P$ if $x \preceq b$ for every $x \in T$, and $m \in T$ is a *maximal element* for T if $x \in T$ and $m \preceq x$ implies that $x = m$.

Zorn's Lemma If P is a non-empty partially ordered set in which every chain has an upper bound, then P contains at least one maximal element.

We will need some other terminology and minor results for the proof.

An *initial segment* of a chain S is a subset $T \subseteq S$ such that if $u, v \in S$ with $u \preceq v$ and $v \in T$, then $u \in T$. We will write $T \lhd S$.

A *well-ordered* set is a totally ordered set in which every non-empty subset has a least element (i.e. every non-empty subset A contains an element s such that $s \preceq a$ for every $a \in A$). Such a least element is unique, since if $s_1, s_2 \in A$

are both least elements of A, then $s_1 \preceq s_2$ and $s_2 \preceq s_1$, which implies that $s_1 = s_2$.

Fact 1 If S is a well-ordered subset of a partially ordered set P and $t \notin S$ is an upper bound for S in P, then $S \cup \{t\}$ is well ordered.

Proof If $a, b \in S \cup \{t\}$, then there are three possibilities: (i) $a, b \in S$ so are comparable; (ii) $a \in S$, $b = t$ so $a \preceq t$; (iii) $a = b = t$ so $a \preceq b$ and $b \preceq a$. Any subset of $S \cup \{t\}$ contains a least element: $\{t\}$ has least element t; a subset of S contains a least element; and for $A \subset S$ the set $A \cup \{t\}$ has the same least element as A.   □

Fact 2 If $\mathcal{Z}$ is a set of well-ordered subsets of a partially ordered set P, such that for all $X, Y \in \mathcal{Z}$, either $X \lhd Y$ or $Y \lhd X$, then $\cup_{X \in \mathcal{Z}} X$ is well ordered.

Proof First we show that $U = \cup_{X \in \mathcal{Z}} X$ is totally ordered. Given any two elements $a, b \in U$, $a \in X$ and $b \in Y$, where $X, Y \in \mathcal{Z}$; but either $X \lhd Y$ or $Y \lhd X$, so a, b are comparable.

Now we show that any non-empty subset V of U has a least element. Since $V \subset U$, V has a non-empty intersection with some $X \in \mathcal{Z}$, and then $V \cap X$ has a least element s (since X is well ordered). Now suppose that we also have $V \cap Y \neq \varnothing$ for another $Y \in \mathcal{Z}$. Then either

(i) $Y \lhd X$, in which case $Y \subseteq X$ so $V \cap Y \subseteq V \cap X$; since s is the least element of $V \cap X$ it is also the least element of $V \cap Y$; or
(ii) $X \lhd Y$; in this case suppose that there exists $v \in Y$ with $v \preceq s$. Then since $s \in X$ and $X \lhd Y$ it follows that $v \in X$; but then, since s is the least element of $U \cap X$, we also have $s \preceq v$, which shows that $v = s$.

It follows that V has s as its least element, and hence U is totally ordered. □

Theorem A.1 *Zorn's Lemma is equivalent to the Axiom of Choice.*

Proof First we show that the Axiom of Choice implies Zorn's Lemma.

Let P be a non-empty partially ordered set with the property that every chain in P is bounded.

In particular, for any chain C the set of all upper bounds for C is non-empty. Suppose that C does not contain an element that is maximal for P; then C must have upper bounds that do not lie in C. Otherwise, suppose that $b \in C$ is an upper bound for C and $m \in P$ satisfies $b \preceq m$; then m is an upper bound for C and so $m \in S$; therefore $m \preceq b$, whence $m = b$. It follows that b is a maximal element of P.

We denote that set of these upper bounds for C that do not lie in C by $B(C)$, and using the Axiom of Choice for each chain C we choose one element of $B(C)$, and denote it by $\varphi(C)$.

Now we would like to argue as follows: choose some $p_0 \in P$. If this is not maximal, then let $p_1 = \varphi(\{p_0\}) \succeq p_0$. If p_1 is not maximal, then let $p_2 = \varphi(\{p_0, p_1\}) \succeq p_1$. If this process never terminates, then we let

$$p^* = \varphi(\{p_0, p_1, p_2, \ldots\}).$$

If p^* is not maximal in P, then we append p^* to the above chain and continue...

Now, fix an element $p \in P$, and let $\mathcal{Z}$ denote the set of subsets S of P that have the following properties:

(i) S is a well-ordered chain in P;
(ii) p is the least element of S;
(iii) for every proper non-empty initial segment $T \subset S$ the least element of $S \setminus T$ is $\varphi(T)$.

Note that $\mathcal{Z}$ is non-empty, since it contains $\{p\}$.

If S and S' are two members of $\mathcal{Z}$ then one is an initial segment of the other. To see this, let R denote the union of all sets that are initial segments of both S and S' – the 'greatest common initial segment' (R is non-empty since $\{p\}$ is an initial segment of every $S \in \mathcal{Z}$). If R is a proper subset of S and of S', then by (iii) the element $\varphi(R)$ is the least element of both $S \setminus R$ and $S' \setminus R$; this would mean that $R \cup \varphi(R)$ is an initial segment of both S and S', but this contradicts the maximality of R. Therefore $R = S$ or $R = S'$, i.e. one is an initial segment of the other.

By Fact 2, the set U, the union of all members of $\mathcal{Z}$, is well ordered. All members of $\mathcal{Z}$ are initial segments of U (suppose that $u, v \in U$, $u \preceq v$, and $v \in X$ ($X \in \mathcal{Z}$); if $u \in Y$, then either Y is an initial segment of X, in which case $u \in X$ immediately; or X is an initial segment of Y and it follows from this that $u \in X$) and the least element of U is $\{p\}$. Furthermore, U also satisfies (iii): if T is a proper non-empty initial segment of U, then there exists some $u \in U \setminus T$. By construction of U, $u \in S$ for some $S \in \mathcal{Z}$, and so T must be a proper initial segment of S. Hence (iii) ensures that $\varphi(T)$ is the least element of $S \setminus T$, and since S is an initial segment of U, $\varphi(T)$ is also the least element of $U \setminus T$.

Therefore U is a member of $\mathcal{Z}$. If U does not contain a maximal element of P, then $U \cup \{\varphi(U)\}$ will be an element of $\mathcal{Z}$ that is not a subset of U, which contradicts the definition of U.

Now we show that Zorn's Lemma implies the Axiom of Choice.

Let P be the collection of all subsets $\Phi \subset \mathbb{A} \times \cup_{\alpha \in \mathbb{A}} X_\alpha$ with the property that (i) for each $\alpha \in \mathbb{A}$ there is at most one element of the form $(\alpha, \varphi) \in \Phi$ and (ii) if $(\alpha, \varphi) \in \Phi$ then φ is a single element of X_α.

We partially order P by inclusion. P is non-empty because the empty set is a member of P (or, alternatively, $\{(\alpha, \varphi)\} \in P$ for any choice of $\alpha \in \mathbb{A}$ and $\varphi \in X_\alpha$). Now suppose that C is a chain in P; then $U = \cup_{S \in C} S$ is an upper bound for C, since $S \subseteq U$ for every $S \in C$. It follows that P has a maximal element Φ^*.

If there exists an $\alpha \in \mathbb{A}$ such that Φ^* contains no element of the form (α, ξ) with $\xi \in X_\alpha$ we can consider $\Psi := \Phi^* \cup \{(\alpha, \xi)\}$ for some $\xi \in X_\alpha$, and then $\Psi \in P$, any $\Phi \in P$ satisfies $\Phi \preceq \Psi$, but $\Psi \neq \Phi^*$ which contradicts the maximality of Φ^*. $\square$

Appendix B
Lebesgue Integration

In this appendix we give an outline of the construction of the Lebesgue integral in $\mathbb{R}^d$, omitting the proofs. We then use this theory to give intrinsic definitions of the spaces $L^p(\Omega)$ for open subsets Ω in $\mathbb{R}^d$. Finally, we prove the properties of these spaces that we have obtained in other ways in the main part of the book. In Sections B.1 and B.2 we follow quite closely the presentation in Stein and Shakarchi (2005); for Section B.3 we follow Chapter 3 of Rudin (1966) and Chapter 2 of Adams (1975). An alternative construction of the Lebesgue integral based on step functions rather than simple functions can be found in Priestley (1997).

B.1 The Lebesgue Measure on $\mathbb{R}^d$

For full details and proofs of statements in this section see Chapter 1 of Stein and Shakarchi (2005).

A 'cube' $Q \subset \mathbb{R}^d$ is a set of the form $[a_1, b_1] \times \cdots \times [a_d, b_d]$ with

$$b_1 - a_1 = \cdots = b_d - a_d.$$

The volume of a cube Q we write as $|Q|$, i.e. $|Q| = (b_1 - a_1)^d$.

Given any subset E of $\mathbb{R}^d$ we define the outer measure of E to be

$$\mu^*(E) := \inf \sum_{j=1}^{\infty} |Q_j|,$$

where the infimum is taken over all countable collections of cubes $\{Q_j\}$ that cover E. A subset of $\mathbb{R}^d$ is called *measurable* if for any $\varepsilon > 0$ we can find an open subset U of $\mathbb{R}^d$ such that $\mu^*(U \setminus E) < \varepsilon$. If E is measurable, then we define its Lebesgue measure $\mu(E)$ to be $\mu(E) = \mu^*(E)$. The Lebesgue measure is countably additive, i.e. for any countable collection $\{A_j\}_{j=1}^{\infty}$ of disjoint measurable sets we have

$$\mu\left(\bigcup_{j=1}^{\infty} A_j\right) = \sum_{j=1}^{\infty} \mu(A_j).$$

We denote by $\mathcal{M}(\mathbb{R}^d)$ the collection of all measurable sets in $\mathbb{R}^d$, and for any measurable subset Ω of $\mathbb{R}^d$ we let $\mathcal{M}(\Omega)$ be the set of all measurable subsets of Ω (it consists of $E \cap \Omega$ for all $E \in \mathcal{M}(\mathbb{R}^d)$). These collections are both examples of σ-algebras.

Definition B.1 A collection Σ of subsets of A is called a σ-*algebra on* A provided that

(i) $\varnothing \in \Sigma$;
(ii) if $E \in \Sigma$, then $A \setminus E \in \Sigma$;
(iii) if $\{E_j\}_{j=1}^{\infty} \subset \Sigma$, then $\bigcup_{j=1}^{\infty} E_j \in \Sigma$.

Note that as a consequence of the definition $A \in \Sigma$ and whenever $\{E_j\}_{j=1}^{\infty} \subset \Sigma, \bigcap_{j=1}^{\infty} E_j \in \Sigma$.

Theorem B.2 (Lebesgue measure) *The σ-algebra $\mathcal{M} = \mathcal{M}(\mathbb{R}^d)$ of measurable subsets of $\mathbb{R}^d$ and the Lebesgue measure $\mu \colon \mathcal{M} \to [0, \infty]$ satisfy*

(i) *every open subset of $\mathbb{R}^d$ belongs to $\mathcal{M}$, as does every closed subset;*
(ii) *if $A \subset B$ and $B \in \mathcal{M}$ with $\mu(B) = 0$, then $A \in \mathcal{M}$ and $\mu(A) = 0$;*
(iii) *every set A of the form $\{x \in \mathbb{R}^d : a_i \leq x_i \leq b_i\}$ belongs to $\mathcal{M}$ and $\mu(A) = \prod_{i=1}^{d}(b_i - a_i)$;*
(iv) *μ is translation invariant: if $A \in \mathcal{M}$, then $A + x \in \mathcal{M}$ for any $x \in \mathbb{R}^d$ and $\mu(A + x) = \mu(A)$.*

More generally, a signed measure on a σ-algebra Σ is a countably additive function $\mu \colon \Sigma \to \mathbb{R}$. If the range of μ is $[0, \infty]$, then μ is a positive measure; if its range is $\mathbb{C}$, then it is a complex measure.

A function $f \colon \mathbb{R}^d \to \mathbb{R}$ is called *measurable* if $f^{-1}((c, \infty)) \subset \mathcal{M}$ for every $c \in \mathbb{R}$. A measurable function $f \colon \mathbb{R}^d \to \mathbb{R}$ is called *simple* if it takes only finitely many values $\{a_1, \ldots, a_n\}$. Then we can write

$$f(x) = \sum_{j=1}^{n} a_j \chi_{E_j}(x), \tag{B.1}$$

where $E_j = f^{-1}(a_j)$ is measurable and the function χ_E denotes the characteristic function of E, i.e.

$$\chi_E(x) = \begin{cases} 1 & x \in E \\ 0 & x \notin E. \end{cases}$$

(Note that any simple function will in general have many representations in the form of (B.1).) We will write $s(\mathbb{R}^d)$ for the collection of all simple functions on $\mathbb{R}^d$.

Lemma B.3 *Given any non-negative measurable function* $f \in \mathfrak{M}$ *there exists an increasing sequence of non-negative simple functions* (f_n) *that converges pointwise to* f, *i.e.* $f_n(x) \to f(x)$ *for each* $x \in \mathbb{R}^d$.

Measurable functions are 'almost continuous', in a sense made precise by the following theorem.

Theorem B.4 (Lusin's Theorem) *If* f *is measurable, bounded, and non-zero on a set of finite measure, then there exists* $g \in C(\mathbb{R}^d)$ *such that*

$$\mu\{x : f(x) \neq g(x)\} < \varepsilon \qquad and \qquad \|g\|_\infty \leq \sup_{x \in \mathbb{R}^d} |f(x)|.$$

(This is a stronger version of Lusin's Theorem than that given in Stein and Shakarchi (2005); it appears as Theorem 2.23 in Rudin (1966).)

B.2 The Lebesgue Integral in $\mathbb{R}^d$

This section summarises results that can be found in Chapter 2 of Stein and Shakarchi (2005).

Integrals of Simple Functions

For any simple function as in (B.1) we define

$$\int_{\mathbb{R}^d} f(x)\,\mathrm{d}x := \sum_{j=1}^{n} a_j \mu(E_j);$$

since $\chi_E(x)f(x)$ is another simple function whenever E is a measurable set, we can also define

$$\int_E f(x)\,\mathrm{d}x = \int_{\mathbb{R}^d} \chi_E(x)f(x)\,\mathrm{d}x. \tag{B.2}$$

One can check that these definitions are independent of the choice of right-hand side in (B.1). (As we build up the integral we will concentrate on defining the integral over $\mathbb{R}^d$; at each stage we can also define the integral over any measurable subset using an expression like (B.2); we will not write this repeatedly.)

This definition gives an integral that satisfies the following properties.

Lemma B.5 *The integral on $s(\mathbb{R}^d)$ satisfies*

- *linearity: for all $f, g \in s(\mathbb{R}^d)$ and $\alpha, \beta \in \mathbb{R}$*

$$\int_{\mathbb{R}^d} \alpha f(x) + \beta g(x)\, dx = \alpha \int_{\mathbb{R}^d} f(x)\, dx + \beta \int_{\mathbb{R}^d} g(x)\, dx;$$

- *additivity: if E and F are disjoint sets in $\mathcal{M}$ with finite measure, then*

$$\int_{E\cup F} f(x)\, dx = \int_{E} f(x)\, dx + \int_{F} f(x)\, dx, \qquad f \in s(\mathbb{R}^d);$$

- *monotonicity: if $f, g \in s(\mathbb{R}^d)$ with $f(x) \leq g(x)$, then*

$$\int_{\mathbb{R}^d} f(x)\, dx \leq \int_{\mathbb{R}^d} g(x)\, dx;$$

- *triangle inequality: if $f \in s(\mathbb{R}^d)$, then $|f| \in s(\mathbb{R}^d)$ and*

$$\left| \int_{\mathbb{R}^d} f(x)\, dx \right| \leq \int_{\mathbb{R}^d} |f(x)|\, dx.$$

As we extend the definition of the integral to larger classes of functions we will ensure that these four properties still hold.

Integrals of Functions in $b(\mathbb{R}^d)$

Now we extend the definition of integral to the set $b(\mathbb{R}^d)$ of bounded functions that are non-zero only a set of finite measure.

If a property holds everywhere except on a set $A \in \mathcal{M}$ with $\mu(A) = 0$, then it is said to hold *almost everywhere*. This is the key notion that allows us to build up the integral for larger classes of functions: if $f \in b(\mathbb{R}^d)$ and (ϕ_n) is a sequence in $s(\mathbb{R}^d)$ that is non-zero only on some set E with $\mu(E) < \infty$, is bounded by some constant M, and converges almost everywhere to f, then

$$\lim_{n\to\infty} \int_{\mathbb{R}^d} \phi_n \quad \text{exists.} \tag{B.3}$$

Furthermore, if $f = 0$ almost everywhere, then the limit in (B.3) is zero.

This allows us to define

$$\int_{\mathbb{R}^d} f(x)\, dx := \lim_{n\to\infty} \int_{\mathbb{R}^d} \phi_n(x)\, dx$$

whenever f and (ϕ_n) are as in the previous paragraph. One can check that this definition is independent of the particular limiting sequence (ϕ_n), and then that the resulting integral on $b(\mathbb{R}^d)$ satisfies all the same properties as did the integral on $s(\mathbb{R}^d)$ (replace $s(\mathbb{R}^d)$ by $b(\mathbb{R}^d)$ throughout Lemma B.5).

Integrals of Non-negative Measurable Functions

The next extension is to all non-negative measurable functions (which may not be bounded). We simply set

$$\int_{\mathbb{R}^d} f(x)\,dx = \sup_{0 \le g \le f} \int_{\mathbb{R}^d} g(x)\,dx,$$

where the supremum is taken over all $g \in b(\mathbb{R}^d)$ with $0 \le g \le f$, allowing for the value $+\infty$ for $\int_{\mathbb{R}^d} g(x)\,dx$ on the right-hand side. If the supremum on the right-hand side is finite, then we say that f is (Lebesgue) integrable. The integral on this collection of functions is still linear, additive, and monotonic (i.e. Lemma B.5 holds for all functions in this class, although the triangle inequality property is now trivial since $f \ge 0$). In addition,

- if f is integrable and $0 \le g \le f$, then g is integrable;
- if f is integrable, then $f(x) < \infty$ for almost every x; and
- if $\int f = 0$, then $f(x) = 0$ for almost every x.

Two of the main convergence theorems for the Lebesgue integral concern only non-negative integrable functions, and since we have now defined the integral for this class of functions we state them here.

Lemma B.6 (Fatou's Lemma) *If* (f_n) *is a sequence of non-negative measurable functions and* $f(x) = \lim_{n\to\infty} f_n(x)$ *almost everywhere, then*

$$\int_{\mathbb{R}^d} f(x)\,dx \le \liminf_{n\to\infty} \int_a^b f_n(x)\,dx.$$

Theorem B.7 (Monotone Convergence Theorem) *If* (f_n) *is an increasing sequence of non-negative measurable functions, i.e.*

$$f_{n+1}(x) \ge f_n(x) \text{ almost everywhere}$$

and $f(x) = \lim_{n\to\infty} f_n(x)$ *almost everywhere, then*

$$\int_{\mathbb{R}^d} f(x)\,dx = \lim_{n\to\infty} \int_{\mathbb{R}^d} f_n(x)\,dx.$$

We are now in a position to the define the Lebesgue integral.

Integrals of Integrable Functions

Finally, a measurable function f is (Lebesgue) integrable if $|f|$ is integrable using the definition above; we then define $\int_{\mathbb{R}^d} f(x)\,dx$ by splitting f into[1] $f^+(x) = \max(f(x), 0)$ and $f^-(x) = \max(-f(x), 0)$ and setting

$$\int_{\mathbb{R}^d} f(x)\,dx := \int_{\mathbb{R}^d} f^+(x)\,dx - \int_{\mathbb{R}^d} f^-(x)\,dx.$$

This gives the set $\mathcal{L}^1(\mathbb{R}^d)$ of all integrable functions on $\mathbb{R}^d$. Defined on this collection of functions the integral is linear, additive, monotonic, and satisfies the triangle inequality, i.e. Lemma B.5 holds with $s(\mathbb{R}^d)$ replaced by $\mathcal{L}^1(\mathbb{R}^d)$ throughout.

The Dominated Convergence Theorem gives a compelling reason for preferring the Lebesgue integration to the Riemann integral.

Theorem B.8 (Dominated Convergence Theorem) *Let (f_n) be a sequence of measurable functions on $\mathbb{R}^d$ such that $f_n(x) \to f(x)$ almost everywhere. If there exists $g \in \mathcal{L}^1(\mathbb{R}^d)$ such that $|f_n(x)| \le g(x)$ almost everywhere, then $f \in \mathcal{L}^1(\mathbb{R}^d)$ and*

$$\int_{\mathbb{R}^d} f(x)\,dx = \lim_{n\to\infty} \int_{\mathbb{R}^d} f_n(x)\,dx.$$

We end this section on the Lebesgue integral with a very useful theorem that allows us to change the order of integration in double integrals.

Theorem B.9 (Fubini–Tonelli Theorem) *If*

$$\int_{\mathbb{R}^{d_1} \times \mathbb{R}^{d_2}} |f(x, y)|\,dx\,dy < \infty$$

then for almost every $y \in \mathbb{R}^{d_2}$ the function $f_y(x) := f(x, y)$ is integrable on $\mathbb{R}^{d_1}$ and the function $F(y) := \int_{\mathbb{R}}^{d_1} f_y(x)\,dx$ is integrable on $\mathbb{R}^{d_2}$. Furthermore,

$$\int_{\mathbb{R}^{d_2}} \left(\int_{\mathbb{R}^{d_1}} f(x, y)\,dx \right) dy = \int_{\mathbb{R}^{d_1} \times \mathbb{R}^{d_2}} f = \int_{\mathbb{R}^{d_1}} \left(\int_{\mathbb{R}^{d_2}} f(x, y)\,dy \right) dx.$$

(The result as stated combines Fubini's Theorem, which gives the same conclusions under the assumption that f is integrable on $\mathbb{R}^{d_1} \times \mathbb{R}^{d_2}$, and Tonelli's Theorem, which gives the same conclusions for a non-negative measurable function on $\mathbb{R}^{d_1} \times \mathbb{R}^{d_2}$.)

---

[1] A similar approach (splitting f into positive and negative real and complex parts) allows us to define the integral when f is complex valued.

B.3 The Lebesgue Spaces $L^p(\Omega)$

We now give an intrinsic definition of the L^p spaces, based on the theory of Lebesgue integration outlined in the previous section. We show that they are complete, and that $C(\overline{\Omega})$ is dense in $L^p(\Omega)$ for $1 \le p < \infty$; this shows that the definition here agrees with the 'definition by completion' we used in Chapter 7. We follow Rudin (1966).

Definition B.10 Let Ω be an open subset of $\mathbb{R}^d$. We let $L^p(\Omega)$ denote the space of all measurable functions on $\Omega \subset \mathbb{R}^d$ such that

$$\|f\|_{L^p} := \left(\int_\Omega |f(x)|^p \, dx \right)^{1/p} < \infty,$$

identifying functions that agree almost everywhere.

(Recall that $\int_\Omega f(x) \, dx := \int_{\mathbb{R}^d} f(x) \chi_\Omega(x) \, dx$; we therefore take

$$\mathcal{S}(\Omega) := \{ f \chi_\Omega : \ f \in \mathcal{S}(\mathbb{R}^d) \},$$

for the various spaces $\mathcal{S}(\mathbb{R}^d)$ defined in the previous two sections.)

We checked the triangle inequality for the L^p norm in Example 3.13, and since we identify functions that agree almost everywhere we have ensured that $\|f\|_{L^p} = 0$ if and only if $f = 0$ in L^p.

The space $L^\infty(\Omega)$ consists of measurable functions on Ω such that

$$\|f\|_{L^\infty} := \inf\{M : \ |f(x)| \le M \text{ almost everywhere}\} < \infty;$$

again, we identify functions that agree almost everywhere.

We now want to show, directly from these definitions, that $L^p(\Omega)$ is complete and that for $1 \le p < \infty$ the space $C(\overline{\Omega})$ is dense in $L^p(\Omega)$.

Theorem B.11 *The space $L^p(\Omega)$ is complete (with the L^p norm) for every $1 \le p \le \infty$.*

Proof First we take $1 \le p < \infty$ and let (f_n) be a Cauchy sequence in L^p. Then we can find a subsequence f_{n_j} such that

$$\|f_{n_{j+1}} - f_{n_j}\|_{L^p} < 2^{-j}$$

(cf. the proof of Lemma 4.14). It follows that

$$\left\| \sum_{j=1}^k |f_{n_{j+1}} - f_{n_j}| \right\|_{L^p} \le \sum_{j=1}^k \|f_{n_{j+1}} - f_{n_j}\|_{L^p} < \sum_{j=1}^k 2^{-j} < 1.$$

If we define

$$g(x) := \lim_{k \to \infty} \sum_{j=1}^{k} |f_{n_{j+1}}(x) - f_{n_j}(x)|,$$

allowing $g(x) = \infty$, we can use Fatou's Lemma (Lemma B.6) to deduce that $\|v\|_{L^p} \leq 1$, and hence that $v(x) < \infty$ almost everywhere. In particular, the series

$$f_{n_1}(x) + \sum_{j=1}^{k} \left(f_{n_{j+1}}(x) - f_{n_j}(x) \right)$$

converges absolutely for almost every $x \in \Omega$. Since the partial sums of this series are just

$$f_{n_1}(x) + \sum_{j=1}^{k} \left(f_{n_{j+1}}(x) - f_{n_j}(x) \right) = f_{n_{k+1}}(x),$$

it follows that for almost every $x \in \Omega$ we can define

$$f(x) := \lim_{j \to \infty} f_{n_j}(x),$$

and we can set f to be whatever we like elsewhere. In this way we have found a function f such that $f_{n_j} \to f$ almost everywhere in Ω.

To complete the proof we show that $f \in L^p(\Omega)$ and that $\|f_n - f\|_{L^p} \to 0$ as $n \to \infty$. Choosing N such that

$$\|f_n - f_m\|_{L^p} < \varepsilon/2, \qquad n, m \geq N$$

it follows that once j is large enough that $n_j \geq N$ we have

$$\|f_n - f_{n_j}\|_{L^p} < \varepsilon/2 \qquad n \geq N,$$

and then taking $j \to \infty$ and using Fatou's Lemma once again we obtain

$$\|f_n - f\|_{L^p} \leq \liminf_{j \to \infty} \|f - f_{n_j}\|_{L^p} < \varepsilon/2 \qquad n \geq N. \qquad \text{(B.4)}$$

It follows that $f_N - f \in L^p$, and since $f = (f - f_N) + f_N$ and L^p is a vector space we have $f \in L^p$. That $f_n \to f$ in L^p is now immediate from (B.4).

For the case $p = \infty$, we observe that if (f_n) is a Cauchy sequence in L^∞ then there exist sets A_n and $B_{m,n}$ of zero measure such that

$$\sup_{\Omega \backslash A_n} |f_n(x)| \leq \|f_n\|_{L^\infty} \quad \text{and} \quad \sup_{\Omega \backslash B_{n,m}} |f_n(x) - f_m(x)| \leq \|f_n - f_m\|_{L^\infty}.$$

If we take $E = \cup_n A_n \cup \cup_{n,m} B_{n,m}$ then E still has zero measure, and (f_n) converges uniformly to some f on $\Omega \backslash E$. Setting $f(x) = 0$ for $x \in E$ it follows that $f \in L^\infty(\Omega)$ and $f_n \to f$ in $L^\infty(\Omega)$. $\qquad \square$

As part of this proof we have obtained the following useful result.

Corollary B.12 *If (f_n) is Cauchy in $L^p(\Omega)$ then it has a subsequence that converges almost everywhere in Ω.*

We now want to show that $C(\overline{\Omega})$ is dense in $L^p(\Omega)$.

Theorem B.13 *If $\Omega \subset \mathbb{R}^d$ then $C(\overline{\Omega})$ is dense in $L^p(\Omega)$, $1 \le p < \infty$.*

Proof Let $S(\Omega)$ denote the set of all $f \in s(\Omega)$ such that

$$\mu\{x : f(x) \ne 0\} < \infty.$$

First we show that $S(\Omega)$ is dense in $L^p(\Omega)$, and then use Lusin's Theorem to show that $C(\overline{\Omega})$ is dense in $S(\Omega)$.

First, note that $S(\Omega) \subset L^p(\Omega)$. Take a non-negative $f \in L^p(\Omega)$ and let (s_n) be a sequence of non-negative simple functions that converges pointwise to f (Lemma B.3); since $0 \le s_n \le f$ it follows that $s_n \in L^p(\Omega)$. Note that since s_n is simple this implies that $s_n \in S(\Omega)$ (otherwise its L^p norm would be infinite). Since $|f(x) - s_n(x)|^p \le |f(x)|^p$ for every $x \in \Omega$, the Dominated Convergence Theorem implies that $\|s_n - f\|_{L^p} \to 0$. For a general $f \in L^p$ we write $f = f^+ - f^-$ with f^+, f^- both non-negative (as in the construction of the integral); then $f^+, f^- \in L^p$, and both can be approximated by elements of $S(\Omega)$. This shows that $S(\Omega)$ is dense in $L^p(\Omega)$.

Now if $s \in S(\Omega)$, given any $\varepsilon > 0$, Lusin's Theorem (Theorem B.4) guarantees that there exists $g \in C(\overline{\Omega})$ such that $g(x) = s(x)$ except on a set E with $\mu(E) < \varepsilon$, and that $\|g\|_\infty \le \|s\|_{L^\infty}$. It follows that

$$\|g-s\|_{L^p} = \left(\int_E |g(x) - s(x)|^p \, dx\right)^{1/p} \le \mu(E)^{1/p}\|g-s\|_\infty \le 2\|s\|_\infty \varepsilon^{1/p}.$$

Since $S(\Omega)$ is dense in $L^p(\Omega)$, the result now follows. $\qquad\square$

Since $L^p(\Omega)$ is complete and $C(\overline{\Omega})$ is a dense subset, it follows (see Lemma 7.2) that $L^p(\Omega)$ is the completion of $C(\overline{\Omega})$ in the L^p norm, which was the definition we gave in Chapter 7.

Lemma B.14 *For any $\Omega \subset \mathbb{R}^d$ and $1 \le p < \infty$ the space $L^p(\Omega)$ is separable.*

Proof Suppose first that Ω is bounded; then $\Omega \subset [-M, M]^d =: Q_M$ for some $M > 0$ and $|\Omega| < \infty$. If we use an argument similar to that of Exercise 6.7 we can show that the collection of functions

$$A := \left\{ \sum_{i_1,\ldots,i_n=1}^{k} c_{i_1,i_2,\ldots,i_n} x_1^{i_1} \cdots x_n^{i_n} : n \in \mathbb{N}, \ c_{i_1,\ldots,i_n} \in \mathbb{R} \right\}$$

is uniformly dense in $C(Q_M)$, and hence uniformly dense in $C(\overline{\Omega})$. Indeed, A is a subalgebra of $C(Q_M)$ that separates points, and hence $\overline{A} = C(Q_M)$ by the Stone–Weierstrass Theorem. Since

$$\|f\|_{L^p(\Omega)}^p = \int_{\Omega} |f(x)|^p \le |\Omega| \|f\|_\infty^p,$$

A is also dense in $C(Q_M)$ in the L^p norm. Since A is the linear span of the countable set

$$A' := \{x_1^{i_1} \cdots x_n^{i_n} : i_j \in \mathbb{N}, \ j = 1, \ldots, n\},$$

it follows from Exercise 3.14 that the linear span of A' is also dense in $L^p(\Omega)$, and so $L^p(\Omega)$ is separable.

If Ω is unbounded, then, given any $f \in L^p(\Omega)$ and $\varepsilon > 0$, there exists $N \in \mathbb{N}$ such that

$$\int_{\Omega \setminus Q_N} |f|^p < \varepsilon/2;$$

this is a consequence of the Monotone Convergence Theorem applied to the sequence of functions $f \chi_{Q_n}$. We have just shown that for any $j \in \mathbb{N}$ the function $f|_{Q_j}$ can be approximated to within $\varepsilon/2$ by elements of the linear span of $\{x^k|_{Q_j}\}$. Since the collection

$$\bigcup_{k,j \in \mathbb{N}} \{x^k|_{Q_j}\}$$

is countable, this shows that $L^p(\Omega)$ is separable. $\qquad\square$

B.4 Dual Spaces of L^p, $1 \le p < \infty$

In the final section of this appendix we show that for $1 \le p < \infty$ we have $(L^p)^* \equiv L^q$ with (p, q) conjugate.

In this section we follow Rudin (1966) and give the standard measure-theoretic argument based on the Radon–Nikodym Theorem. We also give an alternative proof that is based on the uniform convexity of L^p taken from Adams (1975).

Given any $f \in L^1(\Omega)$ the function $\lambda \colon \mathcal{M}(\Omega) \to \mathbb{R}$ defined by setting

$$\lambda(A) = \int_A f(x) \, dx$$

defines a signed measure on $\mathcal{M}(\Omega)$. That this can be reversed is the content of the Radon–Nikodym Theorem.

Theorem B.15 (Radon–Nikodym Theorem) *If $\lambda \colon \mathcal{M}(\Omega) \to \mathbb{R}$ is a measure on $\mathcal{M}(\Omega)$ such that $\lambda(A) = 0$ whenever $\mu(A) = 0$, then there exists $f \in L^1(\Omega)$ such that*

$$\lambda(A) = \int_A f(x)\,dx \qquad A \in \mathcal{M},$$

and f is unique (up to sets of measure zero).

We now use this to find the dual spaces of $L^p(\Omega)$.

Theorem B.16 *For $1 \le p < \infty$ we have $(L^p(\Omega))^* \equiv L^q(\Omega)$, where p and q are conjugate.*

Proof We showed in the proof of Theorem 18.8 given any $g \in L^q(\Omega)$ the linear functional

$$\Phi_g(f) := \int_\Omega f(x)g(x)\,dx, \qquad f \in L^p(\Omega) \tag{B.5}$$

is an element of $(L^p)^*$ with $\|\Phi_g\|_{(L^p)^*} = \|g\|_{L^q}$. What remains is to show that any $\Phi \in (L^p)^*$ can be written in the form (B.5) for some $g \in L^q(\Omega)$.

We give the proof when $\mu(\Omega) < \infty$. For the steps required to prove the result when $\mu(\Omega) = \infty$ see the conclusion of Theorem 6.16 in Rudin (1966).

Given $\Phi \in (L^p)^*$, for any $E \in \mathcal{M}(\Omega)$ define

$$\lambda(E) = \Phi(\chi_E).$$

Clearly λ is additive, since if $A, B \in \mathcal{M}(\Omega)$ are disjoint $\chi_{A \cup B} = \chi_A + \chi_B$, and so

$$\lambda(A \cup B) = \phi(\chi_{A \cup B}) = \phi(\chi_A + \chi_B) = \Phi(\chi_A) + \Phi(\chi_B) = \lambda(A) + \lambda(B).$$

If $E = \cup_{j=1}^\infty A_j$, where the $\{A_j\}$ are disjoint subsets of $\mathcal{M}(\Omega)$, then, if we set $E_k = \cup_{j=1}^k A_j$, we have

$$\|\chi_E - \chi_{E_k}\|_{L^p} = \|\chi_{E \setminus E_k}\|_{L^p} = \mu(E \setminus E_k)^{1/p},$$

which tends to zero as $k \to \infty$ since $\mu(E) = \sum_{j=1}^\infty \mu(A_j) < \infty$. Since Φ is continuous, it follows that

$$\sum_{j=1}^k \lambda(A_j) = \lambda(E_k)\Phi(\chi_{E_k}) \to \Phi(\chi_E) = \lambda(E),$$

and so $\lambda(E) = \sum_{j=1}^{\infty} \lambda(A_j)$ and λ is countably additive. Furthermore, if $\mu(E) = 0$, then $\lambda(E) = \int_E f(x)\,dx = 0$.

We can therefore apply the Radon–Nikodym Theorem to guarantee the existence of $g \in L^1(\Omega)$ such that

$$\Phi(\chi_E) = \int_E g\,dx = \int_{\Omega} \chi_E g\,dx \qquad \text{for every } E \in \mathcal{M}(\Omega). \tag{B.6}$$

Since simple measurable functions are linear combinations of characteristic functions, it follows by linearity of the integral that

$$\Phi(f) = \int f g\,dx \qquad \text{for every } f \in s(\Omega).$$

Since every $f \in L^\infty(\Omega)$ is the uniform limit of simple functions,[2] it follows that

$$\Phi(f) = \int f g\,dx \qquad \text{for every } f \in L^\infty(\Omega). \tag{B.7}$$

We now need to show that $g \in L^q$ and that $\Phi(f) = \int f g$ for every $f \in L^p$. If $p = 1$, then, from (B.6), it follows that

$$\left| \int_E g\,dx \right| = |\Phi(\chi_E)| \leq \|\Phi\|_{(L^1)^*} \|\chi_E\|_{L^1} = \|\Phi\|_{(L^1)^*} \mu(E),$$

which shows that $|g(x)| \leq \|\Phi\|_{(L^1)^*}$ almost everywhere (since otherwise there would be a set E of positive measure for which the inequality was violated) and so $\|g\|_{L^\infty} \leq \|\Phi\|_{(L^1)^*}$.

If $1 < p < \infty$, then for each $k \in \mathbb{N}$ define

$$g_k(x) = \begin{cases} g(x) & |g(x)| \leq k \\ 0 & \text{otherwise.} \end{cases}$$

We apply (B.7) to the function $f_k \in L^\infty(\Omega)$ defined by setting

$$f_k(x) = \begin{cases} |g_k(x)|^{q-2} g_k(x) & g_k(x) \neq 0 \\ 0 & \text{otherwise} \end{cases}$$

and obtain

$$\int_{\Omega} |g_k(x)|^q\,dx = \int_{\Omega} g f_k\,dx = \Phi(f_k) \leq \|\Phi\|_{(L^p)^*} \left(\int_{\Omega} |g_k|^q \right)^{1/p},$$

---

[2] Set $M = \|f\|_{L^\infty}$ and for each $n \in \mathbb{N}$ and $-n \leq j \leq n-1$, let $a_{j,n} = jM/n$ and $E_{j,n} = f^{-1}(jM/n, (j+1)M/n)$; then take $s_n = \sum_{j=-n}^{n-1} a_{j,n} \chi_{E_{j,n}}(x)$.

which implies that

$$\left(\int_\Omega |g_k(x)|^q \, dx \right)^{1/q} \le \|\Phi\|_{(L^p)^*}.$$

Since $|g_{k+1}(x)| \ge |g_k(x)|$ and $g_k(x) \to g(x)$ for almost every x, we can apply the Monotone Convergence Theorem to deduce that $g \in L^q(\Omega)$ with $\|g\|_{L^q} \le \|\Phi\|_{(L^p)^*}$.

It only remains to show that $\Phi(f) = \int_\Omega fg \, dx$ for every $f \in L^p(\Omega)$; we currently only have this for $f \in L^\infty(\Omega)$ (in (B.7)). Since we now know that $f \in L^q(\Omega)$, both $\Phi(f)$ and $\int_\Omega fg \, dx$ define continuous linear functionals on $L^p(\Omega)$ that agree for $f \in L^\infty(\Omega)$. To finish the proof, note that $L^\infty(\Omega)$ is dense in $L^p(\Omega)$, since $L^\infty(\Omega) \supset C^0(\overline{\Omega})$ and (as we showed in Theorem B.13) $C(\overline{\Omega})$ is dense in $L^p(\Omega)$. It follows that $\Phi(f) = \int_\Omega fg \, dx$ for every $f \in L^p(\Omega)$ and we have finished. $\square$

The proof of reflexivity of L^p, which is very similar to the proof that ℓ^p is reflexive once we have identified the dual spaces, is given in Chapter 26 (Proposition 26.6).

We end this appendix with an alternative proof of the 'onto' part of Theorem B.16 that does not use the Radon–Nikodym Theorem. This proof, which can be found in Adams (1975), uses the uniform convexity of L^p ($1 < p < \infty$). The proof as presented here also uses many results that are proved in various exercises in this book. The only part we have not proved is the uniform convexity of L^p for $1 < p < 2$, which uses Clarkson's second inequality. However, this case was treated without the Radon–Nikodym Theorem in Exercise 18.6.

Alternative proof that $(L^p)^* \equiv L^q$ for $1 < p < \infty$. We use the uniform convexity of L^p, which we proved in Exercise 10.6.

Suppose that we have $\Phi \in (L^p)^*$ with $\|\Phi\|_{(L^p)^*} = 1$. Then we can find a unique $\hat{g} \in L^p$ such that $\|\hat{g}\|_{L^p} = 1$ and $\Phi(\hat{g}) = 1$. Indeed, we showed in Exercise 10.7 that any closed convex subset of a uniformly convex Banach space contains a unique element of minimum norm, so the set

$$G := \{ f \in L^p : \Phi(f) = 1 \},$$

which is closed (it is the preimage of 1 under the continuous map ℓ) and convex, contains a unique element $\hat{g}$ of minimum norm. Since for any $f \in L^p$ we have

$$1 = |\Phi(f)| \le \|\Phi\|_{(L^p)^*} \|f\|_{L^p}$$

it follows that $\|f\|_{L^p} \geq 1$ for every $f \in G$, while from the definition of $\|\Phi\|_{(L^p)^*}$ we know that for every $\delta > 0$ there exists an $f \in G$ such that $\|f\|_{L^p} < 1 + \delta$. It follows that $\|\hat{g}\|_{L^p} = 1$ as claimed.

We now use the facts that any uniformly convex space is strictly convex (Exercise 10.4), and that in a strictly convex space for each $x \in X$ with $x \neq 0$ there is a *unique* linear functional $\phi \in X^*$ such that

$$\|\phi\|_{X^*} = 1 \text{ and } \phi(x) = \|x\|_X$$

(Exercise 20.5). Set

$$g(x) = \begin{cases} |\hat{g}(x)|^{p-2}\overline{\hat{g}(x)} & \hat{g}(x) \neq 0 \\ 0 & \text{otherwise.} \end{cases}$$

Then

$$\|g\|_{L^q} = \left(\int_\Omega |\hat{g}(x)|^{(p-1)q} \, dx \right)^{1/q} = \left(\int_\Omega |\hat{g}(x)|^p \, dx \right)^{1/q} = 1,$$

and so the linear map Φ_g as defined in (B.5) is an element of $(L^p)^*$ with $\|\Phi_g\|_{(L^p)^*} = 1$. Furthermore, we have

$$\Phi_g(\hat{g}) = \int_\Omega g(x)\hat{g}(x) \, dx = \int_\Omega |\hat{g}(x)|^p \, dx = 1.$$

It follows from the uniqueness result from the beginning of this paragraph that $\Phi_g = \Phi$.

If we start with $\Phi \in (L^p)^*$ with $v = \|\Phi\|_{(L^p)^*} \neq 1$, then we can apply this result to $v^{-1}\Phi$, obtain g such that $\Phi_g = v^{-1}\Phi$, and then $\Phi_{vg} = \Phi$. $\square$

For a limiting argument that allows one to obtain the result for $p = 1$, see Theorem 2.34 in Adams (1975).

Appendix C
The Banach–Alaoglu Theorem

This appendix gives a proof of two powerful results: the Tychonoff Theorem and the Banach–Alaoglu Theorem.

The first guarantees that the topological product of any collection of compact topological spaces is compact (if we use the product topology). The second, which uses the Tychonoff Theorem as a key ingredient in the proof, says that if X is a Banach space, then the closed unit ball in X^* is compact in the weak-$*$ topology.

The Banach–Alaoglu Theorem is a more general version of Helly's Theorem (sequential weak-$*$ compactness of the closed unit ball in X^* when X is separable) that we proved as Theorem 27.11. (By finding a metric that gives rise to the weak-$*$ topology on the closed unit ball when X is separable we can derive Theorem 27.11 as a consequence.)

We have not discussed topologies and topological spaces elsewhere in this book, so we first give a quick overview of this material, including the notion of continuity and convergence in the topological setting. We show how to construct a topology from a basis or sub-basis, and define the product topology. We then prove Tychonoff's Theorem, derive the Banach–Alaoglu Theorem as a (non-trivial) consequence, and show that the weak-$*$ topology (on the closed unit ball) is metrisable when X is separable. To complete the proof that Helly's Theorem follows from the Banach–Alaoglu Theorem we show that compactness and sequential compactness are equivalent in a metric space.

For a more leisurely treatment of the essentials of topological spaces, see Sutherland (1975) or Munkres (2000). The presentation here is heavily influenced by David Preiss's lecture notes for the University of Warwick *Metric spaces* module, with the later sections also drawing on Brezis (2011).

C.1 Topologies and Continuity

A *topology* $\mathcal{T}$ on a set T is a collection of subsets of T such that

(T1) $T, \varnothing \in \mathfrak{T}$,
(T2) the union of any collection of elements of $\mathfrak{T}$ is in $\mathfrak{T}$, and
(T3) if $U, V \in \mathfrak{T}$, then $U \cap V \in \mathfrak{T}$.

The pair $(T, \mathfrak{T})$ is a *topological space*. Any set in $\mathfrak{T}$ is called 'open', and any subset U of T with $T \setminus U \in \mathfrak{T}$ is called 'closed'.

For example, the collection of all open sets in a metric space (X, d) forms a topology on X (see Lemma 2.6).

We showed in Lemma 2.13 that continuity of a function between metric spaces can be defined in terms of preimages of open sets; we can use this form of the definition for functions between general topological spaces.

Definition C.1 A map $f : (T, \mathfrak{T}) \rightarrow (S, \mathcal{S})$ is said to be continuous if $f^{-1}(S) \in \mathfrak{T}$ for every $S \in \mathcal{S}$, where

$$f^{-1}(S) = \{x \in T : \ f(x) \in S\}.$$

Lemma 2.9 provides, similarly, a definition of convergence in terms only of open sets, and we adopt this definition in a topological space.

Definition C.2 A sequence $(x_n) \in T$ converges to x in the topology $\mathfrak{T}$ if for every $U \in \mathfrak{T}$ with $x \in U$ there exists N such that $x_n \in U$ for all $n \geq N$.

C.2 Bases and Sub-bases

In a metric space we do not have to specify all the open sets: we can build them up from open balls (see Exercise 2.7). We can do something similar in an abstract topological space.

Definition C.3 A *basis* for a topology $\mathfrak{T}$ on T is a collection $\mathcal{B} \subseteq \mathfrak{T}$ such that every set in $\mathfrak{T}$ is a union of sets from $\mathcal{B}$.

The following lemma is an immediate consequence of the definition of a basis, since $T \in \mathfrak{T}$, and if $B_1, B_2 \in \mathcal{B}$, then $B_1, B_2 \in \mathfrak{T}$ so $B_1 \cap B_2 \in \mathfrak{T}$.

Lemma C.4 *If $\mathcal{B}$ is any basis for $\mathfrak{T}$, then*

(B1) T *is the union of sets from* $\mathcal{B}$;
(B2) *if $B_1, B_2 \in \mathcal{B}$, then $B_1 \cap B_2$ is the union of sets from $\mathcal{B}$.*

However, this can be reversed.

Proposition C.5 *Let $\mathcal{B}$ be a collection of subsets of a set T that satisfy* (B1) *and* (B2)*. Then there is a unique topology of T whose basis is $\mathcal{B}$; its open sets are precisely the unions of sets from $\mathcal{B}$.*

Note that $\mathcal{T}$ is the smallest topology that contains $\mathcal{B}$.

Proof If there is such a topology, then, by the definition of a basis, its sets consist of the unions of sets from $\mathcal{B}$. So we only need check that if $\mathcal{T}$ consists of unions of sets from $\mathcal{B}$, then this is indeed a topology on T. We check properties (T1)–(T3).

(T1): T is the union of sets from $\mathcal{B}$ by (B1).

(T2): if $U, V \in \mathcal{T}$, then $U = \cup_{i \in I} B_i$ and $V = \cup_{j \in I} D_j$, with $B_i, D_j \in \mathcal{B}$, and so

$$U \cap V = \bigcup_{i,j} B_i \cap D_j,$$

which is a union of sets in $\mathcal{B}$ by (B2) and hence an element of $\mathcal{T}$.

(T3): any union of unions of sets from $\mathcal{B}$ is a union of sets from $\mathcal{B}$. □

There is a smaller collection of sets from which we can construct the topology $\mathcal{T}$ by allowing for not only unions (as in (T2)) but also finite intersections (as in (T3)).

Definition C.6 A *sub-basis* for a topology $\mathcal{T}$ on T is a collection $\mathcal{B} \subseteq \mathcal{T}$ such that every set in $\mathcal{T}$ is a union of finite intersections of sets from $\mathcal{B}$.

(Note that if $\mathcal{B}$ is a basis for $\mathcal{T}$, then it is also a sub-basis for $\mathcal{T}$; but in general a sub-basis will be 'smaller'.)

Example: the collection of intervals (a, ∞) and $(-\infty, b)$ (ranging over all $a, b \in \mathbb{R}$) is a sub-basis for the usual topology on $\mathbb{R}$, since intersections give the open intervals (a, b) and these are a basis for the usual topology.

Proposition C.7 *If $\mathcal{B}$ is any collection of subsets of a set T whose union is T, then there is a unique topology $\mathcal{T}$ on T with sub-basis $\mathcal{B}$. Its open sets are precisely the unions of finite intersections of sets from $\mathcal{B}$.*

Proof If $\mathcal{B}$ is a sub-basis for T, then any topology has $\mathcal{D}$, the collection of all finite intersections of elements of $\mathcal{B}$, as a basis. But $\mathcal{D}$ satisfies (B1) and (B2) from Lemma C.4, so by Proposition C.5 there is a unique topology $\mathcal{T}$ with basis $\mathcal{D}$, which is also the unique topology with sub-basis $\mathcal{B}$. □

Note that the topology $\mathcal{T}$ from this proposition is the smallest topology on T that contains $\mathcal{B}$.

To check that a map $f \colon T_1 \to T_2$ is continuous it is enough to check that preimages of a sub-basis for T_2 are open in T_1 (since any basis is also a sub-basis, we could check for a basis if we wanted).

Lemma C.8 *Suppose that $f \colon (T_1, \mathcal{T}_1) \to (T_2, \mathcal{T}_2)$, and that $\mathcal{B}$ is a sub-basis for the topology $\mathcal{T}_2$. Then f is continuous if and only if $f^{-1}(B) \in \mathcal{T}_1$ for every $B \in \mathcal{B}$.*

Proof 'Only if' is clear since every element of the sub-basis is an element of $\mathcal{T}_2$.

Now, any element U of $\mathcal{T}_2$ can be written as $U = \cup_i D_i$, for some $\{D_i\}$ that are finite intersections of elements of $\mathcal{B}$. So

$$f^{-1}(U) = f^{-1}(\cup_i D_i) = \cup_i f^{-1}(D_i),$$

and since for each D_i we have $D_i = \cap_{j=1}^{n(i)} B_j$ with $B_j \in \mathcal{B}$ we have

$$f^{-1}\left(\cap_{j=1}^{n} B_j\right) = \cap_{j=1}^{n} f^{-1}(B_j),$$

which is open by assumption. So $f^{-1}(U)$ is a union of open sets, so open. $\qquad\square$

C.3 The Weak-∗ Topology

The weak-∗ topology on X^* is the smallest topology $\mathcal{T}_*$ on X^* such that every map $\delta_x \colon X^* \to \mathbb{K}$ given by

$$\delta_x(f) := f(x)$$

is continuous. Since the open balls $\{B(z, \varepsilon) : z \in \mathbb{K}, \ \varepsilon > 0\}$ form a basis (and so also a sub-basis) for the topology of $\mathbb{K}$, by Lemma C.8 it is enough to guarantee that $\delta_x^{-1}(B(z, \varepsilon))$ is an element of $\mathcal{T}_*$ for every $z \in K, \varepsilon > 0$. In other words we should take $\mathcal{T}_*$ to be the topology with sub-basis

$$\{\phi \in X^* : \phi(x) \in B(z, \varepsilon)\} \qquad x \in X, \ z \in \mathbb{K}, \ \varepsilon > 0, \qquad \text{(C.1)}$$

i.e. $\mathcal{T}_*$ is formed of all unions of finite intersections of sets of the form (C.1).

We now show that weak-∗ convergence and convergence in the weak-∗ topology coincide.

Proposition C.9 *If $(f_n) \in X^*$, then $f_n \overset{*}{\rightharpoonup} f$ if and only if $f_n \to f$ in $\mathcal{T}_*$.*

Proof Suppose that $f_n \overset{*}{\rightharpoonup} f$, and let $U \in \mathcal{T}_*$ be such that $f \in U$. Then, since U is the union of finite intersections of sets of the form in (C.1), f must belong to one such intersection that itself is a subset of U; so there exist m, $y_i \in X$, $z_i \in \mathbb{K}$, $\varepsilon > 0$ such that

$$f \in \bigcap_{i=1}^{m} \{\phi \in X^* : \phi(y_i) \in B(z_i, \varepsilon)\} \subset U.$$

It follows that, in particular, $f(y_i) \in B(z_i, \varepsilon)$. Since $f_n \overset{*}{\rightharpoonup} f$, for n sufficiently large we have $f_n(y_i) \in B(z_i, \varepsilon)$ for all $i = 1, \ldots, m$, i.e. $f_n \in U$. So $f_n \to f$ in $\mathcal{T}_*$.

Conversely, if $f_n \to f$ in $\mathcal{T}_*$ and we choose some $x \in X$ and $\varepsilon > 0$, then f is contained in the open set

$$U := \{\phi \in X^* : \phi(x) \in B(f(x), \varepsilon)\}$$

and so $f_n \in U$ for all $n \geq N$, i.e. $|f_n(x) - f(x)| < \varepsilon$, so $f_n \overset{*}{\rightharpoonup} f$. $\qquad\square$

C.4 Compactness and Sequential Compactness

A topological space $(T, \mathcal{T})$ is *compact* if every open cover of T has a finite subcover, i.e. if

$$T = \bigcup_{\alpha \mathbb{A}} U_\alpha, \qquad U_\alpha \in \mathcal{T},$$

then there exist $U_{\alpha_1}, \ldots, U_{\alpha_n}$ such that

$$T = \bigcup_{i=1}^{n} U_{\alpha_i}.$$

A topological space is *sequentially compact* if any sequence $(x_n) \in T$ has a subsequence (x_{n_j}) that converges to a limit (that lies in T).

Two relatively simple results will enable us to prove the equivalence of compactness and sequential compactness in any metric space.

Theorem C.10 *Let $\mathcal{F}$ be a collection of non-empty closed subsets of a compact space T such that every finite subcollection of $\mathcal{F}$ has a non-empty intersection. Then the intersection of all the sets from $\mathcal{F}$ is non-empty.*

Proof Suppose that the intersection of all the sets from $\mathcal{F}$ is empty, and let $\mathcal{U}$ be the collection of their complements,

$$\mathcal{U} := \{T \setminus F : F \in \mathcal{F}\}.$$

Then $\mathcal{U}$ is an open cover of T, since

$$T \setminus \bigcup_{U \in \mathcal{U}} U = \bigcap_{U \in U} (T \setminus U) = \bigcap_{F \in \mathcal{F}} F = \varnothing.$$

Thus $\mathcal{U}$ has a finite subcover $U_1, \ldots, U_n$, which implies that

$$\varnothing = T \setminus \bigcup_{i=1}^{n} U_i = \bigcap_{i=1}^{n} F_i,$$

a contradiction. $\qquad\square$

Lemma C.11 *Any closed subset K of a compact space T is compact.*

Proof Let $\{U_\alpha\}_{\alpha \in \mathbb{A}}$ be an open cover of K. Then $\{U_\alpha\}_{\alpha \in \mathbb{A}} \cup [T \setminus K]$ is an open cover of T, so it has a finite subcover

$$\{U_{\alpha_j}\}_{j=1}^{n} \qquad \text{or} \qquad \{U_{\alpha_j}\}_{j=1}^{n} \cup [T \setminus K].$$

Either way, $\{U_{\alpha_j}\}_{j=1}^{n}$ is a finite subcover of K, so K is compact. $\qquad\square$

Corollary C.12 *Let $F_1 \supset F_2 \supset F_3 \supset \cdots$ be non-empty closed subsets of a compact space T. Then $\bigcap_{j=1}^{\infty} F_j \neq \varnothing$.*

Lemma C.13 *If K is a sequentially compact subset of a metric space, then, given any open cover $\mathcal{U}$ of K, there exists $\delta > 0$ such that for any $x \in X$ there exists $U \in \mathcal{U}$ such that $B(x, \delta) \subset U$.*

(Such a δ is called a *Lebesgue number* for the cover $\mathcal{U}$.)

Proof Suppose that $\mathcal{U}$ is an open cover of K for which no such δ exists. Then for every $\varepsilon > 0$ there exists $x \in X$ such that $B(x, \varepsilon)$ is not contained in any element of $\mathcal{U}$.

Choose x_n such that $B(x_n, 1/n)$ is not contained in any element of $\mathcal{U}$.

Then x_n has a convergent subsequence, $x_{n_j} \to x$. Since $\mathcal{U}$ covers K, $x \in U$ for some element $U \in \mathcal{U}$. Since U is open, $B(x, \varepsilon) \subset U$ for some $\varepsilon > 0$.

But now take n sufficiently large that $d(x_n, x) < \varepsilon/2$ and $1/n < \varepsilon/2$. Then $B(x_n, 1/n) \subset B(x, \varepsilon) \subset U$, contradicting the definition of x_n. $\qquad\square$

Theorem C.14 *A subset K of a metric space (X, d) is sequentially compact if and only if it is compact.*

Proof Step 1. Compactness implies sequential compactness.

Let (x_j) be a sequence in a compact set K. Consider the sets F_n defined by setting

$$F_n = \overline{\{x_n, x_{n+1}, \dots\}}.$$

The sets F_n are a decreasing sequence of closed subsets of K, so we can find

$$x \in \bigcap_{j=1}^{\infty} F_j.$$

We now show that there is a subsequence that converges to x:

- since $x \in \overline{\{x_j : j \geq 1\}}$ there exists j_1 such that $d(x_{j_1}, x) < 1$;
- since $x \in \overline{\{x_j : j > j_1\}}$ there exists $j_2 > j_1$ such that $d(x_{j_2}, x) < 1/2$;
- continue in this way to find $j_k > j_{k-1}$ such that $d(x_{j_k}, x) < 1/k$.

Then x_{j_k} is a subsequence of (x_j) that converges to x.

Step 2. Sequential compactness implies compactness.

First we show that for every $\varepsilon > 0$ there is a cover of K by a finite number of sets of the form $B(x_j, \varepsilon)$ for some $x_j \in K$.

Suppose that this is not true, and that we have found $\{x_1, \dots, x_n\}$ such that $d(x_i, x_j) \geq \varepsilon$ for all $i, j = 1, \dots, n$. Since the collection $B(x_j, \varepsilon)$ does not cover K, there exists $x_{n+1} \in K$ such that $\{x_1, \dots, x_{n+1}\}$ all satisfy $d(x_i, x_j) \geq \varepsilon$ for all $i, j = 1, \dots, n + 1$. Therefore the sequence (x_j) has no Cauchy subsequence, so no convergent subsequence, which contradicts the sequential compactness of K.

Now, given any open cover $\mathcal{U}$ of K, consider the finite points $y_1, \dots, y_N$ such that $B(y_i, \delta)$ cover K, where δ is the δ from Lemma C.13 for the original cover. Then $B(y_i, \delta) \subset U_i$ for some $U_i \in \mathcal{U}$, and we have

$$K \subset \bigcup_{i=1}^{N} B(y_i, \delta) \subset \bigcup_{i=1}^{N} U_i,$$

so we have found a finite subcover. $\qquad\square$

A topology $\mathcal{T}$ on T is called *metrisable* if it coincides with the open sets for some metric d defined on T. Since compactness is defined only in terms of open sets, and the notion of convergence used for sequential compactness can be defined solely in terms of open sets, it follows that in any metrisable topological space compactness and sequential compactness of equivalent.

C.5 Tychonoff's Theorem

Suppose that $\{T_\alpha \ : \ \alpha \in \mathbb{A}\}$ is a collection of topological spaces. Then the product of the $\{T_\alpha\}$,

$$T = \prod_{\alpha \in \mathbb{A}} T_\alpha,$$

is the collection of all functions $x \colon \mathbb{A} \to \cup_{\alpha \in \mathbb{A}} T_\alpha$ such that $x(\alpha) \in T_\alpha$ for each $\alpha \in \mathbb{A}$. The product topology on T is the smallest topology such that for each $\alpha \in \mathbb{A}$ the map $\pi_\alpha \colon T \to T_\alpha$ defined by setting $\pi_\alpha(x) := x(\alpha)$ is continuous. This is the topology with sub-basis

$$\{x \in T : \ x(\alpha) \in \pi_\alpha^{-1}(U), \ U \in T_\alpha\}, \qquad \alpha \in \mathbb{A}.$$

The topology is therefore formed of unions of sets of the form

$$\{x \in T : \ x(\alpha_i) \in \pi_{\alpha_i}^{-1}(U_i), \ U_i \in T_{\alpha_i}\}, \qquad \alpha_1, \ldots, \alpha_n \in \mathbb{A}. \qquad \text{(C.2)}$$

(Note that this basis is *not* all sets of the form $x \in T$ with $x(\alpha) \in T_\alpha$ for each α; this defines the 'box topology' on T, and in general T is not compact in this topology.)

Tychonoff's Theorem guarantees that if all of the T_α are compact, then so is the product T (with the product topology). In order to prove the theorem we will need a preparatory lemma.

Lemma C.15 *Let $(T, \mathcal{T})$ be a topological space. If T is not compact, then it has a maximal open cover $\mathcal{V}$ that has no finite subcover. Moreover, if $(S, \mathcal{S})$ is any topological space, then for any continuous $\pi \colon (T, \mathcal{T}) \to (S, \mathcal{S})$*

(i) the collection

$$\mathcal{U} := \{U \in \mathcal{S} : \ \pi^{-1}(U) \in \mathcal{V}\}$$

contains no finite cover of S; and

(ii) if $U \subset S$ is open and not in $\mathcal{U}$, then $\pi^{-1}(S \setminus U)$ can be covered by finitely many sets from $\mathcal{V}$.

Proof We use Zorn's Lemma to guarantee the existence of such a $\mathcal{V}$.

If T is not compact, then it has an open cover $\mathcal{V}'$ with no finite subcover. Let E be the collection of all open covers of T that contain $\mathcal{V}'$ and have no finite subcover; define a partial order $\preceq$ on E so that $\mathcal{W} \preceq \mathcal{W}'$ if $\mathcal{W} \subseteq \mathcal{W}'$.

Given any chain $\mathscr{C} \subset E$, let

$$\mathcal{Z} := \bigcup_{\mathcal{C} \in \mathscr{C}} \mathcal{C}.$$

Then $\mathcal{Z} \supset \mathcal{C}$ for every $\mathcal{C} \in \mathscr{C}$, and so provided that $\mathcal{Z} \in E$ it is an upper bound for $\mathscr{C}$. Certainly $\mathcal{Z}$ is an open cover of T that contains $\mathcal{V}'$, and it has no finite subcover: if it did, then this subcover would consist of sets $\{U_i : i = 1, \ldots, n\}$ with $U_i \in \mathcal{C}_i$, and since $\mathscr{C}$ is a chain we would have $U_i \in \mathcal{C}$ for all i for some $\mathcal{C} \in \mathscr{C}$. This would yield some $\mathcal{C}$ with a finite subcover, which is not possible.

Since every chain has an upper bound, there exists a maximal element $\mathcal{V}$ in E.

(i) if $\mathcal{U}$ contained a finite cover $\{U_1, \ldots, U_n\}$ of S, then $\pi^{-1}(U_i)$ would give a finite open (as π is continuous) cover of T using sets from $\mathcal{V}$.

If (ii) does not hold, then we claim that $\mathcal{V}' := \pi^{-1}(U) \cup \mathcal{V}$ is an open cover of T that is strictly larger than $\mathcal{V}$ but that still does not have a finite subcover. It is clearly an open cover of T (since $\mathcal{V}$ is), $\pi^{-1}(U) \notin \mathcal{V}$ since $U \notin \mathcal{U}$, and if $\pi^{-1}(S \setminus U)$ cannot be covered by finitely many sets from $\mathcal{V}$ then

$$\pi^{-1}(U) \cup \pi^{-1}(S \setminus U) = T$$

cannot be covered by finitely many sets from $\pi^{-1}(U) \cup \mathcal{V}$, so $\mathcal{V}'$ has no finite subcover, contradicting the maximality of $\mathcal{V}$. $\qquad\square$

We can now prove Tychonoff's Theorem.

Theorem C.16 (Tychonoff's Theorem) *The topological product of compact spaces is compact.*

Proof Let $(T_\alpha)_{\alpha \in \mathbb{A}}$ be a collection of compact topological spaces, and let

$$T = \prod_{\alpha \in \mathbb{A}} T_\alpha$$

be their topological product. Suppose that T is not compact; then it has a maximal cover with no finite sub-cover, which we will call $\mathcal{V}$.

Let $\mathcal{U}_\alpha$ be the collection of open sets U in T_α such that $\pi_\alpha^{-1}(U) \in \mathcal{V}$. Since T_α is compact, if $\mathcal{U}_\alpha$ covers T_α it would have a finite subcover, but this cannot hold using part (i) of Lemma C.15; so $\mathcal{U}_\alpha$ does not cover T_α.

Therefore for each $\alpha \in \mathbb{A}$ we can choose some $x(\alpha) \in T_\alpha$ that is not covered by $\mathcal{U}_\alpha$. It follows, from part (ii) of Lemma C.15, that any open set U that contains $x(\alpha)$ is not an element of $\mathcal{U}_\alpha$, and so $\pi_\alpha^{-1}(T_\alpha \setminus U)$ can be covered by finitely many sets from $\mathcal{V}$.

The point $x = (x(\alpha))_{\alpha \in \mathbb{A}}$ belongs to some set $V \in \mathcal{V}$, since $\mathcal{V}$ is a cover of T. So we can find $\{\alpha_1, \ldots, \alpha_n\}$ and open sets $U_{\alpha_i} \subset T_{\alpha_i}$ with $x(\alpha_i) \in U_{\alpha_i}$ such that

$$\bigcap_{i=1}^{n} \pi^{-1}(U_{\alpha_i})$$

is contained in V (recall that open sets in $\mathcal{T}$ are unions of sets of this form; see (C.2)). But now, using part (ii) of Lemma C.15, since each U_{α_i} is not in $\mathcal{U}_{\alpha_i}$ (as $x(\alpha_i) \in U_{\alpha_i}$) it follows that

$$\pi^{-1}(T_{\alpha_i} \setminus U_{\alpha_i})$$

can be covered by a finite number of sets from $\mathcal{V}$. Thus

$$T = \bigcap_{i=1}^{n} \pi_{\alpha_i}^{-1}(U_{\alpha_i}) \cup \bigcup_{i=1}^{n} \pi_{\alpha_i}^{-1}(T_{\alpha_i} \setminus U_{\alpha_i})$$

is a cover of T by a finite collection of sets from $\mathcal{V}$, a contradiction. $\qquad\square$

C.6 The Banach–Alaoglu Theorem

We can now prove the Banach–Alaoglu Theorem. This is a 'topological' version of Theorem 27.11; unlike that theorem it does *not* say that bounded sequences in X^* always have weakly-∗ convergent subsequences, although we will be able to deduce this sequential result from the topological theorem when X is separable in the next section.

Theorem C.17 (Banach–Alaoglu Theorem) *If X is a Banach space, then the closed unit ball in X^* is compact in the weak-∗ topology.*

Proof Denote the closed unit ball in X^* by $\mathbb{B}_{X^*}$, and consider the topological space

$$Y := \prod_{x \in X} \{\lambda \in \mathbb{K} : |\lambda| \le \|x\|\}$$

equipped with the product topology. By Tychonoff's Theorem (Theorem C.16), Y is a compact topological space.

Elements of Y are maps $f : X \to \mathbb{K}$ such that $|f(x)| \le \|x\|$; the product topology is the smallest topology on Y that makes all the individual evaluation maps $\delta_x : f \mapsto f(x)$ continuous. So restricted to $\mathbb{B}_{X^*} \subset Y$ (the collection of all elements in Y that are also linear maps) the product topology coincides with the weak-∗ topology.

Since $\mathbb{B}_{X^*} \subset Y$, to show that $\mathbb{B}_{X^*}$ is compact, it suffices to show that $\mathbb{B}_{X^*}$ is closed (in the product topology). To do this, observe that $\mathbb{B}_{X^*}$ consists of all elements of Y that are linear, i.e.

$$\mathbb{B}_{X^*} = \{\phi \in Y : \phi(\lambda x + \mu y) = \lambda\phi(x) + \mu\phi(y), \ \lambda, \mu \in \mathbb{K}, \ x, y \in X\}$$

$$= Y \cap \bigcap_{x,y \in X, \ \lambda,\mu \in \mathbb{K}} (\delta_{\lambda x + \mu y} - \lambda\delta_x - \mu\delta_y)^{-1}\{0\}.$$

Since all evaluation maps are continuous, so is $\delta_{\lambda x + \mu y} - \lambda \delta_x - \mu \delta_y$, and so each set in the intersection is the preimage of the closed set $\{0\}$ under a continuous map, and so closed. Therefore $\mathbb{B}_{X^*}$ is the intersection of closed sets, and so closed itself. □

C.7 Metrisability of the Weak-∗ Topology in a Separable Space

We showed in Theorem C.14 that compactness and sequential compactness are equivalent in metrisable topologies. We can therefore obtain Theorem 27.11 from Theorem C.17 as a consequence of the following observation.

Lemma C.18 *If X is separable, then the weak-∗ topology on $\mathbb{B}_{X^*}$, the closed unit ball in X^*, is metrisable; it can be derived from the metric*

$$d(f, g) = \sum_{j=1}^{\infty} \frac{|f(x_j) - g(x_j)|}{2^j (1 + \|x_j\|)} \tag{C.3}$$

where $(x_j)_{j=1}^{\infty}$ is any countable dense subset of X.

Note that (i) open sets in $(\mathbb{B}_{X^*}, \mathcal{T}_*)$ are unions of sets of the form

$$\{\phi \in \mathbb{B}_{X^*} : \phi(y) \in B(z_j, \varepsilon_j) : y \in X, \ z_j \in \mathbb{K}, \ \varepsilon_j > 0, \ j = 1, \ldots, n\}$$

and (ii) that for each j we have (from (C.3))

$$|f(x_j) - g(x_j)| \le 2^j (1 + \|x_j\|) d(f, g).$$

Proof Suppose that $U \in \mathcal{T}_*$ and that $f \in U$. Then there exist $y_i \in X$, for $i = 1, \ldots, n$, and $\varepsilon > 0$ such that

$$f \in \{\phi \in \mathbb{B}_{X^*} : |\phi(y_i) - f(y_i)| < \varepsilon, \ i = 1, \ldots, n\} \subset U.$$

Now for each y_i find $x_{j(i)}$ such that

$$\|y_i - x_{j(i)}\| < \varepsilon/3,$$

and set

$$M = \max_{i=1,\ldots,n} 2^{j(i)} (1 + \|x_{j(i)}\|).$$

Then, if $d(\phi, f) < \varepsilon/3M$, we have

$$|\phi(y_i) - f(y_i)|$$
$$= |\phi(y_i) - \phi(x_{j(i)}) + \phi(x_{j(i)}) - f(x_{j(i)}) + f(x_{j(i)}) - f(y_i)|$$

$$\leq |\phi(y_i) - \phi(x_{j(i)})| + |\phi(x_{j(i)}) - f(x_{j(i)})| + |f(x_{j(i)}) - f(y_i)|$$

$$\leq \|y_i - x_{i(j)}\| + 2^{j(i)}(1 + \|x_{j(i)}\|)\, d(\phi, f) + \|x_{j(i)} - y_i\|$$

$$\leq \frac{\varepsilon}{3} + \frac{\varepsilon}{3} + \frac{\varepsilon}{3} = \varepsilon.$$

So U is open in (X^*, d).

Conversely, consider the open ball in $(\mathbb{B}_{X^*}, d)$ given by

$$U := \{\phi \in B : d(\phi, f) < \varepsilon\}.$$

Given any $g \in U$ set $\delta = d(g, f)$ and $r = \varepsilon - \delta$. Note that for any $\phi \in B$ we have

$$\frac{|\phi(x_j) - g(x_j)|}{2^j(1 + \|x_j\|)} \leq \frac{2\|x_j\|}{2^j(1 + \|x_j\|)} \leq 2^{-(j-1)},$$

so that, taking N such that $2^{-(N-1)} < r/2$ we obtain

$$\sum_{j=N+1}^{\infty} \frac{|\phi(x_j) - g(x_j)|}{2^j(1 + \|x_j\|)} \leq \sum_{j=N+1}^{\infty} 2^{-(j-1)} = 2^{-(N-1)} < r/2.$$

It follows that if

$$\psi \in \bigcap_{j=1}^{N} \{\phi \in \mathbb{B}_{X^*} : |\phi(x_i) - g(x_i)| < r/2\},$$

then

$$d(\psi, g) = \sum_{j=1}^{N} \frac{|\psi(x_j) - g(x_j)|}{2^j(1 + \|x_j\|)} + \sum_{j=N+1}^{\infty} \frac{|\psi(x_j) - g(x_j)|}{2^j(1 + \|x_j\|)}$$

$$\leq \frac{r}{2} \sum_{j=1}^{N} 2^{-j} + \frac{r}{2} < r,$$

and so

$$d(\psi, f) \leq d(\psi, g) + d(g, f) \leq r + \delta = \varepsilon,$$

i.e. $\psi \in U$, so $U \in \mathcal{T}_*$. $\qquad\qquad\qquad\qquad\qquad\qquad\square$

Solutions to Exercises

Chapter 1

1.1 Dividing by b^p and setting $x = a/b$ it suffices to show that

$$(1 + x)^p \le 2^{p-1}(1 + x^p).$$

The maximum of $f(x) := (1 + x)^p/(1 + x^p)$ occurs when

$$f'(x) = \frac{p(1 + x)^{p-1}(1 + x^p) - p(1 + x)^p x^{p-1}}{(1 + x^p)^2} = 0,$$

i.e. when $(1 + x^p) - (1 + x)x^{p-1} = 0$ which happens at $x = 1$. Thus $f(x) \le 2^{p-1}$ which yields the required inequality.

1.2 The only thing that is not immediate is that $f + g \in \tilde{L}^p(0, 1)$ whenever we take $f, g \in \tilde{L}^p(0, 1)$. First we have $f + g \in C(0, 1)$ as the sum of two continuous functions is continuous, and then we use the result of the previous exercise to guarantee that

$$\int_0^1 |f(x) + g(x)|^p \le 2^{p-1} \int_0^1 |f(x)|^p + |g(x)|^p < \infty.$$

1.3 Suppose that $v = \sum_{j=1}^n \alpha_j e_j = \sum_{j=1}^n \beta_j e_j$; then $\sum_{j=1}^n (\alpha_j - \beta_j)e_j = 0$, and since the $\{e_j\}$ are linearly independent it follows that $\alpha_j - \beta_j = 0$ for each j, i.e. $\alpha_j = \beta_j$ for each j.

1.4 Let $E := \{e_j\}_{j=1}^n$ be a basis for V, and $F := \{f_i\}_{i=1}^m$ another basis for V. Suppose that $m > n$; otherwise we can reverse the roles of the two bases.

Since $\mathrm{Span}(F) = V$, we can write

$$e_1 = \sum_{i=1}^m \alpha_i f_i; \tag{S.1}$$

since $e_1 \ne 0$, there is a $k \in \{1, \ldots, m\}$ such that $\alpha_k \ne 0$, and so we have

$$f_k = \alpha_k^{-1} e_1 - \sum_{i \ne k} \alpha_k^{-1} \alpha_i f_i. \tag{S.2}$$

We can therefore replace f_k by e_1 to get a new basis F' for V: F' still spans V because of (S.2), and is still linearly independent, since if

$$\beta_k e_1 + \sum_{i \neq k} \beta_i f_i = 0$$

then, using (S.1),

$$\sum_{i=1}^{m} \beta_k \alpha_i f_i + \sum_{i \neq k} \beta_i f_i = 0,$$

which yields

$$\beta_k \alpha_k f_k + \sum_{i \neq k} (\beta_k \alpha_i + \beta_i) f_i = 0,$$

so $\beta_k = 0$, and then $\beta_i = 0$ for all $i \neq k$.

We can repeat this procedure in turn for each element of the basis E, ending up with a new basis $F^{(n)}$ that still contains m elements but into which we have swapped all of the n elements of E. Since E spans V, $F^{(n)}$ can no longer be linearly independent, contradicting our assumption that F contains more than n elements initially.

1.5 If $x, y \in \mathrm{Ker}(T)$ and $\alpha \in \mathbb{K}$, then

$$T(x+y) = Tx + Ty = 0 + 0 = 0 \qquad \text{and} \qquad T(\alpha x) = \alpha Tx = 0,$$

so $\mathrm{Ker}(T)$ is a vector space. If $\xi, \eta \in \mathrm{Range}(T)$, then $\xi = Tx$ and $\eta = Ty$ for some $x, y \in X$; then $T(x+y) = \xi + \eta$ and $T(\alpha x) = \alpha Tx = \alpha \xi$, and so $\mathrm{Range}(T)$ is also a vector space.

1.6 Since U is a vector space, we have

$$[x] + [y] = x + U + y + U = x + y + U = [x+y]$$

and $\lambda[x] = \lambda x + \lambda U = \lambda x + U = [\lambda x]$. These two properties show immediately that the map $x \mapsto [x]$ is linear.

1.7 This is clearly true if $n = 1$. Suppose true for $n = m$, i.e. $c_j \preceq c_i \preceq c_k$ for $i = 1, \ldots, m$. Then since $\mathcal{C}$ is a chain we have $c_{m+1} \preceq c_j$ or $c_j \preceq c_{m+1}$; in the former case set $j' = m+1$, in the latter set $j' = j$. Similarly we have $c_k \preceq c_{m+1}$ or $c_{m+1} \preceq c_k$; in the former case set $k' = m+1$ and in the latter set $k' = k$. Then we obtain $c_{j'} \preceq c_i \preceq c_{k'}$ for $i = 1, \ldots, m+1$.

1.8 Let P be the collection of all linearly independent subsets of V that contain Z. We define a partial order on P as in the proof of Theorem 1.28. Arguing just as we did there, any chain in P has an upper bound, so P has a maximal element. This is a maximal linearly independent set, so is a Hamel basis for V that contains Z.

Chapter 2

2.1 If $\eta = 0$ the inequality is immediate, so take $\eta \neq 0$. Since $\sum_{j=1}^{n} (\xi_j - \lambda \eta_j)^2 \geq 0$, it follows that

$$0 \le \sum_{j=1}^{n} \xi_j^2 - 2\lambda \xi_j \eta_j + \lambda^2 \eta_j^2 = \sum_{j=1}^{n} \xi_j^2 - 2\lambda \sum_{j=1}^{n} \xi_j \eta_j + \lambda^2 \sum_{j=1}^{n} \eta_j^2.$$

Now choose $\lambda = \left(\sum_j \xi_j \eta_j \right) / \sum_j \eta_j^2$ and rearrange.

2.2 Once we prove that this expression provides a norm on $X_1 \times X_2$ it is enough to proceed by induction. The only thing that is not trivial to check is the triangle inequality in the case $p \ne \infty$, and for this we use (2.10): taking any three points $(x_1, x_2), (y_1, y_2), (z_1, z_2) \in X_1 \times X_2$ we have

$$\begin{aligned}
\varrho_p((x_1, x_2), (y_1, y_2)) &= \left(d_1(x_1, y_1)^p + d_2(x_2, y_2)^p \right)^{1/p} \\
&\le \left([d_1(x_1, z_1) + d_1(z_1, y_1)]^p + [d_2(x_2, z_2) + d_2(z_2, y_2)]^p \right)^{1/p} \\
&\le [d_1(x_1, z_1)^p + d_2(x_2, z_2)^p]^{1/p} + [d_1(z_1, y_1)^p + d_2(z_2, y_2)^p]^{1/p} \\
&= \varrho_p((x_1, x_2), (z_1, z_2)) + \varrho_p((z_1, z_2), (y_1, y_2)).
\end{aligned}$$

To go from the first to second line, we use the triangle inequality in X_1 and in X_2; to go from the second to third line, we use (2.10); to go from the third to the fourth line, we use the definition of ϱ_p.

2.3 Clearly $\hat{d}(x, y) \ge 0$, $\hat{d}(x, y) = 0$ if and only if $d(x, y) = 0$ if and only if $x = y$, and $\hat{d}(x, y) = \hat{d}(y, x)$. It remains only to check the triangle inequality, and this follows from the fact that the map $t \mapsto t/(1 + t)$ is monotonically increasing: since d satisfies the triangle inequality, $d(x, y) \le d(x, z) + d(z, y)$, we have

$$\begin{aligned}
\hat{d}(x, y) = \frac{d(x, y)}{1 + d(x, y)} &\le \frac{d(x, z) + d(z, y)}{1 + d(x, z) + d(z, y)} \\
&\le \frac{d(x, z)}{1 + d(x, z)} + \frac{d(z, y)}{1 + d(z, y)} = \hat{d}(x, z) + \hat{d}(z, y).
\end{aligned}$$

(To see that $t \mapsto t/(1 + t)$ is increasing, note that the function $f(t) = t/(1 + t)$ has derivative $f'(t) = 1/(1 + t)^2$.)

2.4 Clearly $d(x, y) \ge 0$, and if $d(x, y) = 0$, then $|x_j - y_j| = 0$ for every j and so $x = y$. That $d(x, y) = d(y, x)$ is clear. To check the triangle inequality, note that

$$\begin{aligned}
d(x, y) &= \sum_{j=1}^{\infty} 2^{-j} \frac{|x_j - y_j|}{1 + |x_j - y_j|} \\
&\le \sum_{j=1}^{\infty} 2^{-j} \frac{|x_j - z_j| + |z_j - y_j|}{1 + |x_j - z_j| + |z_j - y_j|} \\
&\le \sum_{j=1}^{\infty} 2^{-j} \left[\frac{|x_j - z_j|}{1 + |x_j - z_j|} + \frac{|z_j - y_j|}{1 + |z_j - y_j|} \right] \\
&= d(x, z) + d(z, y),
\end{aligned}$$

using the facts that $|x_j - y_j| \le |x_j - z_j| + |z_j - y_j|$ and the map $t \mapsto t/(1 + t)$ is monotonically increasing in t.

If $d(x^{(n)}, y) \to 0$, then

$$\sum_{j=1}^{\infty} 2^{-j} \frac{|x_j^{(n)} - y_j|}{1 + |x_j^{(n)} - y_j|} \to 0;$$

in particular, each term converges to zero, so we must have $x_j^{(n)} \to y_j$ as $n \to \infty$ for each j. Conversely, given any $\varepsilon > 0$, choose M such that $\sum_{j=M+1}^{\infty} 2^{-j} < \varepsilon/2$, and then choose N such that

$$|x_j^{(n)} - y_j| < \varepsilon/2 \qquad j = 1, \ldots, M$$

for all $n \geq N$. Then for $n \geq N$ we have

$$d(x^{(n)}, y) = \sum_{j=1}^{\infty} 2^{-j} \frac{|x_j^{(n)} - y_j|}{1 + |x_j^{(n)} - y_j|}$$

$$\leq \sum_{j=1}^{M} 2^{-j} \frac{\varepsilon}{2} + \sum_{j=M+1}^{\infty} 2^{-j} \leq \frac{\varepsilon}{2} + \frac{\varepsilon}{2} = \varepsilon,$$

and so $x^{(n)} \to y$.

2.5 If $\{F_j\}_{j=1}^n$ are closed, then $X \setminus F_j$ are open, and

$$X \setminus \bigcup_{j=1}^{n} F_j = \bigcap_{j=1}^{n} X \setminus F_j$$

is open since the finite intersection of open sets is open: this shows that $\cap_{j=1}^n F_j$ is closed. Similarly, if $\{F_\alpha\}_{\alpha \in \mathbb{A}}$ is any family of closed sets, then $\{X \setminus F_\alpha\}_{\alpha \in \mathbb{A}}$ is a family of open sets, so

$$X \setminus \bigcap_{\alpha \in \mathbb{A}} F_\alpha = \bigcup_{\alpha \in \mathbb{A}} X \setminus F_\alpha$$

is open since any union of open sets is open: this shows that $\cap_\alpha F_\alpha$ is closed.

2.6 Take any $y \in B(x, r)$ and let $\varepsilon = r - d(y, x) > 0$. If $z \in B(y, \varepsilon)$, then

$$d(z, x) \leq d(z, y) + d(y, x) < \varepsilon + d(y, x) = r,$$

and so $B(y, \varepsilon) \subseteq B(x, r)$, which shows that $B(x, r)$ is open.

2.7 A set U is open in a metric space (X, d) if for every $x \in U$ there exists $\varepsilon_x > 0$ such that $B(x, \varepsilon_x) \subseteq U$. So we can write

$$U = \bigcup_{x \in U} B(x, \varepsilon_x)$$

(every $x \in U$ is contained in the right-hand side, and all sets in the union on the right-hand side are subsets of U).

2.8 If $f \colon (X, d_X) \to (Y, d_Y)$ is continuous and $x_n \to x$ in X, then since f is continuous at x there exists $\delta > 0$ such that $d_X(x_n, x) < \delta$ implies that $d_Y(f(x_n), f(x)) < \varepsilon$.

Now, since $x_n \to x$ we can choose N such that $d_X(x_n, x) < \delta$ for all $n \geq N$; this implies that $d_Y(f(x_n), f(x)) < \varepsilon$ for all $n \geq N$, so $f(x_n) \to f(x)$ in Y as claimed.

If f is not continuous at x, then there exists $\varepsilon > 0$ such that for every $\delta > 0$ there exists $y \in X$ with $d_X(y, x) < \delta$ by $d_Y(f(y), f(x)) \geq \varepsilon$. For each n we set $\delta = 1/n$ and thereby obtain a sequence $(x_n) \in X$ such that $d_X(x_n, x) < 1/n$ and $d_Y(f(x_n), f(x)) \geq \varepsilon$: so we have found a sequence (x_n) such that $x_n \to x$ but $f(x_n) \not\to f(x)$.

2.9 If A is a countable dense subset of (X, d_X) and B is a countable dense subset of (Y, d_Y), then $A \times B$ is a countable subset of $X \times Y$. To show that it is dense if we use the metric ϱ_p, taking $1 \leq p < \infty$ given $(x, y) \in X \times Y$ and $\varepsilon > 0$ choose $(a, b) \in A \times B$ such that

$$d(a, x) < 2^{-1/p}\varepsilon \qquad \text{and} \qquad d(b, y) < 2^{-1/p}\varepsilon,$$

and then

$$\varrho_p((a, b), (x, y))^p = d(a, x)^p + d(b, y)^p < \varepsilon^p.$$

For $p = \infty$ we find $(a, b) \in A \times B$ with $d(a, x) < \varepsilon$ and $d(b, y) < \varepsilon$.

2.10 Suppose that $\bigcap_{\alpha \in \mathbb{A}} F_\alpha$ is empty. Then

$$X = X \setminus \bigcap_{\alpha \in \mathbb{A}} F_\alpha = \bigcup_{\alpha \in \mathbb{A}} X \setminus F_\alpha.$$

So $\{X \setminus F_\alpha\}_{\alpha \in \mathbb{A}}$ is an open cover of X. Since X is compact, there is a finite subcover,

$$X = \bigcup_{j=1}^{n} X \setminus F_{\alpha_j} = X \setminus \bigcap_{j=1}^{n} F_{\alpha_j};$$

this implies that $\bigcap_{j=1}^{n} F_{\alpha_j} = \varnothing$, but this contradicts the assumption that such an intersection is always non-empty. So $\bigcap_\alpha F_\alpha$ must be non-empty.

2.11 The sets $\{F_j\}$ have the finite intersection property from the previous exercise, since for any finite collection $\bigcap_{j=1}^{k} F_{n_j} \supset F_k$, where $k = \max_j n_j$. It follows that $\bigcap_j F_j \neq \varnothing$.

2.12 If $\alpha = \sup(S)$, then for any $\varepsilon > 0$ there exists $x \in S$ such that $x > \alpha - \varepsilon$. In particular, there exist $x_n \in S$ such that $\alpha - 1/n < x_n \leq \alpha$, and so $x_n \to \alpha$. Since S is closed, $\alpha \in S$.

2.13 Take any closed subset A of X. Then, since A is a closed subset of a compact set, it is compact, and then since f is continuous, $f(A)$ is a compact subset of Y. It follows that $f(A)$ is a closed subset of Y. Now, since f is a bijection $f(A) = (f^{-1})^{-1}(A)$; since $(f^{-1})^{-1}(A)$ is therefore a closed subset of Y it follows using Lemma 2.13 that f^{-1} is continuous.

An alternative, less topological proof, proceeds via contradiction. Suppose that f^{-1} is not continuous: then there exist $\varepsilon > 0$, $y \in Y$, and a sequence $(y_n) \in Y$ such that $d_Y(y_n, y) \to 0$ but

$$d_X(f^{-1}(y_n), f^{-1}(y)) \geq \varepsilon. \tag{S.3}$$

Since X is compact, there is a subsequence of $f^{-1}(y_n)$ such that $f^{-1}(y_{n_j}) \to x$. Since f is continuous, it follows that $y_{n_j} \to f(x)$, so $y = f(x)$. Since f is injective, it follows that $f^{-1}(y) = x$, but then we have $f^{-1}(y_{n_j}) \to f^{-1}(y)$ contradicting (S.3).

2.14 For each n cover X by open balls of radius $1/n$. Since X is compact, this cover has a finite subcover $\{B(y_i^{(n)}, 1/n)\}_{i=1}^{N_n}$. Take the centres of the balls in each of these finite subcovers, adding them successively (for each n) to our sequence (x_j). (More explicitly, if we set $s_0 = 0$ and $s_n = \sum_{j=1}^{n} N_j$, then we let $x_{s_n+i} = y_i^{(n)}$ for all $1 \le i \le N_n$.) Now for any $x \in X$ and $\varepsilon > 0$ there exists an n such that $1/n < \varepsilon$; if we set $M(\varepsilon) = s_n$, then there is an x_j with $1 \le j \le M(\varepsilon)$ such that $d(x_j, x) < 1/n < \varepsilon$.

Chapter 3

3.1 We have

$$d(x + z, y + z) = \|(x + z) - (y + z)\| = \|x - y\| = d(x, y)$$

and

$$d(\alpha x, \alpha y) = \|\alpha x - \alpha y\| = |\alpha| \|x - y\| = |\alpha| d(x, y).$$

The metric from Exercise 2.4 does not satisfy the second of these two properties.

3.2 If $x_1, x_2 \in A + B$, then $x_i = a_i + b_i$, with $a_i \in A$ and $b_i \in B$, and for any $\lambda \in (0, 1)$

$$\lambda x_1 + (1 - \lambda)x_2 = \lambda(a_1 + b_1) + (1 - \lambda)(a_2 + b_2)$$
$$= (\lambda a_1 + (1 - \lambda)a_2) + (\lambda b_1 + (1 - \lambda)b_2) \in A + B.$$

3.3 A simple inductive argument shows that if $a, b \in C$, then $\lambda a + (1 - \lambda b) \in C$ for any λ of the form $a2^{-k}$ with $a = 0, \ldots, 2^k$. Given any $\lambda \in (0, 1)$ we can find n_k with $0 \le n_k \le 2^k$ such that $\frac{n_k}{2^k} \to \lambda$; since C is closed it follows that

$$\lambda a + (1 - \lambda b) = \lim_{k \to \infty} \left[\frac{n_k}{2^k} a + \left(1 - \frac{n_k}{2^k}\right) b \right] \in C.$$

3.4 If $f'' \ge 0$ on (a, b), then f' is increasing on $[a, b]$. Take $a < c < b$, where $c = (1 - t)a + tb$. Then, by the Mean Value Theorem, there exists $\xi \in (a, c)$ and $\eta \in (c, b)$ such that

$$\frac{f(c) - f(a)}{c - a} = f'(\xi) \quad \text{and} \quad \frac{f(b) - f(c)}{b - c} = f'(\eta).$$

Since f' is increasing,

$$\frac{f(c) - f(a)}{c - a} \le \frac{f(b) - f(c)}{b - c}.$$

Since $c = (1 - t)a + tb$, we have $b - c = (1 - t)(b - a)$ and $c - a = t(b - a)$; it follows that

$$(1 - t)(f(c) - f(a)) \le t(f(b) - f(c)),$$

and so f is convex.

Convexity of $f(x) = e^x$ is now immediate, since $f''(x) = e^x > 0$.

For convexity of $f(t) = |t|^q$ we first consider $t \geq 0$, where we have $f(t) = t^q$. This is twice differentiable, with $f''(t) = q(q-1)t^{q-2} \geq 0$ on $[0, \infty)$, so f is convex on $[0, \infty)$. A similar argument shows that f is convex on $(-\infty, 0]$. So it only remains to check the required convexity inequality if $x < 0$ and $y > 0$, and then, using convexity on $[0, \infty)$,

$$|tx + (1-t)y|^q \leq |t|x| + (q-t)|y||^q \leq t|x|^q + (1-t)|y|^q.$$

3.5 The inequality is almost immediate, since $m = |x_i|$ for some i, and $|x_j| \leq m$ for every $j = 1, \ldots, n$. Taking the pth root yields

$$m \leq \left(\sum_{j=1}^{n} |x_j|^p \right)^{1/p} \leq n^{1/p} m,$$

and $n^{1/p} \to 1$ as $p \to \infty$. So $\|x\|_{\ell^p} \to m = \|x\|_{\ell^\infty}$ as $p \to \infty$.

3.6 If $x \in \ell^1$, then $x \in \ell^p$ for every $p \in [1, \infty]$ so $\|x\|_{\ell^p} < \infty$ for every $p \in [1, \infty]$.

Fix $\varepsilon > 0$. Since $x \in \ell^1$, there exists N such that

$$\sum_{j=n+1}^{\infty} |x_j| < \varepsilon \qquad \text{for every } n \geq N.$$

In particular,

$$\|x - (x_1, \ldots, x_n, 0, 0, 0, \ldots)\|_{\ell^p} \leq \|x - (x_1, \ldots, x_n, 0, 0, 0, \ldots)\|_{\ell^1} < \varepsilon$$

for every $n \geq N$. It follows that for every $p \in [1, \infty)$ we have

$$\|x\|_{\ell^p} - \varepsilon \leq \|(x_1, \ldots, x_n, 0, 0, 0, \ldots)\|_{\ell^p} = \|(x_1, \ldots, x_n)\|_{\ell^p} \leq \|x\|_{\ell^p}. \quad \text{(S.4)}$$

In ℓ^∞ the inequality is immediate from the definition of the supremum: there exists an N' such that for $n \geq N'$

$$\|x\|_{\ell^\infty} - \varepsilon \leq \max_{j=1,\ldots,n} |x_j| = \|(x_1, \ldots, x_n)\|_{\ell^\infty} \leq \|x\|_{\ell^\infty}. \quad \text{(S.5)}$$

Now fix some $n \geq \max(N, N')$, and using Exercise 3.5 choose p_0 such that

$$\|(x_1, \ldots, x_n)\|_{\ell^\infty} \leq \|(x_1, \ldots, x_n)\|_{\ell^p} \leq \|(x_1, \ldots, x_n)\|_{\ell^\infty} + \varepsilon \quad \text{(S.6)}$$

for all $p \geq p_0$.

Combining (S.4), (S.5), and (S.6) we obtain

$$\|x\|_{\ell^\infty} - \varepsilon \leq \|(x_1, \ldots, x_n)\|_{\ell^p} \leq \|x\|_{\ell^p}$$

and

$$\|x\|_{\ell^p} - \varepsilon \leq \|(x_1, \ldots, x_n)\|_{\ell^p} \leq \|(x_1, \ldots, x_n)\|_{\ell^\infty} + \varepsilon \leq \|x\|_{\ell^\infty} + \varepsilon,$$

and so

$$\|x\|_{\ell^\infty} - \varepsilon \leq \|x\|_{\ell^p} \leq \|x\|_{\ell^\infty} + 2\varepsilon$$

for all $p \geq p_0$, i.e. $\|x\|_{\ell^p} \to \|x\|_{\ell^\infty}$ as $p \to \infty$.

3.7 Set $x_n = [1/n(\log n)^2]^{1/p}$. Then

$$\sum_{n=1}^{\infty} |x_n|^p = \sum_{n=1}^{\infty} \frac{1}{n(\log n)^2} < \infty$$

so $\boldsymbol{x} \in \ell^p$ but

$$\sum_{n=1}^{\infty} |x_n|^q = \sum_{n=1}^{\infty} \frac{1}{(n(\log n)^2)^{q/p}} = \infty$$

so $\boldsymbol{y} \notin \ell^q$ for all $q < p$.

3.8 Take $x \in X$. Since U is a subspace, $0 \in U$; since U is open, $B_X(0, \varepsilon) \subset U$ for some $\varepsilon > 0$. We have already observed that $0 \in U$, while if $x \neq 0$ we have $\varepsilon x / 2\|x\| \in B_X(0\varepsilon) \subset U$, and then, since U is a subspace,

$$x = \frac{2\|x\|}{\varepsilon} \frac{\varepsilon x}{2\|x\|} \in U.$$

3.9 Suppose that $x, y \in \overline{U}$ and $\alpha \in \mathbb{K}$. Then there exist $(x_n) \in U$ and $(y_n) \in U$ such that $x_n \to x$ and $y_n \to y$. It follows that $x_n + y_n \to x + y$, so $x + y \in \overline{U}$, and $\alpha x_n \to \alpha x$, so $\alpha x \in \overline{U}$. This shows that $\overline{U}$ is also a linear subspace, and it is closed since $\overline{A}$ is closed for any set A.

3.10 If $\boldsymbol{x} = (x_1, \ldots, x_n) \in \mathbb{R}^n$, then

$$\max_{j=1,\ldots,n} |x_j| \le \sum_{j=1}^{n} |x_j| \le n \max_{j=1,\ldots,n} |x_j|,$$

i.e. $\|\boldsymbol{x}\|_{\ell^\infty} \le \|\boldsymbol{x}\|_{\ell^1} \le n\|\boldsymbol{x}\|_{\ell^\infty}$; and

$$\max_{j=1,\ldots,n} |x_j| \le \left(\sum_{j=1}^{n} |x_j|^2\right)^{1/2} \le [n(\max_{j=1,\ldots,n} |x_j|)^2]^{1/2} = n^{1/2} \max_{j=1,\ldots,n} |x_j|,$$

i.e. $\|\boldsymbol{x}\|_{\ell^\infty} \le \|\boldsymbol{x}\|_{\ell^2} \le n^{1/2}\|\boldsymbol{x}\|_{\ell^\infty}$. Combining these two 'equivalences' yields $n^{-1/2}\|\boldsymbol{x}\|_{\ell^2} \le \|\boldsymbol{x}\|_{\ell^1} \le n\|\boldsymbol{x}\|_{\ell^2}$.

3.11 If $\|f_n - f\|_\infty \to 0$, then

$$\|f_n - f\|_{L^p}^p = \int_0^1 |f_n(x) - f(x)|^p \, \mathrm{d}x \le \|f_n - f\|_\infty^p \to 0,$$

so $\|f_n - f\|_{L^p} \to 0$; pointwise convergence follows almost immediately, since for each $x \in [0, 1]$

$$|f_n(x) - f(x)| \le \|f_n - f\|_\infty.$$

3.12 The linear span of the elements $\{e^{(j)}\}_{j=1}^{\infty}$ is dense in $c_0(\mathbb{K})$. Given any $\varepsilon > 0$ there exists N such that $|x_n| < \varepsilon$ for all $n \ge N$, and then

$$\left\| \boldsymbol{x} - \sum_{j=1}^{n} x_j e^{(j)} \right\|_{\ell^\infty} = \sup_{j>n} |x_j| < \varepsilon.$$

This shows that $c_0(\mathbb{K})$ is separable using Lemma 3.23 (iii).

3.13 Let A be a countable dense subset of $(X, \|\cdot\|_X)$, and let $\phi\colon X \to Y$ be an isomorphism. Then $\phi(A) = \{\phi(a) : a \in A\}$ is a countable subset of Y that is dense in Y: given any $y \in Y$ and $\varepsilon > 0$, we know that $y = \phi(x)$ for some $x \in X$; since A is dense in X we can find $a \in A$ such that $\|x - a\|_X < \varepsilon$, and then, since ϕ is an isomorphism, $\|\phi(x) - \phi(a)\|_Y \le c_2\|x - a\|_X < c_2\varepsilon$.

3.14 Since $A \subset B$, we have $\mathrm{clin}(A) \subseteq \mathrm{clin}(B)$. Given any $x \in \mathrm{clin}(B)$, we have

$$\left\| x - \sum_{j=1}^n \alpha_j b_j \right\| < \varepsilon/2 \qquad \text{for some } n \in \mathbb{N}, \ \alpha_j \in \mathbb{K}, \ b_j \in B.$$

Since A is dense in B, for each b_j there exists $a_j \in A$ with $\|a_j - b_j\| < \varepsilon/2n|\alpha_j|$, and then

$$\left\| x - \sum_{j=1}^n \alpha_j a_j \right\| \le \left\| x - \sum_{j=1}^n \alpha_j b_j \right\| + \sum_{j=1}^n |\alpha_j|\|b_j - a_j\| < \varepsilon,$$

which shows that $x \in \mathrm{clin}(A)$ and therefore $\mathrm{clin}(B) \subseteq \mathrm{clin}(A)$. So if A is dense in B then $\mathrm{clin}(A) = \mathrm{clin}(B)$. Now B is dense in X if $\mathrm{clin}(B) = X$, and so if A is dense in B, then $\mathrm{clin}(A) = \mathrm{clin}(B) = X$.

3.15 We use Cantor's diagonal argument. Suppose that $\mathfrak{b}$ is countable; then there is a bijection $\phi\colon \mathbb{N} \to \mathfrak{b}$. Consider the sequence $(x^{(j)}) \in \ell^\infty$ with $x^{(j)} := \phi(j)$; every element of $\mathfrak{b}$ must occur as some $x^{(j)}$ since ϕ is a bijection. However, if we set

$$x_j = \begin{cases} 1 & x_j^{(j)} = 0 \\ 0 & x_j^{(j)} = 1, \end{cases}$$

then this defines an element $x \in \mathfrak{b}$, but $x \ne x^{(j)}$ for any j, since $x_j \ne x_j^{(j)}$. So $\mathfrak{b}$ must be uncountable.

3.16 Let $A = \{a_j\}_{j=1}^\infty$ be a countable set whose linear span is dense in X and let $X_n = \mathrm{Span}(a_1, \ldots, a_n)$. Then X_n has dimension at most n, and if $x \in X$ we can find elements y_n in the linear span of A such that $y_n \to x$ as $n \to \infty$. Since each element of the linear span of A is contained in one of the spaces X_n, it follows that $\overline{\cup X_n} = X$.

On the other hand, if $X = \overline{\cup X_n}$, with each X_n finite-dimensional, then for each n find a basis E_n for X_n, which will have a finite number of elements. Let $A = \cup_n E_n$; then A is a countable set whose linear span is dense in X.

3.17 Suppose that X is the closed linear span of a compact set K. Since K is compact, it has a countable dense subset A (see Exercise 2.14). Using Exercise 3.14 we have $\mathrm{clin}(A) = \mathrm{clin}(K) = X$, so X is separable.

If X is separable, then $X = \mathrm{clin}(\{x_j\}_{j=1}^\infty)$ for some countable set $\{x_j\}_{j=1}^\infty$. But this is the same as the closed linear span of $(x_n/n\|x_n\|)_{n=1}^\infty \cup \{0\}$, which is compact.

Chapter 4

4.1 (i) Taking $\varepsilon = 1$ in the definition of a Cauchy sequence there exists N such that $\|x_n - x_m\| < 1$ for every $n, m \geq N$. In particular, $\|x_n - x_N\| < 1$ for every $n \geq N$. It follows that for every n we have

$$\|x_n\| < \max\left(\|x_1\|, \ldots, \|x_{N-1}\|, \|x_N\| + 1\right)$$

and so (x_n) is bounded.

(ii) Given $\varepsilon > 0$ choose N such that

$$\|x_i - x_k\| < \varepsilon/2 \qquad i, k \geq N;$$

now choose J such that $n_J \geq N$ and

$$\|x_{n_j} - x\| < \varepsilon/2 \qquad j \geq J.$$

Then, for all $i \geq n_J$, we have

$$\|x_i - x\| \leq \|x_i - x_{n_J}\| + \|x_{n_J} - x\| < \varepsilon$$

and so $x_i \to x$ as claimed.

4.2 Suppose that $(x^{(k)})$ with $x^{(k)} = (x_1^{(k)}, x_2^{(k)}, \cdots)$ is a Cauchy sequence in $\ell^\infty(\mathbb{K})$. Then for every $\varepsilon > 0$ there exists an N_ε such that

$$\|x^{(n)} - x^{(m)}\|_{\ell^\infty} = \sup_j |x_j^{(n)} - x_j^{(m)}| < \varepsilon \qquad \text{for all} \qquad n, m \geq N_\varepsilon.$$

In particular, for every $j \in \mathbb{N}$ we have

$$|x_j^{(n)} - x_j^{(m)}| < \varepsilon \qquad \text{for all} \qquad n, m \geq N_\varepsilon \qquad \text{(S.7)}$$

and so $(x_j^{(n)})_{n=1}^\infty$ is a Cauchy sequence in $\mathbb{K}$ for every fixed j. Since $\mathbb{K}$ is complete, it follows that for each $j \in \mathbb{N}$

$$x_j^{(n)} \to a_j \qquad \text{as} \qquad n \to \infty$$

for some $a_j \in \mathbb{K}$. Set $a = (a_1, a_2, \ldots)$.

We need to show that $a \in \ell^\infty$ and that $\|x^{(n)} - a\|_{\ell^\infty} \to 0$ as $n \to \infty$. Letting $m \to \infty$ in (S.7) we obtain $|x_j^{(n)} - a_j| \leq \varepsilon$ for all $n \geq N_\varepsilon$, since $x_j^{(m)} \to a_j$ as $m \to \infty$. Since this holds for every j, it follows that

$$\|x^{(n)} - a\|_{\ell^\infty} = \sup_j |x_j^{(n)} - a_j| \leq \varepsilon \qquad \text{for all} \qquad n \geq N_\varepsilon.$$

This shows that $a \in \ell^\infty$ and $x^{(n)} \to a$ in ℓ^∞ as $n \to \infty$.

4.3 Suppose that $\mathbb{K}^{j-1}$ is complete. Since $\mathbb{K}^j = \mathbb{K}^{j-1} \times \mathbb{K}$ with the norm

$$\|(x, x')\|_{\mathbb{K}^{j-1} \times \mathbb{K}}^2 = \|x\|_{\mathbb{K}^{j-1}}^2 + \|x'\|_{\mathbb{K}}^2, \qquad x \in \mathbb{K}^{j-1}, \ x' \in \mathbb{K},$$

and we know that $\mathbb{K}^{j-1}$ and $\mathbb{K}$ are complete it follows from Lemma 4.6 that $\mathbb{K}^j$ is complete. Completeness of $\mathbb{K}^n$ for any n now follows by induction.

4.4 Given $(x^{(n)}) \in c_0(\mathbb{K})$ with $x^{(n)} \to x$ in ℓ^∞, we need to show that $x \in c_0(\mathbb{K})$. Take $\varepsilon > 0$; then there exists an N such that $\|x^{(n)} - x\|_{\ell^\infty} < \varepsilon/2$ for all $n \geq N$. In particular, $\|x^{(N)} - x\|_{\ell^\infty} < \varepsilon/2$. Now since $x^{(N)} \in c_0(\mathbb{K})$ there exists J such that $|x_j^{(N)}| < \varepsilon/2$ for all $j \geq J$. It follows that for all $j \geq J$ we have

$$|x_j| < |x_j - x_j^{(N)}| + |x_j^{(N)}| < \varepsilon/2 + \varepsilon/2 = \varepsilon,$$

and so $x_j \to 0$ as $j \to \infty$, i.e. $x \in c_0(\mathbb{K})$.

4.5 Take $x, y \in X$; then for any $u, v \in U$ we have

$$\|[x + y]\| \leq \|x + y + u + v\| \leq \|x + u\| + \|y + v\|,$$

so

$$\|[x + y]\| \leq \inf_{u \in U} \|x + u\| + \inf_{v \in U} \|y + v\| = \|[x]\| + \|[y]\|.$$

Given any $\alpha \in \mathbb{K}$ we have

$$\|[\alpha x]\| = \inf_{u \in U} \|\alpha x + u\| = \inf_{u \in U} |\alpha| \|x + u\| = \alpha \|[x]\|.$$

All that remains is to show that $\|[x]\| = 0$ implies that $[x] = 0$; but $\|[x]\| = 0$ means that

$$\inf_{u \in U} \|x + u\| = 0,$$

i.e. there exist $(u_n) \in U$ such that $u_n \to x$. Since U is closed, $x \in U$, and so $[x] = [0]$.

Now assume that X is complete. To show that X/U is complete we show that if $([x_n]) \in X/U$ with $\sum_{n=1}^\infty \|[x_n]\|_{X/U} < \infty$, then $\sum_{n=1}^\infty [x_n]$ converges to some element $[y] \in X/U$. For each n, choose $\xi_n \in [x_n]$ such that

$$\|\xi_n\|_X < \|[x_n]\|_{X/U} + 2^{-n}.$$

Then $\sum_{n=1}^\infty \|\xi_n\| < \infty$, and since X is complete, it follows that $\sum_{n=1}^\infty \xi_n$ converges to some $\eta \in X$. Noting that

$$\left(\sum_{n=1}^N [x_n]\right) - [\eta] = \left(\sum_{n=1}^N \xi_n\right) - \eta + U$$

it follows that

$$\left\|\sum_{n=1}^N [x_n] - [\eta]\right\|_{X/U} \leq \left\|\left(\sum_{n=1}^N \xi_n\right) - \eta\right\|_X$$

and so $\sum_{n=1}^\infty [x_n]$ converges to $[\eta]$, and hence X/U is complete.

4.6 Let $X = \mathbb{R}$ and $K = [1, \infty)$ and define $f(x) = x + 1/x$. Then

$$|f(x) - f(y)| = |x - y|(1 - (xy)^{-1}) < |x - y|$$

but there is no $x \in [1, \infty)$ such that $f(x) = x$.

If K is compact and $f : K \to K$ has no fixed point, then $\|f(x) - x\| > 0$ for every $x \in X$. Since the function $x \mapsto \|f(x) - x\|$ is continuous and K is compact, $\|f(x) - x\|$ attains its lower bound on K, i.e.

$$\inf_{x \in K} \|f(x) - x\| = \|f(y) - y\| \tag{S.8}$$

for some $y \in K$. But then $f(y) \in X$ and $\|f(f(y)) - f(y)\| < \|f(y) - y\|$, contradicting (S.8), so f must have a fixed point in K. The uniqueness follows as in the standard theorem.

4.7 Integrating (4.10) from 0 to t we obtain (4.11). Setting $t = 0$ we obtain $x(0) = x_0$; the Fundamental Theorem of Calculus ensures that x is differentiable with derivative $\dot{x}(t) = f(x(t))$.

4.8 The space $C([0, T])$ is complete (with the supremum norm), so we can apply the Contraction Mapping Theorem in this space. We have

$$|(\mathcal{J}x)(t) - (\mathcal{J}y)(t)| = \left| \int_0^t f(x(s)) \, ds - \int_0^t f(y(s)) \, ds \right|$$

$$\leq \int_0^t |f(x(s)) - f(y(s))| \, ds$$

$$\leq L \int_0^t |x(s) - y(s)| \, ds \leq L \int_0^t \|x - y\|_\infty \, ds \leq LT \|x - y\|_\infty.$$

Since this holds for every $t \in [0, T]$, it follows that

$$\|\mathcal{J}x - \mathcal{J}y\|_\infty \leq LT \|x - y\|_\infty,$$

and so $\mathcal{J}$ is a contraction on $[0, T]$ for any $T < 1/L$.

Now observe that the time of existence given by this argument does not depend on $x(0)$; we obtain existence on $[0, 1/2L]$ for any $x(0)$. Set $x_0(t) = x(t)$ for all $t \in [0, T]$. To extend the existence time, observe that if $x_n(t)$ solves $\dot{x}_n = f(x_n)$ with $x_n(0) = x_{n-1}(1/2L)$, then

$$\hat{x}(t) = x_k(t), \qquad (k-1)/2L \leq t < 2kL,$$

solves $\dot{\hat{x}} = f(\hat{x})$ with $\hat{x}(0) = x(0)$ for all $t \geq 0$.

4.9 If (f_n) is Cauchy, then for any $\varepsilon > 0$ there exists N such that

$$d(f_i, f_j) = \sum_{n=1}^\infty \frac{1}{2^n} \frac{\|f_i - f_j\|_{[-n,n]}}{1 + \|f_i - f_j\|_{[-n,n]}} < \varepsilon \qquad \text{for } i, j \geq N.$$

In particular, for each n we have

$$\|f_i - f_j\|_{[-n,n]} < \varepsilon \qquad \text{for } i, j \geq N,$$

so $f_i|_{[-n,n]}$ is a Cauchy sequence in $C^0([-n, n])$. It follows that for each n there exists $g_n \in C^0([-n, n])$ such that $f_i|_{[-n,n]}$ converges uniformly to g_n on $[-n, n]$.

Noting that if $m > n$ we must have $g_n = g_m|_{[-n,n]}$ we can unambiguously define $g \in C(\mathbb{R})$ by setting

$$g(x) = g_n(x), \qquad x \in [-n, n],$$

and then f_i converges uniformly to g on each interval $[-n, n]$, which makes $C(\mathbb{R})$ complete using the metric.

Just as in Exercise 3.1 this metric does not come from a norm, since it does not have the homogeneity property $d(\lambda x, \lambda y) = \lambda d(x, y)$.

Chapter 5

5.1 Note that since K is non-empty there exists some $x \in K$; since K is symmetric $-x \in K$, and since K is convex it follows that $\frac{1}{2}(x + (-x)) = 0 \in K$. Since K is open, it follows that K contains some ball $B(0, r)$, using the usual Euclidean norm on $\mathbb{R}^n$; since K is bounded $K \subset B(0, R)$.

Hence for every non-zero $x \in \mathbb{R}^n$ we have $\frac{r}{2\|x\|}x \in K$ and $\frac{2R}{\|x\|}x \notin K$, so

$$\frac{1}{2R}\|x\| < N(x) < \frac{2}{r}\|x\|$$

and $N(x) = 0$ if and only if $x = 0$.

Taking any $\lambda \in \mathbb{R}$ we have

$$\begin{aligned}
N(\lambda x) &= \inf\{M > 0 : M^{-1}\lambda x \in K\} \\
&= \inf\{M > 0 : M^{-1}|\lambda|x \in K\} \\
&= |\lambda|\inf\{(M/|\lambda|) > 0 : M^{-1}|\lambda|x \in K\} \\
&= |\lambda|\inf\{L : L^{-1}x \in K\} = |\lambda|N(x),
\end{aligned}$$

where we have used the fact that K is symmetric.

Finally, we show that the 'closed unit ball' $\mathbb{B} = \{x : N(x) \le 1\}$ is convex. If $x, y \in \mathbb{B}$, then for any $\alpha > 1$ we have $\alpha^{-1}x \in K$ and $\alpha^{-1}y \in K$. Since K is convex,

$$\alpha^{-1}(\lambda x + (1 - \lambda)y) = \lambda\alpha^{-1}x + (1 - \lambda)\alpha^{-1}y \in K,$$

which shows that $N(\lambda x + (1 - \lambda)y) \le \alpha$. Since this holds for all $\alpha > 1$, it follows that $N(\lambda x + (1 - \lambda)y) \le 1$, so $\mathbb{B}$ is convex and N is a norm on $\mathbb{R}^n$.

5.2 We have $\|x\|_W = \|Tx\|_V \ge 0$, and if $0 = \|x\|_W = \|Tx\|_V$, then $Tx = 0$ (since $\|\cdot\|_V$ is a norm on V), which implies that $x = 0$ since T is an isomorphism (and $T0 = 0$). We also have

$$\|\alpha x\|_W = \|T(\alpha x)\|_V = \|\alpha Tx\|_V = |\alpha|\|Tx\|_V = |\alpha|\|x\|_W,$$

since T is linear. Finally, the triangle inequality follows, since for every $x, y \in W$ we have

$$\begin{aligned}
\|x + y\|_W = \|T(x + y)\|_V = \|Tx + Ty\|_V &\le \|Tx\|_V + \|Ty\|_V \\
&= \|x\|_W + \|y\|_W.
\end{aligned}$$

That T is an isometry is now immediate from the construction, and so T is a bijective linear isometry, i.e. an isometric isomorphism (see Definition 3.19).

5.3 Theorem 5.2 guarantees that any finite-dimensional normed space – and hence any finite-dimensional subspace of a normed space – is complete. It follows from Lemma 4.3 that any finite-dimensional subspace of a Banach space must be closed.

5.4 For any $x \in X \setminus Y$ there exist $(y_n) \in Y$ such that $\|x - y_n\| \to \mathrm{dist}(x, Y)$ as $n \to \infty$. In particular, (y_n) is a bounded sequence, and so, as Y is finite-dimensional, there is a convergent subsequence (y_{n_j}) with $y_{n_j} \to y$, for some $y \in Y$ (we know that Y is closed by the previous exercise). Since $x \mapsto \|x\|$ is continuous (Lemma 3.14), it follows that $\|x - y\| = \mathrm{dist}(x, Y)$.

5.5 For any $y \in Y$

$$\|\alpha x - y\| = |\alpha| \left\| x - \frac{y}{\alpha} \right\|;$$

since Y is a linear space $Y/\alpha = Y$, so

$$\text{dist}(\alpha x, Y) = \inf_{y \in Y} \|\alpha x - y\| = |\alpha| \inf_{z \in Y} \|x - z\| = |\alpha| \text{dist}(x, Y).$$

Similarly for any $w \in Y$ and $y \in Y$ we have

$$\|(x + w) - y\| = \|x - (y - w)\|;$$

since $Y - w = Y$ it follows that

$$\text{dist}(x + w, Y) = \inf_{y \in Y} \|(x + w) - y\| = \inf_{z \in Y} \|x - z\| = \text{dist}(x, Y).$$

5.6 First choose any $w \in X \setminus Y$ and let $\delta := \text{dist}(w, Y) > 0$. Using Exercise 5.4 there exists $z \in Y$ such that $\|w - z\| = \delta$. Using the result of the previous exercise if we set $w' = rw/\delta$ and $z' = rz/\delta$ (which is still in Y), then

$$\|w' - z'\| = \text{dist}(w', Y) = r.$$

Finally, we let $x = w' - z' + y$; since $y - z' \in Y$ we have

$$\|x - y\| = \|w' - z'\| = \text{dist}(x, Y) = r,$$

as required.

5.7 Suppose that X is a normed space with a countably infinite Hamel basis $\{e_j\}_{j=1}^{\infty}$. Set $X_n = \text{Span}(e_1, \ldots, e_n)$; since $\{e_j\}$ is a Hamel basis every element of X be written as a finite linear combination of the $\{e_j\}$, so $X = \bigcup_{j=1}^{\infty} X_j$.

Starting with $y_1 = e_1/\|e_1\|$, for each n use the result of the previous exercise to find $y_n \in X_n$ such that

$$\|y_n - y_{n-1}\| = \text{dist}(y_n, X_{n-1}) = 3^{-n}.$$

Then (y_n) is a Cauchy sequence, since if $m > n$, we have

$$\|y_n - y_m\| \leq \sum_{k=n+1}^{m} \|y_k - y_{k-1}\| = \sum_{k=n+1}^{m} 3^{-k} < \frac{3^{-k}}{2}.$$

However, note that if $m > n$, then

$$\text{dist}(y_m, X_n) \geq \text{dist}(y_{n+1}, X_n) - \sum_{j=n+2}^{m} \|y_j - y_{j-1}\|$$

$$= 3^{-(n+1)} - \sum_{j=n+2}^{m} 3^{-j} \geq 3^{-(n+1)} - \frac{3^{-(n+1)}}{2} = \frac{3^{-(n+1)}}{2} > 0,$$

so (y_n) cannot have a limit lying in X_n. Since this holds for every n, the sequence (y_n) does not converge.

Chapter 6

6.1 We have

$$\binom{2k}{k} = \frac{(2k)!}{(k!)^2} \leq \frac{e(2k)^{2k+1/2}e^{-2k}}{2\pi k^{2k+1}e^{-2k}} = \frac{e}{\sqrt{2\pi}}\frac{2^{2k}}{\sqrt{k}}.$$

6.2 We calculate the derivatives of $f(x) = (1-x)^{1/2}$. We have

$$f'(x) = -\frac{1}{2}(1-x)^{-1/2}, \quad f''(x) = -\frac{1}{2}\frac{1}{2}(1-x)^{-3/2}, \quad f'''(x) = -\frac{1}{2}\frac{1}{2}\frac{3}{2}(1-x)^{-5/2},$$

and in general

$$f^{(n)}(x) = -\frac{1}{2^n}(2n-3)!!(1-x)^{-(2n-1)/2} = -\frac{(2(n-1))!}{2^{2n-1}(n-1)!}(1-x)^{-(2n-1)/2},$$

using the fact that $(2n-3)!! = 1 \cdot 3 \cdot 5 \cdots (2n-3) = (2(n-2))!/2^{n-1}(n-1)!$.

This gives the n-term Taylor expansion in terms up to x^n as

$$f(x) = \left[1 - 2\sum_{k=1}^{n}\frac{(2(k-1))!}{2^{2(k-1)}(k-1)!k!}x^k\right] - \frac{2(2n)!}{2^{2n}(n!)^2}(1-c)^{-(2n+1)/2}x(x-c)^n$$

for some $c \in (0, x)$. In particular, the error term is bounded by

$$\frac{2(2n)!}{2^{2n}(n!)^2}x\frac{(x-c)^n}{(1-c)^{n+1/2}} \leq \frac{2(2n!)}{2^{2n}(n!)^2}\frac{1}{(1-x)^{n+1/2}} \leq \frac{A(x)}{\sqrt{n}},$$

where $A(x)$ is a constant depending only on x (using the result of the previous exercise to bound the factorial terms), which tends to zero as $n \to \infty$.

6.3 Using the Weierstrass Approximation Theorem, given any $\varepsilon > 0$ we can find a polynomial $p(x) = \sum_{k=0}^{n}c_k x^k$ such that

$$\sup_{x\in[-1,1]}|p(x) - |x|| < \varepsilon.$$

Then, given any $f \in A$ with $\|f\|_\infty \leq 1$, we have

$$\sup_{x\in X}\left|\sum_{k=0}^{n}c_k f^k(x) - |f(x)|\right| < \varepsilon.$$

Since A is an algebra, $\sum_{k=0}^{n}c_k f^k \in A$. It follows that $|f| \in \overline{A}$. The rest of the lemma follows as before.

6.4 Consider

$$E_n := \{x \in [a, b] : f(x) - f_n(x) < \varepsilon\}.$$

Then, since $f - f_n$ is continuous, each E_n is open, since it is the preimage of the open set $(-\infty, \varepsilon)$. Since $f_n \to f$ pointwise, $\cup_n E_n = [a, b]$; this gives an open cover of the compact set $[a, b]$, and so there is a finite subcover. Since the sets E_n are nested, with $E_{n+1} \supseteq E_n$, the largest of the E_j in this subcover is a cover itself, i.e. $E_N = [a, b]$ for some N, which shows that $|f(x) - f_N(x)| < \varepsilon$ for all $n \geq N$, i.e. f_n converges uniformly to f.

6.5 For any fixed $p \in [-1, 1]$ consider the function $f(p) = \frac{1}{2}x^2 + p - \frac{1}{2}p^2$. Then

$$0 \le p \le |x| \le 1 \quad \Rightarrow \quad p \le f(p) \le |x|,$$

and $p = f(p)$ if and only if $p = |x|$. Now for each fixed $x \in [0, a]$ the sequence $(p_n(x))$ satisfies $p_{n+1}(x) = f(p_n(x))$, and is therefore a positive, monotone, bounded sequence. It follows that for each fixed $x \in [-1, 1]$ we have $p_n(x) \to p_x$ for some $p_x \in [-1, 1]$; taking limits as $n \to \infty$ in $p_{n+1}(x) = f(p_n(x))$ yields $p_x = f(p_x)$, and so $p_x = |x|$. Dini's Theorem (the result of the previous exercise) now guarantees that $p_n(x)$ converges uniformly to $|x|$ on $[0, 1]$.

To apply this to give another proof of Lemma 6.7, observe that if $p_n(x)$ is a polynomial in x, then so is $p_{n+1}(x)$; we now approximate $|f(x)|$ by $p_n(f(x))$, which will converge uniformly for all f with $\|f\|_\infty \le 1$.

6.6 The collection A of all functions of the form $\sum_{i=1}^n f_i(x)g_i(y)$ is clearly closed under addition and multiplication. We have $1 \in A$, since we can take $n = 1$ and $f_1(x) = g_1(x) = 1$, so A is an algebra. It is clear that A separates points, since $x \in A$ and $y \in A$. The result now follows immediately from the Stone–Weierstrass Theorem.

6.7 The argument is very similar to that of the previous exercise. We let A be the collection of all functions of the form $\sum_{i=1}^n a_{ij}x^i y^j$. This forms an algebra, and separates points since $x \in A$ and $y \in A$.

6.8 Take $f \in C([a, b] \times [c, d]; \mathbb{R})$ and $\varepsilon > 0$. Then, using the result of the previous exercise, we can find $n \in \mathbb{N}$ and $a_{ij} \in \mathbb{R}$ such that

$$\left\| f(x, y) - \sum_{i,j=1}^n a_{ij}x^i y^j \right\|_\infty < \varepsilon.$$

Therefore

$$\left| \int_a^b \int_c^d f(x, y)\, dy\, dx - \int_a^b \int_c^d \sum_{i,j=1}^n a_{ij}x^i y^j\, dy\, dx \right| < (b - a)(d - c)\varepsilon.$$

Now for any i, j, we have

$$\int_a^b \int_c^d x^i y^j\, dy\, dx = \int_a^b x^i \left[\frac{d^{j+1} - c^{j+1}}{j+1} \right] dx$$

$$= \frac{b^{i+1} - a^{i+1}}{i+1} \frac{d^{i+1} - c^{i+1}}{i+1}$$

$$= \int_c^d \left[\frac{b^{i+1} - a^{i+1}}{i+1} \right] y^j\, dy = \int_c^d \int_a^b x^i y^j\, dx\, dy,$$

and so

$$\left| \int_a^b \int_c^d f(x, y)\, dy\, dx - \int_c^d \int_a^b f(x, y)\, dx\, dy \right| < 2\varepsilon(b - a)(d - c).$$

Since this holds for any $\varepsilon > 0$, we obtain the required result.

6.9 Take $z, z_0 \in S^1$ with $z = e^{ix}$ and $z_0 = e^{ix}$, where $x, x_0 \in [-\pi, \pi)$. Since f is continuous, for any $\varepsilon > 0$ there exists $\delta > 0$ such that (i) $|x - x_0| < \delta$ ensures that $|f(x) - f(x_0)| < \varepsilon$ and (ii) either $|x - \pi| < \delta$ or $|x + \pi| < \delta$ implies that $|f(x) - f(\pi)| < \varepsilon/2$.

Now, observe that

$$|z - z_0|^2 = |e^{ix} - e^{ix_0}|^2 = (e^{ix} - e^{ix_0})(e^{-ix} - e^{-ix_0})$$

$$= 2(1 - \cos(x - x_0)).$$

It follows that if $|z - z_0| < \delta/2$, then

$$\cos(x - x_0) > 1 - \frac{\delta^2}{8}.$$

If δ is small, then either $x - x_0$ is close to zero or close to 2π.

Since $\cos\theta \le 1 - \theta^2/2 + \theta^4/4 \le 1 - \theta^2/4$ for $|\theta| \le 1$, it follows in the first case that

$$\frac{|x - x_0|^2}{4} < \frac{\delta^2}{8},$$

and so $|x - x_0| < \delta$, which ensures that $|g(e^{ix}) - g(e^{ix_0})| = |f(x) - f(x_0)| < \varepsilon$.

In the second case, when $x - x_0$ is close to 2π, we have $x = -\pi + \xi$ and $x_0 = \pi - \xi_0$ (or vice versa) with $\xi > 0$ and $\xi_0 > 0$ small. Then

$$\cos(x - x_0) = \cos(\xi_0 + \xi - 2\pi) = \cos(\xi_0 + \xi)$$

and the same argument as before guarantees that $|\xi_0 + \xi| < \delta$, from which it follows that

$$|f(x) - f(x_0)| \le |f(x) - f(-\pi)| + |f(-\pi) - f(\pi)| + |f(\pi) - f(x_0)| < \frac{\varepsilon}{2} + 0 + \frac{\varepsilon}{2} = \varepsilon,$$

and so once again $|g(e^{ix}) - g(e^{ix_0})| = |f(x) - f(x_0)| < \varepsilon$.

6.10 (i) Given $\varepsilon > 0$ let $\{x_1, \ldots, x_n\}$ be such that $A \subset \bigcup_{i=1}^n B(x_i, \varepsilon/2)$. Then we have $\overline{A} \subset \bigcup_{i=1}^n B(x_i, \varepsilon)$, since, given any $y \in \overline{A}$, there exists $x \in A$ with $\|x - y\| < \varepsilon/2$ and x_i such that $\|x - x_i\| < \varepsilon/2$, so $\|y - x_i\| < \varepsilon$.

(ii) Let (x_n) be a sequence in A. Find a cover of A by finitely many balls of radius 1, and choose one of these, $B(y_1, 1)$, such that $x_n \in B(y_1, 1)$ for infinitely many n, giving a subsequence $(x_{n_{1,j}})_j \in B(y_1, 1)$. Now cover A by finitely many balls of radius $1/2$, and from these choose $B(y_2, 1/2)$ that contains infinitely many of the $x_{n_{1,j}}$: these give a further subsequence $(x_{n_{2,j}})_j$ that is contained in $B(y_2, 1/2)$. Continue in this way to find successive subsequences $(x_{n_{k,j}})_j \in B(y_k, 2^{-k})$. The sequence $x_{n_{k,k}}$ is a Cauchy sequence, since for $j > k$ we know that $x_{n_{j,j}}$ is an element of the sequence $(x_{n_{k,j}})_j$, and all elements of this sequence are in $B(y_k, 2^{-k})$ so no more than $2^{-(k-1)}$ apart.

(iii) If A is a subset of a complete normed space that is totally bounded and closed, then any sequence in A has a Cauchy subsequence (since A is totally bounded) that converges to a limit (since X is complete) that must lie in A (since A is closed), i.e. A is compact. Conversely, if $\overline{A}$ is compact, then it is totally bounded, and since $A \subset \overline{A}$ it must be totally bounded too.

6.11 Suppose that (f_n) is bounded and equicontinuous on $\mathbb{R}$. Then it is bounded and equicontinuous on each compact interval $[-N, N]$. We can use the Arzelà–Ascoli Theorem to extract a subsequence $(f_{n_{1,j}})_j$ such that $f_{n_{1,j}}$ converges uniformly on $[-1, 1]$; now take a further subsequence to ensure that $f_{n_{2,j}}$ converges uniformly on $[-2, 2]$, and continue in this way, finding subsequences such that $(f_{n_{k,j}})_j$ converges uniformly on $[-k, k]$. As in the proof of the Arzelà–Ascoli Theorem itself take the 'diagonal' subsequence $(f_{n_{k,k}})_k$ to obtain a subsequence of the initial (f_n) that converges uniformly on every interval of the form $[-k, k]$ for $k \in \mathbb{N}$, and so on any bounded interval in $\mathbb{R}$.

6.12 First it is simple to show that f_δ is bounded, since

$$|f_\delta(x)| \le \frac{1}{2\delta} \int_{x-\delta}^{x+\delta} |f(x)| \, dx \le \frac{1}{2\delta} \int_{x-\delta}^{x+\delta} \|f\|_\infty \, dx = \|f\|_\infty.$$

Consider $x, z \in \mathbb{R}$. If $|x - z| \ge \delta$, then $(x - \delta, x + \delta) \cap (z - \delta, z + \delta) = \varnothing$ and

$$|f_\delta(x) - f_\delta(y)| \le \frac{1}{2\delta} \int_{x-\delta}^{x+\delta} |f(y)| \, dy + \frac{1}{2\delta} \int_{z-\delta}^{z+\delta} |f(y)| \, dy$$
$$\le 2\|f\|_\infty \le \frac{2\|f\|_\infty}{\delta} |x - z|.$$

If $|x - z| < \delta$ and $z > x$, then

$$(x - \delta, x + \delta) \cap (z - \delta, z + \delta) = (z - \delta, x + \delta),$$

and so

$$|f_\delta(x) - f_\delta(y)| \le \frac{1}{2\delta} \int_{x-\delta}^{z-\delta} |f(y)| \, dy + \frac{1}{2\delta} \int_{x+\delta}^{z+\delta} |f(y)| \, dy \le \frac{\|f\|_\infty}{\delta} |z - x|.$$

This shows that $|f_\delta(x) - f_\delta(z)| \le (2\|f\|_\infty/\delta)|x - z|$ and so f is Lipschitz.

To show convergence on $[-R, R]$, first note that since f is continuous it is uniformly continuous on any closed bounded interval (a compact subset of $\mathbb{R}$): so, given any $\varepsilon > 0$ that there exists $\delta > 0$ such that $x, y \in [-(R+1), R+1]$ with $|x - y| < \delta$ implies that $|f(x) - f(y)| < \varepsilon$. It follows that for any $x \in [-R, R]$, if $\delta < 1$ we have

$$|f_\delta(x) - f(x)| = \left| \frac{1}{2\delta} \int_{x-\delta}^{x+\delta} f(y) \, dy - \frac{1}{2\delta} \int_{x-\delta}^{x+\delta} f(x) \, dy \right|$$
$$= \frac{1}{2\delta} \int_{x-\delta}^{x+\delta} |f(y) - f(x)| \, dy < \varepsilon,$$

which gives the required convergence.

6.13 Since x_n satisfies (6.9), it follows that

$$|x_n(t)| \le |x_0| + \int_0^t \|f_n\|_\infty \, ds \le |x_0| + T\|f_n\|_\infty \le B := |x_0| + MT,$$

so (x_n) is a bounded sequence in $C([0, T])$. It is also equicontinuous, since for each n and any $t, t' \in [0, T]$ with $t' > t$ we have

$$|x_n(t') - x_n(t)| = \left| \int_t^{t'} f_n(x_n(s)) \, ds \right| \le M(t' - t).$$

The Arzelà–Ascoli Theorem guarantees that (x_n) has a subsequence that converges uniformly on $[0, T]$ to some limiting function $x \in C([0, T])$. Since $f_n \to f$ uniformly on $[-B, B]$, it follows that $f_n(x_n(\cdot)) \to f(x(\cdot))$ uniformly on $[0, T]$, and so taking limits in (6.9) we obtain (6.10).

6.14 Approximate f by Lipschitz functions (f_n) using Exercise 6.12. Exercise 4.8 guarantees that the ODE $\dot{x}_n = f(x_n)$, $x_n(0) = x_0$, has a solution on $[0, T]$. From Exercise 4.7 this is also a solution of the integral equation

$$x_n(t) = x_0 + \int_0^t f_n(x_n(s))\, ds \qquad t \in [0, T].$$

Using the previous exercise there exists $x \in C([0, T])$ that satisfies the limit equation (6.10), and by Exercise 4.7 this is the required solution of the ODE $\dot{x} = f(x)$ with $x(0) = x_0$.

Chapter 7

7.1 This is immediate from the estimate

$$\|f_n - f\|_{L^1} = \int_a^b |f_n(x) - f(x)|\, dx \le \int_a^b \|f_n - f\|_\infty = (b-a)\|f_n - f\|_\infty.$$

7.2 We have

$$\int_0^1 |f_n(x)|\, dx = \int_0^{1/n} 1 - nx\, dx = \frac{1}{2n} \to 0,$$

but $f_n(0) = 1$ for every n, so f_n does not converge pointwise to zero.

7.3 We have $g_n(0) = 0$ for every n, and for each $x \in (0, 1]$ we have $g_n(x) = 0$ for all $n > 2/x$. However,

$$\int_0^1 |g_n(x)|\, dx = \int_0^{1/n} n^2 x\, dx + \int_{1/n}^{2/n} n(2 - nx)\, dx = 1.$$

7.4 Suppose that $\xi \in \mathscr{X}$; then there exists a sequence $(x_n) \in X$ such that $i(x_n) \to \xi$, and so, in particular,

$$\|\xi\|_{\mathscr{X}} = \lim_{n\to\infty} \|i(x_n)\|_{\mathscr{X}} = \lim_{n\to\infty} \|x_n\|_X,$$

since i is an isometry. The convergence of $i(x_n)$ to ξ means that $(i(x_n))$ must be Cauchy in $\mathscr{X}$, and using once again the fact that i is an isometry it follows that (x_n) is Cauchy in X. Since i' is another isometry, $(i'(x_n))$ is Cauchy in $\mathscr{X}'$, and since $\mathscr{X}'$ is complete $i'(x_n) \to \xi' \in \mathscr{X}'$, from which it follows that

$$\|\xi'\|_{\mathscr{X}'} = \lim_{n\to\infty} \|i'(x_n)\|_{\mathscr{X}'} = \lim_{n\to\infty} \|x_n\|_X = \|\xi\|_{\mathscr{X}}. \tag{S.9}$$

We now define $j: \mathscr{X} \to \mathscr{X}'$ by setting $j(\xi) = \xi'$. This map is well defined, and (S.9) shows that j is an isometry.

7.5 First note that it follows from the triangle inequality that

$$|[i(x)]y| = |d(y, x) - d(y, x_0)| \le d(x, x_0),$$

so $i(x) \in \mathcal{F}_b(X; \mathbb{R})$ as claimed. To show that i is an isometry, we use the triangle inequality again to obtain

$$[i(x)](y) - [i(x')](y) = [d(y, x) - d(y, x_0)] - [d(y, x') - d(y, x_0)]$$
$$= d(y, x) - d(y, x') \le d(x, x'),$$

so $\|i(x) - i(x')\|_{\ell^\infty} \le d(x, x')$. For the opposite inequality, choose $y = x'$, and then

$$[i(x)](x') - [i(x')](x') = d(x', x),$$

which shows that $\|i(x) - i(x')\|_{\ell^\infty} \ge d(x, x')$. We therefore obtain

$$\|i(x) - i(x')\|_{\ell^\infty} = d(x, x')$$

and i is an isometry as claimed.

7.6 We certainly have $\|f\|_{L^\infty} \le \|f\|_\infty$, since the supremum can be taken over a smaller set for the L^∞ norm (excluding a set of measure zero). Suppose that $E \subset \Omega$ is a set of measure zero such that

$$|f(x)| \le \|f\|_{L^\infty} \qquad x \in \Omega \setminus E.$$

Since E has measure zero, given any $x \in E$ there exists a sequence $(x_n) \in \Omega \setminus E$ such that $x_n \to x$; otherwise, there would be an open ball $B(x, \delta) \subset E$, and then E would not have measure zero. But then $|f(x_n)| \le \|f\|_{L^\infty}$, and since f is continuous it follows that $|f(x)| = \lim_{n \to \infty} |f(x_n)| \le \|f\|_{L^\infty}$, which shows that $\|f\|_\infty = \|f\|_{L^\infty}$ as claimed.

Chapter 8

8.1 We check properties (i)–(iv) from Definition 8.1.

(i) We have $(f, f) = \int |f|^2 = \|f\|_{L^2}^2 \ge 0$, and if $(f, f) = 0$, then $f = 0$ almost everywhere, i.e $f = 0$ in L^2.

(ii) We have

$$(f + g, h) = \int (f + g)\overline{h} = \int f\overline{h} + \int g\overline{h} = (f, h) + (g, h)$$

using the linearity of the integral.

(iii) Again, the linearity of the integral yields

$$(\alpha f, g) = \int \alpha f\overline{g} = \alpha \int f\overline{g} = \alpha(f, g).$$

(iv) Finally, we have $(f, g) = \int f\overline{g} = \overline{\int \overline{f}g} = \overline{(g, f)}$.

8.2 We use the Cauchy–Schwarz inequality in L^2 to write

$$\int_\Omega |f(x)| \, dx \le \left(\int_\Omega |f(x)|^2 \, dx \right)^{1/2} \left(\int_\Omega 1 \, dx \right)^{1/2} \le |\Omega| \|f\|_{L^\infty}$$

which immediately yields (8.11).

8.3 First we check that this really does defined an inner product.

(i) $((x, \xi), (x, \xi))_{H \times K} = (x, x)_H + (\xi, \xi)_K \geq 0$ and if $(x, x)_H + (\xi, \xi)_K = 0$, then, since $(x, x)_H \geq 0$ and $(\xi, \xi)_K \geq 0$, we must have $x = \xi = 0$, and so $(x, \xi) = 0$.

(ii) $((x, \xi) + (x', \xi'), (y, \eta))_{H \times K} = (x + x', y)_H + (\xi + \xi', \eta)_K$
$$= [(x, y)_H + (\xi, \eta)_K] + [(x', y)_H + (\xi', \eta)_K]$$
$$= ((x, \xi), (y, \eta))_{H \times K} + ((x', \xi'), (y, \eta))_{H \times K}.$$

(iii) $(\alpha(x, \xi), (y, \eta))_{H \times K} = (\alpha x, y)_H + (\alpha \xi, \eta)_K = \alpha(x, y)_H + \alpha(\xi, \eta)_K$
$$= \alpha((x, \xi), (y, \eta))_{H \times K}.$$

(iv) $((x, \xi), (y, \eta))_{H \times K} = (x, y)_H + (\xi, \eta)_K = \overline{(y, x)_H} + \overline{(\eta, \xi)_K}$
$$= \overline{((y, \eta), (x, \xi))_{H \times K}}.$$

It remains to check that $H \times K$ is complete with the induced norm. But from (i) the induced norm satisfies
$$\|(x, \xi)\|_{H \times K}^2 = \|x\|_H^2 + \|\xi\|_K^2,$$
and $H \times K$ is complete with this norm (see comments around (4.2)).

8.4 If (8.12) holds, then putting $x = y$ yields $\|Tx\|_K^2 = \|x\|_H^2$. For the reverse implication we use the polarisation identity (8.8) to obtain
$$4(Tx, Ty)_K = \|Tx + Ty\|_K^2 - \|Tx - Ty\|_K^2$$
$$= \|T(x + y)\|_K^2 - \|T(x - y)\|_K^2$$
$$= \|x + y\|_H^2 - \|x - y\|_H^2 = 4(x, y)_H.$$

(The proof in the complex case is essentially the same, but notationally more involved since we have to use the complex polarisation identity in (8.9).)

8.5 Take $x = (1, 0, 0, \ldots)$ and $y = (0, 1, 0, \ldots)$. Then $x + y = (1, 1, 0, \ldots)$ and $x - y = (1, -1, 0, \ldots)$. So $\|x\|_{\ell^p} = \|y\|_{\ell^p} = 1$ and
$$\|x + y\|_{\ell^p} = \|x - y\|_{\ell^p} = 2^{1/p};$$
but
$$\|x + y\|_{\ell^p}^2 + \|x - y\|_{\ell^p}^2 = 2^{1+2/p} \neq 4 = 2(\|x\|_{\ell^p}^2 + \|y\|_{\ell^p}^2)$$
unless $p = 2$.

8.6 Consider $f(x) = x$ and $g(x) = 1 - x$, with
$$\|f\|_{L^p} = \|g\|_{L^p} = \left(\int_0^1 x^p \, dx \right)^{1/p} = \frac{1}{(p + 1)^{1/p}}.$$
Now $(f + g)(x) = 1$ so $\|f + g\|_{L^p} = 1$, and $(f - g)(x) = 2x - 1$, so
$$\|f - g\|_{L^p} = \left(\int_0^1 |2x - 1|^p \, dx \right)^{1/p} = \frac{1}{(p + 1)^{1/p}}.$$

Now

$$\|f + g\|_{L^p}^2 + \|f - g\|_{L^p}^2 = 1 + \frac{1}{(p+1)^{2/p}} \neq \frac{2}{(p+1)^{2/p}} = 2(\|f\|_{L^p}^2 + \|g\|_{L^p}^2)$$

unless $p = 2$.

8.7 Using the parallelogram identity we have

$$
\begin{aligned}
\|z - x\|^2 + \|z - y\|^2 &= \left\| \left[z - \frac{x+y}{2} \right] + \frac{y-x}{2} \right\|^2 + \left\| \left[z - \frac{x+y}{2} \right] - \frac{y-x}{2} \right\|^2 \\
&= 2 \left\| z - \frac{x+y}{2} \right\|^2 + 2 \left\| \frac{y-x}{2} \right\|^2 \\
&= \frac{1}{2} \|x - y\|^2 + 2 \left\| z - \frac{x+y}{2} \right\|^2.
\end{aligned}
$$

8.8 To prove (ii), i.e. $\langle x, z \rangle + \langle y, z \rangle = \langle x + y, z \rangle$, we write

$$
\begin{aligned}
4[\langle x, z \rangle + \langle y, z \rangle] &= \|x + z\|^2 - \|x - z\|^2 + \|y + z\|^2 - \|y - z\|^2 \\
&= \left\| \left[z + \frac{x+y}{2} \right] + \frac{x-y}{2} \right\|^2 + \left\| \left[z + \frac{x+y}{2} \right] - \frac{x-y}{2} \right\|^2 \\
&\quad - \left\| \left[z - \frac{x+y}{2} \right] - \frac{x-y}{2} \right\|^2 - \left\| \left[z - \frac{x+y}{2} \right] + \frac{x-y}{2} \right\|^2 \\
&= 2 \left\| z + \frac{x+y}{2} \right\|^2 + 2 \left\| \frac{x-y}{2} \right\|^2 - 2 \left\| z - \frac{x+y}{2} \right\|^2 - 2 \left\| \frac{x-y}{2} \right\|^2 \\
&= 2 \left\| z + \frac{x+y}{2} \right\|^2 - 2 \left\| z - \frac{x+y}{2} \right\|^2 \\
&= 2 \left\langle \frac{x+y}{2}, z \right\rangle.
\end{aligned}
\tag{S.10}
$$

Setting $y = 0$ shows that $2\langle x, z \rangle = 2\langle \frac{x}{2}, z \rangle$, from which it follows from (S.10) that

$$\langle x, z \rangle + \langle y, z \rangle = \langle x + y, z \rangle, \tag{S.11}$$

as required.

To prove (iii), i.e. $\langle \alpha x, z \rangle = \alpha \langle x, z \rangle$, observe that we can now use (S.11) to deduce that for any $n \in \mathbb{N}$ we have

$$\langle nx, z \rangle = n \langle x, z \rangle,$$

and so also $\langle x, z \rangle = \langle m \langle x/m \rangle, z \rangle = m \langle x/m, z \rangle$, which shows that

$$\langle x/m, z \rangle = \frac{1}{m} \langle x, z \rangle.$$

We also have $\langle x, z \rangle + \langle -x, z \rangle = 0$, so $\langle -x, z \rangle = -\langle x, z \rangle$. Combining these shows that $\langle \alpha x, z \rangle = \alpha \langle x, z \rangle$ for any $\alpha \in \mathbb{Q}$. That this holds for every $\alpha \in \mathbb{R}$ now follows from the fact that $x \mapsto \|x\|$ is continuous: given any $\alpha \in \mathbb{R}$, find $\alpha_n \in \mathbb{Q}$ such that $\alpha_n \to \alpha$, and then

$$\alpha \langle x, z \rangle = \lim_{n \to \infty} \alpha_n \langle x, z \rangle = \lim_{n \to \infty} \langle \alpha_n x, z \rangle$$

$$= \lim_{n \to \infty} \frac{1}{4} \left(\| \alpha_n x + z \|^2 - \| \alpha_n x - z \|^2 \right)$$

$$= \frac{1}{4} \left(\| \alpha x + z \|^2 - \| \alpha x - z \|^2 \right) = \langle \alpha x, z \rangle.$$

8.9 Recall that $\| [x] \|_{H/F} = \inf_{u \in U} \| x + u \|$. We show that this norm satisfies the parallelogram identity and therefore we can define an inner product on H/F using the polarisation identity.

Note that $\| [x] + [y] \|^2 + \| [x] - [y] \|^2 = \| [x + y] \|^2 + \| [x - y] \|^2$ for every $x, y \in X$, and observe that for any $\xi, \eta \in U$ we have

$$\| x + y + \xi \|^2 + \| x - y + \eta \|^2$$

$$= \| [x + \tfrac{1}{2}(\xi + \eta)] + [y + \tfrac{1}{2}(\xi - \eta)] \|^2 + \| [x + \tfrac{1}{2}(\xi + \eta)] - [y + \tfrac{1}{2}(\xi - \eta)] \|^2$$

$$= 2 \left(\| x + \tfrac{1}{2}(\xi + \eta) \|^2 + \| y + \tfrac{1}{2}(\xi - \eta) \|^2 \right). \tag{S.12}$$

Taking the infimum over $\xi \in U$ we obtain

$$\| [x + y] \|^2 + \| x - y + \eta \|^2 \ge 2(\| [x] \|^2 + \| [y] \|^2),$$

and then taking the infimum over $\eta \in U$ gives

$$\| [x + y] \|^2 + \| [x - y] \|^2 \ge 2(\| [x] \|^2 + \| [y] \|^2). \tag{S.13}$$

Returning to (S.12) and setting $\xi + \eta = 2\alpha$ and $\xi - \eta = 2\beta$ we have

$$\| x + y + (\alpha + \beta) \|^2 + \| x - y + (\alpha - \beta) \|^2 = 2(\| x + \alpha \|^2 + \| y + \beta \|^2).$$

Now we take first the infimum over $\alpha \in U$,

$$\| [x + y] \|^2 + \| [x - y] \|^2 \le 2(\| [x] \|^2 + \| y + \beta \|^2),$$

and then the infimum over $\beta \in U$ to give

$$\| [x + y] \|^2 + \| [x - y] \|^2 \le 2(\| [x] \|^2 + \| [y] \|^2). \tag{S.14}$$

Combining the two inequalities (S.13) and (S.14) shows that the norm on H/U satisfies the parallelogram identity, and hence comes from an inner product. So H/U is a Hilbert space.

8.10 From the parallelogram identity, if $\| x \| = \| y \| = 1$ and $\| x - y \| > \varepsilon$ then

$$4 = 2(\| x \|^2 + \| y \|^2) = \| x + y \|^2 + \| x - y \|^2 > \| x + y \|^2 + \varepsilon^2$$

which implies that $\| x + y \|^2 < 4 - \varepsilon^2$ so

$$\left\| \frac{x + y}{2} \right\| < \left(1 - \frac{\varepsilon^2}{4} \right)^{1/2} = 1 - \delta$$

if we set $\delta = 1 - \sqrt{1 - \varepsilon^2/4} > 0$.

Chapter 9

9.1 First we check that all the elements of E have norm 1. This is clear for $1/\sqrt{2\pi}$, and for the other elements we have

$$\int_{-\pi}^{\pi} \cos^2 nt \, dt = \frac{1}{2} \int_{-\pi}^{\pi} 1 + \cos 2nt \, dt = \pi$$

and similarly for $\sin^2 nt$.

Now we check the orthogonality properties: for any n, m

$$\int_{-\pi}^{\pi} \cos nt \, dt = \int_{-\pi}^{\pi} \sin nt \, dt = \int_{-\pi}^{\pi} \sin nt \cos mt \, dt = 0;$$

and for any $n \neq m$

$$\int_{-\pi}^{\pi} \cos nt \cos mt \, dt = \int_{-\pi}^{\pi} \sin nt \sin mt \, dt = \int_{-\pi}^{\pi} \sin nt \cos mt \, dt = 0.$$

9.2 Throughout we use the equality

$$\|x + \alpha y\|^2 = (x + \alpha y, x + \alpha y) = \|x\|^2 + \alpha(y, x) + \overline{\alpha}(x, y) + |\alpha|^2 \|y\|^2$$
$$= \|x\|^2 + \text{Re}\,[\alpha(y, x)] + |\alpha|^2 \|y\|^2.$$

The 'only if' parts of (i) and (ii) now follow immediately.

For part (i), if $\|x + \alpha y\| \geq \|x\|$ for every $\alpha \in \mathbb{K}$ then taking $\alpha \in \mathbb{R}$ yields

$$2\alpha \,\text{Re}(y, x) + |\alpha|^2 \|y\|^2 \geq 0,$$

which implies that $\text{Re}(y, x) = 0$, for otherwise we could invalidate the inequality by taking α sufficiently small; similarly taking $\alpha = i\beta$ with $\beta \in \mathbb{R}$ yields

$$2\beta \,\text{Im}(x, y) + |\beta|^2 \|y\|^2 \geq 0,$$

which implies that $\text{Im}(x, y) = 0$. So $(x, y) = 0$.

For part (ii), if $\|x + \alpha y\| = \|x - \alpha y\|$, then we obtain

$$\alpha(y, x) + \overline{\alpha}(x, y) = 0$$

for every $\alpha \in \mathbb{K}$. We take $\alpha \in \mathbb{R}$ and then $\alpha = i\beta$ with $\beta \in \mathbb{R}$ to deduce that $(x, y) = 0$.

9.3 Bessel's inequality guarantees that

$$|\{j : |(x, e_j)|^2 > M\}|M^2 < \sum_{j=1}^{\infty} |(x, e_j)|^2 \leq \|x\|^2,$$

which yields the required inequality.

9.4 For each fixed $m \in \mathbb{N}$, consider the set

$$E_m = \{e \in E : |(x, e)| \geq 1/m\}.$$

Then this set can have no more than $m^2 \|x\|^2$ elements. Indeed, if E_m has $N > m^2 \|x\|^2$ elements, one can select N elements $\{e_1, \ldots, e_N\}$ from E_m, and then

$$\sum_{j=1}^{N} |(x, e_j)|^2 \geq N \times \frac{1}{m^2} > \|x\|^2.$$

But this contradicts Bessel's inequality. Thus each E_m contains only a finite number of elements, and hence

$$\bigcup_{m=1}^{\infty} E_m = \{e \in E : (x, e) \neq 0\}$$

contains at most a countable number of elements.

9.5 We write $u = \sum_{j=1}^{\infty}(u, e_j)e_j$ and $v = \sum_{j=1}^{\infty}(v, e_j)e_j$. Then by the continuity of the inner product we have

$$(u, v) = \lim_{n \to \infty} \left(\sum_{j=1}^{n}(u, e_j)e_j, \sum_{k=1}^{n}(v, e_k)e_k \right)$$

$$\lim_{n \to \infty} \sum_{j,k=1}^{n}(u, e_j)(e_k, v)\delta_{jk}$$

$$= \lim_{n \to \infty} \sum_{j=1}^{\infty}(u, e_j)(e_j, v) = \sum_{j=1}^{\infty}(u, e_j)(e_j, v).$$

(Note that the Cauchy–Schwarz inequality guarantees that the sum converges.)

9.6 Take any sequence $(x^{(n)}) \in Q$. Then, if we let

$$x^{(n)} = \sum_{j=1}^{\infty} \alpha_j^{(n)} e_j,$$

we know that each sequence $(\alpha_j^{(n)})$ is bounded in $\mathbb{K}$. We can use a diagonal argument as in the proof of the Arzelà–Ascoli Theorem to find a subsequence $x^{(n_k)}$ such that $\alpha_j^{(n_k)} \to \alpha_j^*$ as $k \to \infty$ for each $j \in \mathbb{N}$. Note that $|\alpha_j^*| \leq 1/j$. Now we define $x^* = \sum_{j=1}^{\infty} \alpha_j^* e_j$; we want to show that $x^{(n_k)} \to x^*$ as $k \to \infty$. Given $\varepsilon > 0$ find M such that

$$\sum_{j=M+1}^{\infty} \frac{1}{j^2} < \varepsilon^2/8.$$

Now find N such that for all $n \geq N$ we have

$$|\alpha_j^{(n_k)} - \alpha_j^*|^2 < \varepsilon^2/2M \qquad j = 1, \ldots, M;$$

then for all $n \geq N$,

$$\|x^{(n_k)} - x^*\|^2 = \sum_{j=1}^{\infty} |\alpha_j^{(n_k)} - \alpha_j^*|^2 = \sum_{j=1}^{M} |\alpha_j^{(n_k)} - \alpha_j^*|^2 + \sum_{j=M+1}^{\infty} |\alpha_j^{(n_k)} - \alpha_j^*|^2$$

$$\leq M \frac{\varepsilon^2}{2M} + \sum_{j=M+1}^{\infty} \frac{4}{j^2} < \varepsilon^2,$$

which shows that $\|x^{(n_k)} - x^*\| < \varepsilon$, and so $x^{(n_k)} \to x^*$ as claimed.

9.7 Since $\hat{H} := \text{clin}(E)$ is a closed linear subspace of H, it is a Hilbert space itself. Now we can use Proposition 9.14 to deduce $[(\text{e}) \Rightarrow (\text{b})]$ that any element of $\text{clin}(E)$ can be written as $\sum_{j=1}^{\infty} \alpha_j e_j$ for some $(\alpha_j) \in \mathbb{K}$.

9.8 Since $\{e_j\}$ is an orthonormal basis for H, the linear span of the $\{e_j\}$ is dense in H. Noting that e_j can be written as a linear combination of $f_1, \ldots, f_j$, it follows that the linear span of the (f_j) is also dense in H.

Now consider $x \in H$ with $x = \sum_{j=1}^{\infty} j^{-1} e_j$. There is no expansion for x of the form $x = \sum_{j=1}^{\infty} \alpha_j f_j$, since taking the inner product of both sides with e_k would yield

$$\frac{1}{k} = (x, e_k) = \sum_{j=1}^{\infty} \alpha_j (f_j, e_k) = \frac{1}{k} \sum_{j=k}^{\infty} \alpha_j,$$

i.e. $\sum_{j=k}^{\infty} \alpha_j = 1$ for every $k \in \mathbb{N}$; but this is impossible.

9.9 We have $(x - y, z) = 0$ for every z in a dense subset of H. So we can find z_n such that $z_n \to x - y$, and then

$$\|x - y\|^2 = (x - y, x - y) = \left(x - y, \lim_{n \to \infty} z_n \right) = \lim_{n \to \infty} (x - y, z_n) = 0,$$

using the continuity of the inner product. It follows that $x = y$.

9.10 By Lemma 7.7 the set $\mathcal{P}(a, b)$ of polynomials restricted to (a, b) forms a dense subset of $L^2(a, b)$. The given equality shows that

$$(p, f) = \int_a^b p(x) f(x) \, dx = \int_a^b p(x) g(x) \, dx = (p, g)$$

for every $p \in \mathcal{P}$. Since $\mathcal{P}$ is dense in L^2, the result of the previous exercise now guarantees that $f = g$.

9.11 The functions $\{1\} \cup \{\cos k\pi x\}_{k=1}^{\infty}$ are orthonormal in $L^2(0, 1)$ by Example 9.11. We need only show that their linear span is dense in $L^2(0, 1)$. Corollary 6.5 shows that the linear span of these functions is uniformly dense in $C([0, 1])$, and we know that $C([0, 1])$ is dense in $L^2(0, 1)$, so, given any $f \in L^2(0, 1)$ and $\varepsilon > 0$, first find $g \in C([0, 1])$ such that

$$\|f - g\|_{L^2} < \varepsilon/2,$$

and then approximate g by an expression

$$h(x) = \sum_{k=0}^{n} c_k \cos(k\pi x)$$

such that $\|g - h\|_{\infty} < \varepsilon/2$. Since $\|h - g\|_{L^1} = \int_0^1 |h - g| \le \|h - g\|_{\infty}$, it follows that

$$\|f - h\|_{L^2} \le \|f - g\|_{L^2} + \|g - h\|_{L^2} < \varepsilon.$$

The coefficients in the resulting expansion

$$f(x) = \sum_{k=0}^{\infty} a_k \cos k\pi x$$

can be found by taking the inner product with $\cos n\pi x$:

$$a_0 = \int_0^1 f(x)\,dx = a_0 \quad \text{and} \quad a_n = 2\int_0^1 f(x)\cos n\pi x\,dx, \ n \neq 0.$$

9.12 We can write $x = \sum_{k=-\infty}^{\infty} c_k e^{ikx}$ where the Fourier coefficients (c_k) are given by (9.7):

$$c_k = \frac{1}{2\pi}\int_{-\pi}^{\pi} x e^{ikx}\,dx.$$

If $k = 0$ this yields $c_0 = 0$, while for $k \neq 0$ we obtain

$$c_k = \frac{1}{2\pi}\left[\frac{x}{ik}e^{ikx}\right]_{-\pi}^{\pi} - \int_{-\pi}^{\pi}\frac{e^{ikx}}{ik}\,dx = \frac{(-1)^k}{ik}.$$

To use the Parseval identity, recall that the functions $\sqrt{\frac{1}{2\pi}}e^{ikx}$ are orthonormal, so we have

$$x = \sum_{k\neq 0}\frac{(-1)^k\sqrt{2\pi}}{ik}\left[\frac{1}{\sqrt{2\pi}}e^{ikx}\right];$$

therefore

$$4\pi\sum_{k=1}^{\infty}\frac{1}{k^2} = \int_{-\pi}^{\pi}x^2\,dx = \frac{2\pi^3}{3},$$

which yields $\sum_{k=1}^{\infty}k^{-2} = \pi^2/6$.

9.13 Suppose that H contains an orthonormal set $E_k = \{e_j\}_{j=1}^{k}$. Then E_k does not form a basis for H, since H is infinite-dimensional. It follows that there exists a non-zero $u_k \in H$ such that

$$(u_k, e_j) = 0 \quad \text{for all} \quad j = 1,\ldots,k,$$

for otherwise, by characterisation (c) of Proposition 9.14, E_k would be a basis. Now define $e_{k+1} = u_k/\|u_k\|$ to obtain an orthonormal set $E_{k+1} = \{e_1,\ldots,e_k\}$. The result follows by induction, starting with $e_1 = x/\|x\|$ for some non-zero $x \in H$.

9.14 The closed unit ball is both closed and bounded. When H is finite-dimensional this is equivalent to compactness by Theorem 5.3. If H is infinite-dimensional, then by the previous exercise it contains a countable orthonormal set $\{e_j\}_{j=1}^{\infty}$, and we have $\|e_i - e_j\|^2 = 2$ for $i \neq j$. The (e_j) therefore form a sequence in the unit ball that can have no convergent subsequence.

9.15 An application of Zorn's Lemma shows that there is a maximal orthonormal set: let P be the set of all orthonormal subsets of H order by inclusion. Then, for any chain C, the set

$$E = \bigcup_{A\in C} A$$

is an orthonormal set that provides an upper bound for C, and so P has a maximal element $E = (e_\alpha)_{\alpha\in\mathbb{A}}$.

Now suppose that there exists $y \in H$ that cannot be written as $\sum_{j=1}^{\infty} a_j e_{\alpha_j}$ for any choice of $(\alpha_j) \in \mathbb{A}$ and $a_j \in \mathbb{K}$. Set $a_\alpha = (y, e_\alpha)$; then, by Exercise 9.4, a_α is only non-zero for a countable collection $\{\alpha_j\} \in \mathbb{A}$. It then follows from Corollary 9.13 that $x = \sum_{\alpha \in \mathbb{A}} a_\alpha e_\alpha$ converges, with $x \in H$. By construction

$$(y - x, e_\alpha) = (y, e_\alpha) - (x, e_\alpha) = a_\alpha - a_\alpha = 0,$$

and so $y - x$ is orthogonal to every element of E. Since $y - x$ is non-zero by assumption, so is $e' := (y - x)/\|y - x\|$; then $E \cup \{e'\}$ is an orthonormal set this is larger than E. This contradicts the maximality of E, so E is indeed a basis for H.

Chapter 10

10.1 Proposition 10.1 is equivalent to finding the element of

$$A - x = \{a - x : a \in A\}$$

with minimum norm; since $x \notin A$, $0 \notin A - x$, and $A - x$ is a closed convex set whenever A is.

10.2 For any $\delta > 0$ define $\alpha_\delta \in C([-1, 1])$ by

$$\alpha_\delta(t) := \begin{cases} -(1+\delta) & -1 \leq x < -2\delta/(1+\delta) \\ \frac{(1+\delta)^2}{2\delta} x & -2\delta/(1+\delta) \leq x \leq 2\delta/(1+\delta) \\ (1+\delta) & 2\delta/(1+\delta) < x \leq 1. \end{cases}$$

Observe that $\|\alpha_\delta\|_\infty = 1 + \delta$ and that

$$\int_{-1}^0 \alpha_\delta(t)\, dt = -1 \qquad \text{and} \qquad \int_0^1 \alpha_\delta(t)\, dt = 1.$$

It follows that $f(t) := g(t) + \alpha_\delta(t)$ is an element of U with $\|f - g\|_\infty < 1 + \delta$.

Now, for any $f \in U$ we have

$$\int_{-1}^0 g(t) - f(t) = 1 \qquad \text{and} \qquad \int_0^1 g(t) - f(t) = -1.$$

Since g and f are both continuous,

$$\max_{t \in [-1, 0]} g(t) - f(t) \geq 1, \tag{S.15}$$

with equality holding if and only if $g(t) - f(t) = 1$ for all $t \in [-1, 0]$; similarly

$$\min_{t \in [0, 1]} g(t) - f(t) \leq -1, \tag{S.16}$$

with equality holing if and only if $g(t) - f(t) = -1$ for all $t \in [0, 1]$. We cannot have both $g(0) - f(0) = 1$ and $g(0) = f(0) = -1$, so the inequality in either (S.15) or (S.16) must be strict. Hence $\|g - f\|_\infty > 1$ for every $f \in U$.

Combining this with our previous upper bound for a particular choice of f shows that $\text{dist}(g, U) = 1$, but we have also just shown that $\|g - f\|_\infty > 1$ for every $f \in U$.

10.3 Suppose that $x \notin U$, and $u, v \in U$ with $u \neq v$ such that

$$\|x - u\| = \|x - v\| = \text{dist}(x, U).$$

Now, if we set $w = (u + v)/2 \in U$, then by strict convexity

$$\|x - u + x - v\| < 2\,\text{dist}(x, U) \quad \Rightarrow \quad \|x - w\| < \text{dist}(x, U),$$

contradicting the definition of $\text{dist}(x, U)$.

10.4 If $x, y \in X$ with $\|x\| = \|y\| = 1$ and $x \neq y$, then certainly $\|x - y\| > \varepsilon$ for some $\varepsilon > 0$. If X is uniformly convex, then there exists some $\delta > 0$ such that

$$\left\|\frac{x + y}{2}\right\| < 1 - \delta \quad \Rightarrow \quad \|x + y\| < 2(1 - \delta) < 2,$$

and so X is strictly convex.

10.5 First divide the inequality

$$\alpha^p + \beta^p \le (\alpha^2 + \beta^2)^{p/2}$$

by β^p to see that it is sufficient to prove that

$$f(x) := (x^2 + 1)^{p/2} - x^p - 1 \ge 0,$$

which follows since $f(0) = 0$ and

$$f'(x) = xp(x^2 + 1)^{p/2-1} - px^{p-1} \ge 0.$$

Now set $\alpha = |\frac{1}{2}(a + b)|$ and $\beta = |\frac{1}{2}(a - b)|$ to yield

$$\left|\frac{a + b}{2}\right|^p + \left|\frac{a - b}{2}\right|^p \le \left(\left|\frac{a + b}{2}\right|^2 + \left|\frac{a - b}{2}\right|^2\right)^{p/2}$$

$$= \left(\frac{|a|^2}{2} + \frac{|b|^2}{2}\right)^{p/2} \le \frac{1}{2}\left(|a|^p + |b|^p\right),$$

since $t \mapsto |t|^{p/2}$ is convex for $p \ge 2$. Now for the L^p inequality just integrate

$$\left|\frac{f(x) + g(x)}{2}\right|^p + \left|\frac{f(x) - g(x)}{2}\right|^p \le \frac{1}{2}(|f(x)|^p + |g(x)|^p)$$

over Ω to obtain (10.3). For ℓ^p we sum the inequalities

$$\left|\frac{x_j + y_j}{2}\right|^p + \left|\frac{x_j - y_j}{2}\right|^p \le \frac{1}{2}(|x_j|^p + |y_j|^p).$$

10.6 Take $f, g \in L^p$ with $\|f\|_{L^p} = \|g\|_{L^p} = 1$ and $\|f - g\|_{L^p} > \varepsilon$. Then

$$\left\|\frac{f + g}{2}\right\|_{L^p}^p \le 1 - (\varepsilon/2)^p,$$

so $\|\frac{1}{2}(f + g)\|_{L^p} < 1 - \delta$ with $\delta = 1 - [1 - (\varepsilon/2)^p]^{1/p} > 0$, which shows that L^p is uniformly convex. For the case $1 < p \le 2$ we again take $f, g \in L^p$ with $\|f\|_{L^p} = \|g\|_{L^p} = 1$ and $\|f - g\|_{L^p} > \varepsilon$. Clarkson's second inequality gives

$$\left\|\frac{f + g}{2}\right\|_{L^p}^q + \le 1 - (\varepsilon/2)^q,$$

and the argument concludes similarly. (The proof in ℓ^p is more or less identical.)

10.7 (i) Let $d := \inf\{\|x\| : x \in K\} > 0$. Then there exist $k_n \in K$ such that $\|k_n\| \to \delta$. Set $x_n = k_n/\|k_n\|$ so that $\|x_n\| = 1$. Now

$$\frac{x_n + x_m}{2} = \frac{1}{2\|k_n\|}k_n + \frac{1}{2\|k_m\|}k_m$$

$$= \left(\frac{1}{2\|k_n\|} + \frac{1}{2\|k_m\|}\right)(c_n k_n + c_m k_m);$$

note that $c_n + c_m = 1$, and so since K is convex $c_n k_n + c_m k_m \in K$. It follows that $\|c_n k_n + c_m k_m\| \geq d$ and so

$$\left\|\frac{x_n + x_m}{2}\right\| \geq \frac{d}{2\|k_n\|} + \frac{d}{2\|k_m\|}.$$

(ii) Since $\|k_n\| \to d$, we can find N such that for all $n, m \geq N$

$$\left\|\frac{x_n + x_m}{2}\right\| > 1 - \delta. \tag{S.17}$$

(iii) Now fix $\varepsilon > 0$. Since X is uniformly convex, there exists $\delta > 0$ such that $\|x\| = \|x'\| = 1$ and $\|x - x'\| > \varepsilon$ implies that $\|(x + x')/2\| < 1 - \delta$. This implies that if $\|x\| = \|x'\| = 1$ and $\|(x + x')/2\| \geq 1 - \delta$ then $\|x - x'\| \leq \varepsilon$. It follows from (S.17) that

$$\|x_n - x_m\| < \varepsilon$$

for all $n, m \geq N$, so (x_n) is Cauchy.

(iv) We now deduce that (k_n) in also Cauchy. Indeed, since $k_n = \|k_n\|x_n$ and $\|k_n\| \to d$ we can write

$$\|k_n - k_m\| = \big\|\|k_n\|x_n - \|k_m\|x_m\big\|$$

$$\leq \big|\|k_n\| - \|k_m\|\big|\|x_n\| + \|k_m\|\|x_n - x_m\|$$

$$= \big|\|k_n\| - \|k_m\|\big| + M\|x_n - x_m\|,$$

where $\|k_m\| \leq M$ for some $M > 0$ since $\|k_m\| \to \delta$. Give any $\eta > 0$ choose N' such that $\big|\|k_n\| - d\big| < \eta$ for all $n \geq N'$ and $\|x_n - x_m\| < \eta/3$ for all $m, n \geq N'$; then, for $m, n \geq N'$, we have $\|k_n - k_m\| < \eta$, which shows that (k_n) is Cauchy. Since X is complete, it follows that $k_n \to k$; since K is closed we must have $k \in K$, and we have $\|k\| = \lim_{n \to \infty}\|k_n\| = d$.

10.8 If $u \in (X + Y)^\perp$ then

$$(u, x + y) = 0 \qquad x, y \in X, Y.$$

Choosing $y = 0 \in Y$ shows that $u \in X^\perp$; choosing $x = 0 \in X^\perp$ shows that $u \in Y^\perp$, so $u \in X^\perp \cap Y^\perp$. For the reverse inclusion, if $x \in X^\perp \cap Y^\perp$, then

$$(u, x + y) = (u, x) + (u, y) = 0.$$

10.9 Since $\text{Span}(E) \subseteq \text{clin}(E)$, we have $(\text{Span}(E))^\perp \supseteq (\text{clin}(E))^\perp$. To show equality, take $y \in (\text{Span}(E))^\perp$ and $u \in \text{clin}(E)$: we want to show that we have $(y, u) = 0$ so that $u \in (\text{clin}(E))^\perp$. Now, since $u \in \text{clin}(E)$, there exists a sequence $x_n \in \text{Span}(E)$ such that $x_n \to 0$. Therefore

$$(y, u) = (y, \lim_{n \to \infty} x_n) = \lim_{n \to \infty} (y, x_n) = 0,$$

as required.

10.10 Noting that

$$T([x]) = P^\perp(x + M) = P^\perp x$$

it is clear that Range$(T) \subset M^\perp$, and that T is linear. Since for any $m \in M$ we have

$$\|x + m\|^2 = \|P_M(x + m)\|^2 + \|P^\perp x\|^2,$$

it follows that

$$\|[x]\|^2 = \inf_{m \in M} \|x + m\|^2 = \|P^\perp x\|^2,$$

i.e. $\|[x]\| = \|P^\perp x\|$, which in particular shows that T is an isometry. It remains only to show that T is onto, but this is almost immediate, since given $x \in M^\perp$ we have $[x] = x + M$ and $P^\perp(x + M) = x$.

10.11 We have

$$e_4' = x^3 - \left(x^3, \sqrt{\frac{5}{8}}(3x^2 - 1)\right)\sqrt{\frac{5}{8}}(3x^2 - 1) - \left(x^3, \sqrt{\frac{3}{2}}x\right)\sqrt{\frac{3}{2}}x$$

$$- \left(x^3, \frac{1}{\sqrt{2}}\right)\frac{1}{\sqrt{2}}$$

$$= x^3 - \frac{5}{8}(3x^2 - 1)\int_{-1}^{1} 3t^5 - t^3 \, dt - \frac{3}{2}x\int_{-1}^{1} t^4 \, dt - \frac{1}{2}\int_{-1}^{1} t^3 \, dt$$

$$= x^3 - \frac{3}{5}x$$

and

$$\|e_4'\|_{L^2}^2 = \int_{-1}^{1} x^6 - \frac{6}{5}x^4 + \frac{9}{25}x^2 \, dx = 2\left(\frac{1}{7} - \frac{6}{25} + \frac{3}{25}\right) = \frac{8}{175}.$$

It follows that

$$e_4 = \sqrt{\frac{175}{8}}\frac{1}{5}(5x^3 - 3) = \sqrt{\frac{7}{8}}(5x^3 - 3).$$

10.12 We have, using repeated integration by parts,

$$\int_{-1}^{1} x^k u_n(x) \, dx = \int_{-1}^{1} x^k u_n^{(n)}(x) \, dx$$

$$= \left[x^k u^{(n-1)}(x)\right]_{-1}^{1} - k\int_{-1}^{1} x^{k-1} u_n^{(n-1)}(x) \, dx$$

$$= -k\int_{-1}^{1} x^{k-1} u_n^{(n-1)}(x) \, dx$$

$$= \cdots$$

$$= (-1)^k k! \int_{-1}^{1} u_n^{(n-k)}(x)\, dx$$

$$= (-1)^k k! \left[u_n^{(n-k-1)}(x) \right]_{-1}^{1} = 0.$$

Now observe that if $m < n$, then P_m is a polynomial of order m, and P_n is proportional to $u^{(n)}(x)$. It follows that $(P_m, P_n) = 0$ as claimed.

Chapter 11

11.1 Since T is linear, for any $x \in X$ with $\|x\| < 1$ we have

$$\|Tx\|_Y = \|x\|_X \left\| T \frac{x}{\|x\|_X} \right\|_Y,$$

and so

$$\sup_{\|x\|_X \leq 1} \|Tx\|_Y \leq \sup_{\|x\|_X = 1} \|Tx\|_Y.$$

The reverse inequality follows immediately since $\{\|x\|_X = 1\} \subset \{\|x\|_X \leq 1\}$; now use Lemma 11.6.

For the second expression, note that if $x \neq 0$, then

$$\frac{\|Tx\|_Y}{\|x\|_X} = \left\| T \frac{x}{\|x\|} \right\|_Y;$$

therefore

$$\sup_{x \neq 0} \frac{\|Tx\|_Y}{\|x\|_X} \leq \sup_{\|z\|_X = 1} \|Tz\|_Y,$$

and the reverse inequality follows by restricting the supremum to those x for which $\|x\|_X = 1$ on the left-hand side.

11.2 This is clearly a linear mapping, since the integral is linear.

We check that $Tf \in C_b([0, \infty))$. That Tf is continuous at each $x > 0$ is clear, since $\int_0^x f(s)\, ds$ is continuous in x, and $x \mapsto 1/x$ is continuous at each point in $(0, 1)$. So we only need to check that Tf is continuous at $x = 0$ and that it is bounded on $\mathbb{R}$.

To see that Tf is continuous at zero, consider

$$|[Tf](x) - f(0)| = \left| \frac{1}{x} \int_0^x f(s)\, ds - f(0) \right|$$

$$= \left| \frac{1}{x} \int_0^x f(s) - f(0)\, ds \right|$$

$$\leq \frac{1}{x} \int_0^x |f(s) - f(0)|\, ds.$$

Since f is continuous at 0, for any $\varepsilon > 0$ there exists $\delta > 0$ such that $|f(s) - f(0)| < \varepsilon$ for all $0 \leq s < \delta$. If we take $0 \leq x < \delta$, then we obtain

$$|[Tf](x) - f(0)| \leq \frac{1}{x} \int_0^x \varepsilon\, ds = \varepsilon,$$

which shows that $Tf \in C([0, \infty))$.

The function Tf is bounded since $|[Tf](0)| = |f(0)| \le \|f\|_\infty$ and for every $x > 0$ we have

$$\|[Tf](x)\| \le \frac{1}{x}\int_0^x \|f\|_\infty \, dx \le \|f\|_\infty;$$

it follows that $\|Tf\|_\infty \le \|f\|_\infty$, which also shows $\|T\|_{B(X)} \le 1$. The choice $f(x) = 1$ for all x yields $[Tf](x) = 1$ for all x, and so $\|Tf\|_\infty = \|f\|_\infty$ for this f, which shows that $\|T\|_{B(X)} = 1$.

11.3 If T is bounded, then it is continuous, in which case

$$T\left(\sum_{j=1}^\infty x_j\right) = T\left(\lim_{n\to\infty}\sum_{j=1}^n x_j\right) = \lim_{n\to\infty} T\left(\sum_{j=1}^n x_j\right)$$

$$= \lim_{n\to\infty}\sum_{j=1}^n Tx_j = \sum_{j=1}^\infty Tx_j.$$

For the converse, suppose that $y_k \to y$. Then set $y_0 = 0$ and put $x_k = y_k - y_{k-1}$. Then

$$\sum_{j=1}^n x_j = y_n \qquad \text{and} \qquad T\left(\sum_{j=1}^n x_j\right) = \sum_{j=1}^n Ty_j - Ty_{j-1} = Ty_n.$$

The assumed equality means that

$$Ty = T\left(\lim_{n\to\infty}\sum_{j=1}^n x_j\right) = \lim_{n\to\infty}\sum_{j=1}^n Tx_j = \lim_{n\to\infty} Ty_n,$$

i.e. $Ty_n \to Ty$ whenever $y_n \to y$, which shows that T is continuous, and therefore bounded.

11.4 We have

$$\|S_nT_n - ST\|_{B(X,Z)} = \|S_n(T_n - T) + T(S_n - S)\|_{B(X,Z)}$$
$$\le \|S_n\|_{B(Y,Z)}\|T_n - T\|_{B(X,Y)} + \|T\|_{B(X,Y)}\|S_n - S\|_{B(Y,Z)}$$
$$\le M\|T_n - T\|_{B(X,Y)} + \|T\|_{B(X,Y)}\|S_n - S\|_{B(Y,Z)},$$

where we use that fact that $\|S_n\|_{B(Y,Z)} \le M$ for some M since S_n converges in $B(Y,Z)$. Convergence of S_nT_n to ST now follows.

11.5 Since $\sum_{j=1}^\infty \|T^j\| < \infty$ we know that the partial sums are Cauchy; in particular, for any $\varepsilon > 0$ there exists N such that for $n > m \ge N$

$$\sum_{j=m+1}^n \|T^j\| < \varepsilon$$

(as a particular case we have $\|T^j\| \to 0$ as $j \to \infty$). If we consider

$$V_n = I + T + \cdots + T^n,$$

then for $n > m \geq N$ we have

$$\|V_n - V_m\| = \|T^{m+1} + \cdots + T^{n-1} + T^n\|$$

$$\leq \|T^{m+1}\| + \cdots + \|T^{n-1}\| + \|T^n\| = \sum_{j=m+1}^{n} \|T^j\| < \varepsilon.$$

It follows that (V_n) is Cauchy in $B(X)$, and so (since $B(X)$ is complete by Theorem 11.11), V_n converges to some $V \in B(X)$ with $\|V\| \leq \sum_{j=1}^{\infty} \|T^j\|$.

For any finite n we have

$$(I + T + \cdots + T^n)(I - T) = I - T^{n+1} = (I - T)(I + T + \cdots + T^n);$$

taking $n \to \infty$ we can use the fact that $\sum_{j=1}^{n} T^j \to V$ and $T^n \to 0$ as $n \to \infty$ to deduce (using the result of the previous exercise) that

$$V(I - T) = I = (I - T)V,$$

so $V = (I - T)^{-1}$.

If $\|T\| < 1$, then $\|T^n\| \leq \|T\|^n$, and then

$$\sum_{j=1}^{\infty} \|T^j\| \leq \sum_{j=1}^{\infty} \|T\|^j = \frac{1}{1 - \|T\|} < \infty.$$

11.6 Let $P = T^{-1}(T + S) = I + T^{-1}S$. Then, since $\|T^{-1}\|\|S\| < 1$, it follows from the previous exercise that P is invertible with

$$\|P\|^{-1} \leq \frac{1}{1 - \|T^{-1}\|\|S\|}.$$

Using the definition of P we have

$$T^{-1}(T + S)P^{-1} = P^{-1}T^{-1}(T + S) = I;$$

from the first of these identities we have, acting with T on the left and then T^{-1} on the right,

$$(T + S)P^{-1}T^{-1} = I$$

and so $(T + S)^{-1} = P^{-1}T^{-1}$ and (11.15) follows.

11.7 We argue by induction. If $|T^n f(x)|_\infty \leq M^n \|f\|_\infty (x - a)^n/n!$, then

$$\left|T^{n+1} f(x)\right| = |T(T^n f(x))|$$

$$= \left|\int_a^x K(x, y)[T^n f(y)] \, dy\right| \leq \int_a^x |K(x, y)||T^n f(y)| \, dy$$

$$\leq \frac{M^{n+1}}{n!}\|f\|_\infty \int_a^x (y - a)^n \, dy = \frac{M^{n+1}}{(n + 1)!}\|f\|_\infty (x - a)^{n+1}.$$

In particular,

$$\|T^n f\|_\infty \leq M^n \|f\|_\infty \frac{(b - a)^n}{n!} \quad \Rightarrow \quad \|T^n\|_{B(X)} \leq M^n \frac{(b - a)^n}{n!}. \qquad \text{(S.18)}$$

We rewrite the integral equation (11.16) as

$$(I - \lambda T)f = g.$$

The bounds on $\|T^n\|$ above imply that

$$\sum_{j=1}^{\infty} \|(\lambda T)^n\| \le \sum_{j=1}^{\infty} [\lambda M(b-a)]^n / n! < \infty.$$

The previous exercise then shows that $(I - \lambda T)$ is invertible with inverse $\sum_{j=0}^{\infty} (\lambda T)^j$, and so $f = \sum_{j=0}^{\infty} \lambda^j T^j g$.

11.8 That $\|T\|_{B(X,Y)} = \|T^{-1}\|_{B(Y,X)} = 1$ when T is an isometry is immediate. For the opposite implication, we have

$$\|Tx\|_Y \le \|x\|_X \qquad \text{and} \qquad \|T^{-1}y\|_X \le \|y\|_Y;$$

setting $y = Tx$ in the second inequality yields $\|x\|_X \le \|Tx\|_Y$ and shows that $\|Tx\|_Y = \|x\|_X$ as claimed.

11.9 Suppose that $y_n \in \text{Range}(T)$ with $y_n \to y \in Y$; then $y_n = Tx_n$ for some $x_n \in X$. Since y_n converges, it is Cauchy in Y, and since

$$\alpha \|x_n - x_m\|_X \le \|T(x_n - x_m)\|_Y = \|y_n - y_m\|_Y,$$

it follows that (x_n) is Cauchy in X. Because X is complete, $x_n \to x$ for some $x \in X$, and now since T is continuous $y_n = Tx_n \to Tx = y$. So $\text{Range}(T)$ is closed.

11.10 We have

$$\int_{-1}^{1} \int_{-1}^{1} |K(t,s)|^2 \, dt \, ds = \int_{-1}^{1} \int_{-1}^{1} \left(1 + 12ts + 36t^2 s^2\right) dt \, ds$$

$$= \int_{-1}^{1} 2 + 24s^2 \, ds = 4 + 16 = 20,$$

while $Tx(t) = \int_a^b K(t,s)x(s) \, ds = 2e_1(t)(x, e_1) + 4e_1(t)(x, e_2)$.
Since e_1 and e_2 are orthonormal,

$$\|Tx\|_{L^2}^2 = 4|(x, e_1)|^2 + 16|(x, e_2)|^2 \le 16(|(x, e_1)|^2 + |(x, e_2)|^2) \le 16\|x\|_{L^2}^2,$$

and so $\|T\| \le 4$. In fact, since $Te_2 = 4e_2$, we have $\|T\| = 4 < \sqrt{20}$.

11.11 Since $\|T^k\| \le \|T\|^k$, it follows that

$$\sum_{k=0}^{\infty} \|(T^k)/k!\| \le \sum_{k=0}^{\infty} \|T\|^k / k! < \infty.$$

Lemma 4.13 now ensures that $\sum_{k=0}^{\infty} T^k / k!$ converges to an element of $B(X)$.

Chapter 12

12.1 Let X be any infinite-dimensional normed space, let $\Xi := \{x_n\}_{n=1}^{\infty}$ be a linearly independent set with $\|x_n\| = 1$ and let E be a Hamel basis containing this set (see Exercise 1.8). Define $L: X \to \mathbb{R}$ by setting $L(x_n) = n$ for each n and $L(y) = 0$ for all

$y \in E \setminus \Xi$. The operator $L : \mathscr{X} \to \mathbb{K}$ defined by setting $L(\sum_j \alpha_j x_j) = \sum_j j\alpha_j$ is linear but unbounded.

12.2 Let $\{e_1, \dots, e_n\}$ be a basis for V and take any $f \in V^*$; then for any $(\alpha_j)_{j=1}^n \in \mathbb{K}$ we have, since f is linear

$$f\left(\sum_{j=1}^n \alpha_j e_j\right) = \sum_{j=1}^n \alpha_j f(e_j). \tag{S.19}$$

Define linear functionals $\{\phi_j\}_{j=1}^n \in V^*$ by setting $\phi_j(e_i) = \delta_{ij}$. These form a basis for V^*: they are linearly independent since if

$$\sum_{j=1}^n \alpha_j \phi_j = 0$$

we can apply both sides to e_k to show that $\alpha_k = 0$ for each k; and they span V^*, since we can write any $f \in V^*$ as $f = \sum_{i=1}^n f(e_i)\phi_i$. Indeed, using (S.19) we have

$$\left[\sum_{i=1}^n f(e_i)\phi_i\right]\left(\sum_{j=1}^n \alpha_j e_j\right) = \sum_{i,j} f(e_i)\alpha_j \phi_i(e_j)$$

$$= \sum_{i,j} f(e_i)\alpha_j \delta_{ij} = \sum_{j=1}^n \alpha_j f(e_j) = f\left(\sum_{j=1}^n \alpha_j e_j\right).$$

12.3 If u minimises F, then $F(u) \le F(u + tv)$ for every $t \in \mathbb{R}$, so

$$\frac{1}{2}B(u, u) - f(u) \le F(u + tv) = \frac{1}{2}B(u + tv, u + tv) - f(u + tv)$$

$$= \frac{1}{2}\left(B(u, u) + tB(u, v) + tB(v, u) + t^2 B(v, v)\right) - f(u) - tf(v)$$

$$= \frac{1}{2}\left(B(u, u) + 2tB(u, v) + t^2 B(v, v)\right) - f(u) - tf(v), \tag{S.20}$$

which yields

$$t[B(u, v) - f(v)] + \frac{t^2}{t}B(v, v) \ge 0.$$

Taking $t > 0$ and sufficiently small shows that we must have $B(u, v) - f(v) \ge 0$; taking $t < 0$ and sufficiently small shows that we also have $B(u, v) - f(v) \le 0$, and so $B(u, v) = f(v)$ for every $v \in V$.

For the contrary, if $B(u, v) = f(v)$ for every $v \in V$, then the expression in (S.20) gives

$$F(u + tv) = \frac{1}{2}B(u, u) - f(u) + \frac{t^2}{2}B(v, v) \ge F(u).$$

12.4 (i) By assumption for each fixed $x \in H$ the map $x \mapsto B(x, y)$ is a linear map from H into $\mathbb{R}$, so we only need check that it is bounded. But this follows immediately from assumption (ii), namely $|B(x, y)| \leq c\|x\|\|y\|$. The Riesz Representation Theorem guarantees that there exists $w \in H$ such that $B(x, y) = (x, w)$ and $\|w\| \leq c\|y\|$.

(ii) If we set $Ay = w$, then, by definition, $B(x, y) = (x, Ay)$. To show that A is linear, we use the linearity of B to write

$$(x, A(\alpha x + \beta z)) = B(x, \alpha y + \lambda z) = \alpha B(x, y) + \lambda B(x, z)$$
$$= \alpha(x, Ay) + \beta(x, Az) = (x, \alpha Ay + \beta Az);$$

since this holds for every $x \in H$, it follows that A is linear. To see that A is bounded, either use the fact that $\|Ay\| = \|w\| \leq c\|y\|$ from part (i), or note that

$$\|Au\|^2 = (Au, Au) = B(Au, u) \leq c\|u\|\|Au\|,$$

and so $\|Au\| \leq c\|u\|$.

(iii) We have

$$\|Ty - Ty'\|^2 = \|(y - y') - \varrho A(y - y')\|^2$$
$$= \|y - y'\|^2 - 2\varrho(y - y', A(y - y')) + \varrho^2\|A(y - y')\|^2$$
$$= \|y - y'\|^2 - 2\varrho B(y - y', y - y') + \varrho^2\|A(y - y')\|^2$$
$$\leq \|y - y'\|^2[1 - 2\varrho b + \varrho^2 c^2],$$

and by choosing ϱ sufficiently small we can ensure that T is a contraction.

Chapter 13

13.1 We have

$$(Tf, g) = \int_0^1 \left(\int_0^t K(t, s) f(s) \, ds \right) g(t) \, dt$$
$$= \int_0^1 \int_0^t K(t, s) f(s) g(t) \, ds \, dt$$
$$= \int_0^1 \int_s^1 K(t, s) f(s) g(t) \, dt \, ds = (f, T^*g),$$

where

$$T^*g(s) = \int_s^1 K(t, s) g(t) \, dt.$$

13.2 Suppose that $T_n \to T$. Since $\|T_n^* - T^*\| = \|(T_n - T)^*\| = \|T_n - T\|$, it follows that

$$\|T - T^*\| \leq \|T - T_n\| + \|T_n - T_n^*\| + \|T_n^* - T\| = 2\|T_n - T\|;$$

since the right-hand side tends to zero it follows that $\|T - T^*\| = 0$, i.e. that $T = T^*$ and so T is self-adjoint.

13.3 Take $x \in \text{Ker}(T)$ and $z \in \text{Range}(T^*)$, so that $z = T^*y$ for some $y \in H$. Then

$$(x, z) = (x, T^*y) = (Tx, y) = 0;$$

it follows that $\text{Ker}(T) \subset \text{Range}(T^*)^\perp$.

Now suppose that $z \in \text{Range}(T^*)^\perp$; since $T^*(Tz) \in \text{Range}(T^*)$ it follows that

$$\|Tz\|^2 = (Tz, Tz) = (z, T^*(Tz)) = 0,$$

and so $Tz = 0$. This shows that $\text{Range}(T^*)^\perp \subset \text{Ker}(T)$, and equality follows.

13.4 Since $\|T\| = \|T^*\|$, we have

$$\|T^*T\| \le \|T^*\|\|T\| = \|T\|^2.$$

But we also have

$$\|Tx\|^2 = (Tx, Tx) = (x, T^*Tx) \le \|x\|\|T^*Tx\| \le \|T^*T\|\|x\|^2,$$

i.e. $\|T\|^2 \le \|T^*T\|$, which gives equality.

13.5 Since T is invertible, we have $T^{-1} \in B(K, H)$ and $TT^{-1} = T^{-1}T = I$. Taking adjoints yields

$$(T^{-1})^*T^* = T^*(T^{-1})^* = I^* = I.$$

This shows that $(T^*)^{-1} = (T^{-1})^*$, and since $\|(T^{-1})^*\| = \|T^{-1}\|$ it follows that T^* is invertible. If T is self-adjoint, then $T^* = T$, and so $(T^{-1})^* = (T^*)^{-1} = T^{-1}$, i.e. T^{-1} is self-adjoint.

Chapter 14

14.1 First we check that $(\cdot, \cdot)_{H_\mathbb{C}}$ is an inner product on $H_\mathbb{C}$.

(i) positivity: $(x + iy, x + iy)_{H_\mathbb{C}} = (x, x) + (y, y) = \|x\|^2 + \|y\|^2 \ge 0$, and if $\|x\|^2 + \|y\|^2 = 0$, then $x = y = 0$;

(ii) linearity:

$$\begin{aligned}
((x + iy) + (x' + iy'), w + iz)_{H_\mathbb{C}} &= ((x + x') + i(y + y'), w + iz)_{H_\mathbb{C}} \\
&= (x + x', w) + i(y + y', w) - i(x + x', z) + (y + y', z) \\
&= [(x, w) + i(y, w) - i(x, z) + (y, z)] \\
&\quad + [(x', w) + i(y', w) - i(x', z) + (y', z)] \\
&= (x + iy, w + iz)_{H_\mathbb{C}} + (x' + iy', w + iz)_{H_\mathbb{C}};
\end{aligned}$$

(iii) scalar multiples: if $\alpha \in \mathbb{C}$ with $\alpha = a + ib$, $a, b \in \mathbb{R}$, then

$$\begin{aligned}
(\alpha(x + iy), w + iz)_{H_\mathbb{C}} &= ((ax - by) + i(bx + ay), w + iz)_{H_\mathbb{C}} \\
&= (ax - by, w) + i(bx + ay, w) \\
&\quad - i(ax - by, z) + (bx + ay, z) \\
&= (a + ib)(x, w) + (ai - b)(y, w) \\
&\quad + (b - ia)(x, z) + (a + ib)(y, z) \\
&= (a + ib)[(x, w) + i(y, w) - i(x, z) + (y, z)] \\
&= \alpha(x + iy, w + iz)_{H_\mathbb{C}};
\end{aligned}$$

(iv) conjugation: using the fact that $(a, b) = (b, a)$ since H is real we have

$$\overline{(x + iy, w + iz)_{H_{\mathbb{C}}}} = \overline{(x, w) + i(y, w) - i(x, z) + (y, z)}$$
$$= (w, x) - i(w, y) + i(z, x) + (z, y)$$
$$= (w + iz, x + iy)_{H_{\mathbb{C}}}.$$

The space $H_{\mathbb{C}}$ is a Hilbert space: it is isomorphic to $H \times H$ with the norm $\|(x, y)\|^2_{H \times H} = \|x\|^2_H + \|y\|^2_H$, and so complete (see Lemma 4.6).

14.2 (i) We have to show that $T_{\mathbb{C}}$ is complex-linear and bounded. If $a, b \in \mathbb{R}$, then

$$T_{\mathbb{C}}((a + ib)(x + iy)) = T_{\mathbb{C}}((ax - by) + i(bx + ay)) = T(ax - by) + iT(bx + ay)$$
$$= (a + ib)Tx + (ia - b)Ty = (a + ib)[Tx + iTy] = (a + ib)T_{\mathbb{C}}(x + iy),$$

so $T_{\mathbb{C}}$ is complex-linear. To show that $T_{\mathbb{C}}$ is bounded, we estimate

$$\|T_{\mathbb{C}}(x + iy)\|^2_{H_{\mathbb{C}}} = \|Tx\|^2 + \|Ty\|^2 \le \|T\|^2_{B(H)}(\|x\|^2 + \|y\|^2)$$
$$= \|T\|^2_{B(H)}\|x + iy\|^2_{H_{\mathbb{C}}},$$

so $\|T_{\mathbb{C}}\| \le \|T\|$. But also $\|Tx\| = \|T_{\mathbb{C}}(x + i0)\|$, and so $\|T\| \le \|T_{\mathbb{C}}\|$, which shows that $\|T_{\mathbb{C}}\| = \|T\|$.

If λ is an eigenvalue of T with eigenvector x, then

$$T_{\mathbb{C}}(x + i0) = Tx = \lambda x = \lambda(x + i0);$$

while if $T_{\mathbb{C}}$ has eigenvalue $\lambda \in \mathbb{R}$ with eigenvector $x + iy$ (with either x or y non-zero), it follows that

$$Tx + iTy = T_{\mathbb{C}}(x + iy) = \lambda(x + iy) = \lambda x + i\lambda y,$$

and so $Tx = \lambda x$ and $Ty = \lambda y$, so since x or y is non-zero, λ is an eigenvalue of T.

14.3 Suppose that $D_{\alpha}x = \lambda x$; then

$$\alpha_j x_j = \lambda x_j \qquad \Rightarrow \qquad (\alpha_j - \lambda)x_j = 0$$

for every j. So either $\lambda = \alpha_j$ or $x_j = 0$. This shows that the only eigenvalues are $\{\alpha_j\}_{j=1}^{\infty}$.

Suppose now that $\lambda \in \mathbb{C}$ with $\lambda \notin \overline{\sigma_p(D_{\alpha})}$; then $|\lambda - \alpha_j| \ge \delta$ for every $j \in \mathbb{N}$. For such λ note that

$$[(D_{\alpha} - \lambda I)x]_j = (\alpha_j - \lambda)x_j;$$

since $|\alpha_j - \lambda| \ge \delta$ for every j this map is one-to-one and onto, i.e. a bijection. So we can define its inverse by setting

$$[(D_{\alpha} - \lambda I)^{-1}x]_j = (\alpha_j - \lambda)^{-1}x_j,$$

and then

$$|[(D_{\alpha} - \lambda I)^{-1}x]_j| = |(\alpha_j - \lambda)^{-1}x_j| \le \frac{1}{\delta}|x_j|,$$

which shows that $[D_{\alpha} - \lambda I]^{-1}: \ell^{\infty} \to \ell^{\infty}$ is bounded. So $D_{\alpha} - \lambda I$ is invertible and $\lambda \notin \sigma(D_{\alpha})$.

That $\sigma(D_{\alpha}) = \overline{\sigma_p(D_{\alpha})}$ now follows from the fact that the spectrum is closed (and must contain the point spectrum).

Finally, given any compact subset K of $\mathbb{C}$ we can find a set $\{\alpha_j\}_{j=1}^{\infty}$ that is dense in K (see Exercise 2.14).

14.4 We use the Spectral Mapping Theorem (Theorem 14.9) and (14.3)

$$\sigma(T) \subseteq \{\lambda \in \mathbb{C} : |\lambda| \le \|T\|\}.$$

If $\lambda \in \sigma(T)$, then $\lambda^n \in \sigma(T^n)$, so

$$[r_\sigma(T)]^n = r_\sigma(T^n),$$

and since $\sigma(T^n) \subseteq \{\lambda : |\lambda| \le \|T^n\|\}$ it follows that we have

$$r_\sigma(T) \le \|T^n\|^{1/n}$$

for every n, and so $r_\sigma(T) \le \liminf_{n\to\infty} \|T^n\|^{1/n}$.

14.5 This operator is the same as that in Exercise 11.7 for the particular choice $K(x, y) = 1$ and $[a, b] = [0, 1]$. In this case the general bound we obtained in the solution as (S.18) becomes

$$\|T^n\|_{B(X)} \le \frac{1}{n!}.$$

It now follows from the previous exercise that

$$r_\sigma(T) \le \liminf_{n\to\infty} \left(\frac{1}{n!}\right)^n = 0,$$

so no non-zero λ is contained in the spectrum. Since $T: X \to X$ is not invertible we know that $0 \in \sigma(T)$; T is not surjective, since for any $f \in C([0, 1])$ we have $Tf \in C^1([0, 1])$. But 0 is not an eigenvalue, since

$$Tf = 0 \quad \Rightarrow \quad \int_0^x f(s)\,ds = 0 \quad \Rightarrow \quad f(x) = 0, \ x \in [0, 1]$$

using the Fundamental Theorem of Calculus.

14.6 Simply note that $\mathcal{F}^4 = \text{id}$. It follows from the Spectral Mapping Theorem that if $\lambda \in \sigma(T)$, then $\lambda^4 = 1$, so $\sigma(T) \subset \{\pm 1, \pm i\}$.

Chapter 15

15.1 Suppose that T is compact and take any sequence in $T\mathbb{B}_X$, i.e. a sequence (Tx_n) with $(x_n) \in \mathbb{B}_X$. Then (x_n) is a bounded sequence in X, so (Tx_n) has a convergent subsequence.

Conversely, take any bounded sequence $(x_n) \in X$ with $\|x_n\|_X \le R$. The sequence $(x_n/R) \in \mathbb{B}_X$, so $(T(x_n/R)) \in T\mathbb{B}_X$ has a convergent subsequence (Tx_{n_j}/R). Since $T(x_{n_j}/R) = \frac{1}{R}Tx_{n_j}$, Tx_{n_j} converges, which shows that T is compact.

15.2 Let (x_n) be a bounded sequence in X. Suppose that S is compact: then (Tx_n) is a bounded sequence in Y, and so $S(Tx_n) = (S \circ T)x_n$ has a subsequence that converges in Z, i.e. $S \circ T$ is compact. If T is compact, then (Tx_n) has a subsequence that converges in Y, i.e. $Tx_{n_j} \to y \in Y$, and then since S is continuous $(S \circ T)x_{n_j} = S(Tx_{n_j}) \to Sy$, so once again $S \circ T$ is compact.

15.3 Consider the operators $T_n : \ell^2 \to \ell^2$ where

$$(T_n x)_j = \begin{cases} j^{-1} x_j & 1 \le j \le n, \\ 0 & j > n. \end{cases}$$

Then each T_n has finite-dimensional range, so is compact. We also have

$$\|Tx - T_n x\|_{\ell^2}^2 = \sum_{j=n+1}^{\infty} \frac{|x_j|^2}{j^2} \le \frac{1}{(n+1)^2} \|x\|_{\ell^2},$$

so $T_n \to T$ in $B(X)$. That T is compact now follows from Theorem 15.3.

Alternatively we can argue directly, using the compactness of the Hilbert cube Q from Exercise 9.6. Suppose that $(x^{(j)})_j$ is a bounded sequence in ℓ^2, $\|x^{(j)}\|_{\ell^2} \le M$. Then, in particular, $|x_i^{(j)}| \le M$ for all i, j. It follows that $(Tx^{(j)})$ is a sequence in MQ, where Q is the Hilbert cube defined in Exercise 9.6. We showed there that Q is compact, so MQ is also compact, which shows that T is a compact operator.

15.4 Since K is continuous on the compact set $[a, b] \times [a, b]$, it is bounded and uniformly continuous on $[a, b] \times [a, b]$.

Let $\mathbb{B}$ be the closed unit ball in $L^2(a, b)$. We will show that $T(\mathbb{B})$ is a bounded equicontinuous subset of $C([a, b])$, and so precompact in $C([a, b])$ by the Arzelà–Ascoli Theorem (Theorem 6.12). A precompact subset of $C([a, b])$ is also precompact in $L^2(a, b)$, since for any sequence $(f_n) \in C([a, b])$ we have

$$\|f_n - f_m\|_{L^2} \le (b - a)^{1/2} \|f_n - f_m\|_\infty.$$

Using Exercise 15.1 it will follow that $T : L^2 \to L^2$ is compact as claimed.

First, if $u \in \mathbb{B}$, then we have

$$|Tu(x)| \le \left(\int_a^b |K(x, y)|^2 dy \right)^{1/2} \left(\int_a^b |u(y)|^2 dy \right)^{1/2}$$

$$\le (b - a)^{1/2} \|K\|_\infty \|u\|_{L^2}$$

$$\le (b - a)^{1/2} \|K\|_\infty.$$

Since K is uniformly continuous, given any $\varepsilon > 0$, we can choose $\delta > 0$ so that $|K(x, y) - K(x', y)| < \varepsilon$ for all $y \in [a, b]$ whenever $|x - x'| < \delta$. Therefore if $u \in \mathbb{B}$ and $|x - x'| < \delta$ we have

$$|Tu(x) - Tu(x')| = \left| \int_a^b [K(x, y) - K(x', y)] u(y) \, dy \right|$$

$$\le \left(\int_a^b |K(x, y) - K(x', y)|^2 \, dy \right)^{1/2} \|u\|_{L^2}$$

$$\le [(b - a)\varepsilon^2]^{1/2} = (b - a)^{1/2} M\varepsilon.$$

Therefore $T(\mathbb{B})$ is equicontinuous.

15.5 Let $\{e_j\}$ be an orthonormal basis for H; then

$$Tu = T\left(\sum_{j=1}^{\infty}(u, e_j)e_j\right) = \sum_{j=1}^{\infty}(u, e_j)Te_j.$$

Therefore, using the Cauchy–Schwarz inequality,

$$\|Tu\| \le \sum_{j=1}^{\infty}|(u, e_j)|\|Te_j\|$$

$$\le \left(\sum_{j=1}^{\infty}|(u, e_j)|^2\right)^{1/2}\left(\sum_{j=1}^{\infty}\|Te_j\|^2\right)^{1/2} = \|T\|_{\mathrm{HS}}\|u\|,$$

which shows that $\|T\|_{B(H)} \le \|T\|_{\mathrm{HS}}$ as claimed.

15.6 We have

$$\|Te_j\|_{L^2}^2 = \int_a^b |(\kappa_x, e_j)|^2 \, dx = \int_a^b |(K(x, \cdot), e_j(\cdot))|^2 \, dx.$$

Therefore

$$\sum_{j=1}^{\infty}\|Te_j\|_{L^2}^2 = \sum_{j=1}^{\infty}\int_a^b |(\kappa_x, e_j)|^2 \, dx = \int_a^b \left(\sum_{j=1}^{\infty}|(\kappa_x, e_j)|^2\right) dx$$

$$= \int_a^b \|\kappa_x\|_{L^2}^2 \, dx = \int_a^b \int_a^b |K(x, y)|^2 \, dy \, dx < \infty,$$

and so T is Hilbert–Schmidt.

15.7 We show that S is Hilbert–Schmidt. Take the orthonormal basis $\{e^{(j)}\}_j$; then

$$(Se^{(k)})_i = \sum_{i=1}^{\infty} K_{ij}\delta_{kj} = K_{ik},$$

so

$$\|Se^{(k)}\|^2 = \sum_{i=1}^{\infty}|K_{ik}|^2 \quad \text{and} \quad \sum_{k=1}^{\infty}\|Se^{(k)}\|^2 = \sum_{k,i=1}^{\infty}|K_{ik}|^2 < \infty.$$

15.8 If there is no such sequence, then we must have $\|Tx\| \ge \delta$ for all $x \in S_X$. Since X if infinite-dimensional, we can use the argument from Theorem 5.5 to find $(x_j)_{j=1}^{\infty} \in S_X$ with $\|x_i - x_j\|_X \ge 1/2$ whenever $i \ne j$. It follows that

$$\frac{1}{\|x_i - x_j\|}\|Tx_i - Tx_j\| = \left\|T\frac{x_i - x_j}{\|x_i - x_j\|}\right\| \ge \delta,$$

i.e.

$$\|Tx_i - Tx_j\| \ge \frac{\delta}{2}.$$

This shows that (Tx_i) has no convergent subsequence, but this contradicts the compactness of T.

For a counterexample take the map $T : \ell^2 \to \ell^2$ given by

$$T(x_1, x_2, x_3, \ldots) = (x_1, \frac{x_2}{2}, \frac{x_3}{3}, \ldots).$$

This map is compact (see Exercise 15.3), but if $Tx = 0$, then $x_j/j = 0$ for every j, i.e. $x = 0$.

15.9 The operator T' is the composition of the compact operator T with the bounded operator σ_r, so it follows from Exercise 15.2 that T' is compact. To show that it has no eigenvalues, suppose first that $\lambda \neq 0$ is an eigenvalue; then

$$(0, x_1, \frac{x_2}{2}, \frac{x_3}{3}, \cdots) = \lambda(x_1, x_2, x_3, x_4, \ldots),$$

so

$$\lambda x_1 = 0, \qquad \lambda x_2 = x_1, \qquad \lambda x_3 = x_1, \qquad \cdots ;$$

but this implies that $x_j = 0$ for every j, so λ is not an eigenvalue. If we try $\lambda = 0$, then we immediately obtain $x_j = 0$ for every j also. So T' has no eigenvalues.

15.10 To see that T is not compact, consider the sequence $(e^{(2j-1)})_{j=1}^{\infty}$. This is a bounded sequence, but $(Te^{(2j-1)}) = (e^{2j})$ which has no convergent subsequence, so T cannot be compact.

Chapter 16

16.1 $(T(x + iy), x + iy) = (Tx, x) - i(Tx, y) + i(Ty, x) + (Ty, y)$
$$= (Tx, x) + i(y, Tx) - i(x, Ty) + (Ty, y),$$

which shows that

$$(Ty, x) + (x, Ty) = (Tx, y) + (y, Tx). \qquad \text{(S.21)}$$

Also

$$(T(x + y), x + y) = (Tx, x) + (Tx, y) + (Ty, x) + (Ty, y)$$
$$= (Tx, x) + (y, Tx) + (x, Ty) + (Ty, y)$$

which gives

$$(Ty, x) - (x, Ty) = (y, Tx) - (Tx, y). \qquad \text{(S.22)}$$

Adding (S.21) and (S.22) shows that

$$(Ty, x) = (y, Tx) \qquad \text{for every } x, y \in H,$$

and so T is self-adjoint.

16.2 Set $y = Tx$, then $(Tx, Tx) = 0$, so $\|Tx\|^2 = 0$ so $Tx = 0$. If H is complex, then we can use the equalities from the previous exercise: first

$$0 = (T(x + iy), x + iy) = -i(Tx, y) + i(Ty, x) \quad \Rightarrow \quad (Tx, y) - (Ty, x) = 0$$

and then

$$0 = (T(x + y), x + y) = (Tx, y) + (Ty, x).$$

Adding these yields $(Tx, y) = 0$, and we can set $y = Tx$ as before.

16.3 Note first that $((\beta I - T)x, x) \geq 0$ and $(Tx, x) \geq 0$, for every $x \in H$ since $V(T) \subseteq [0, \beta]$. Now, since T is self-adjoint we can write

$$\beta(Tx, x) - \|Tx\|^2 = ((\beta I - T)Tx, x)$$
$$= \frac{1}{\beta}((\beta I - T)Tx, (T + \beta I - T)x)$$
$$= \frac{1}{\beta}[(\beta I - T)Tx, Tx) + (T(\beta I - T)x, (\beta I - T)x)] \geq 0.$$

16.4 We prove the result for $\alpha = \inf V(T)$, since the result for β is more similar to the proof of Theorem 16.3. Find a sequence $(x_n) \in H$ with $\|x_n\| = 1$ such that $(Tx_n, x_n) \to \alpha$. Then $T - \alpha I$ is self-adjoint, $V(T - \alpha I) \subset [0, \beta - \alpha]$, and we have $((T - \alpha I)x, x) \geq 0$ for every $x \in H$, so we can use the result of the previous exercise to obtain

$$\|Tx_n - \alpha x_n\|^2 \leq (\beta - \alpha)((T - \alpha I)x_n, x_n) = (\beta - \alpha)[(Tx_n, x_n) - \alpha].$$

This shows that $Tx_n - \alpha x_n \to 0$ as $n \to \infty$. Now we argue as in the proof of Theorem 16.3 to deduce that α is an eigenvalue of T.

For $\beta = \sup V(T)$ we apply a similar argument to $\beta I - T$, for which we again have $V(\beta I - T) \subset [0, \beta - \alpha]$.

16.5 We have $Te_j = \lambda_j e_j$, so every λ_j is an eigenvalue. If $Tu = \lambda u$, then

$$\sum_{j=1}^{\infty} (\lambda_j - \lambda)(u, e_j)e_j = 0;$$

if $\lambda \neq \lambda_j$ then $(u, e_j) = 0$ for every j and hence $u = 0$. So λ is not an eigenvalue of T.
(i) We have

$$\|Tu\|^2 = \sum_{j=1}^{\infty} |\lambda_j|^2 |(u, e_j)|^2 \leq \left(\sup_j |\lambda_j|^2\right) \|u\|^2,$$

so if $\sup_j |\lambda_j| < \infty$, then T is bounded. If T is bounded, then for $u = e_k$ we have

$$\|Te_k\| = \|\lambda_k e_k\| = |\lambda_k| \leq \|T\|\|e_k\| = \|T\|,$$

so $\sup_k |\lambda_k| \leq \|T\|$.
(ii) If (λ_j) is bounded, then T is self-adjoint: we have

$$(Tu, v) = \left(\sum_{j=1}^{\infty} \lambda_j(u, e_j)e_j, v\right) = \sum_{j=1}^{\infty} \lambda_j(u, e_j)(e_j, v)$$
$$= \sum_{j=1}^{\infty} \lambda_j(u, e_j)\overline{(v, e_j)} = \sum_{j=1}^{\infty} (u, \lambda_j(v, e_j)e_j)$$
$$= \left(u, \sum_{j=1}^{\infty} \lambda_j(v, e_j)e_j\right) = (u, Tv).$$

(iii) Suppose that $\lambda_j \to 0$. Observe that each operator

$$T_n := \sum_{j=1}^{n} \lambda_j (u, e_j) e_j$$

is compact since its range is finite-dimensional (see Example 15.2). To show that T is compact we show that $T_n \to T$ in $B(H)$ and use Theorem 15.3. We have

$$\|(T - T_n)x\|^2 = \left\| \sum_{j=n+1}^{\infty} \lambda_j (u, e_j) e_j \right\|^2 \le \sum_{j=n+1}^{\infty} |\lambda_j|^2 |(u, e_j)|^2$$

$$\le \lambda_{n+1}^2 \sum_{j=n+1}^{\infty} |(u, e_j)|^2 \le \lambda_{n+1}^2 \|u\|^2,$$

which shows that $\|T - T_n\|_{B(H)} \le \lambda_{n+1}$ and $\lambda_{n+1} \to 0$ as $n \to \infty$ by assumption. If $\lambda_j \not\to 0$, then there is a $\delta > 0$ and a subsequence with $|\lambda_{j_k}| > \delta$, and then

$$\|T e_{j_k} - T e_{j_l}\|^2 = \|\lambda_{j_k} e_{j_k} - \lambda_{j_l} e_{j_l}\|^2 > \delta^2,$$

so $(T e_{j_k})$ has no Cauchy subsequence, contradicting the compactness of T.

(iv) Since $\|T e_j\| = |\lambda_j|$, $\sum |\lambda_j|^2 < \infty$ if and only if $\sum \|T e_j\|^2 < \infty$ (recall that the property of being Hilbert–Schmidt does not depend on the choice of orthonormal basis).

16.6 First note that

$$[Tu](x) = \int_a^b K(x, y) u(y) \, dy$$

$$= \int_a^b \sum_{j=1}^{\infty} \lambda_j e_j(x) e_j(y) u(y) \, dy = \sum_{j=1}^{\infty} \lambda_j e_j (e_j, u);$$

if we augment $\{e_j\}$ to an orthonormal basis of H, then this is an operator of the form in Exercise 16.5 if we assign the eigenvalue zero to the additional elements of the basis; the result then follows immediately.

16.7 For (i) we rewrite the kernel as

$$K(t, s) = \cos(t - s) = \cos t \cos s - \sin t \sin s,$$

recalling (see Exercise 9.1) that $\cos t$ and $\sin t$ are orthogonal in $L^2(-\pi, \pi)$. We have

$$T(\sin t) = \int_{-\pi}^{\pi} K(t, s) \sin s \, ds$$

$$= \int_{-\pi}^{\pi} \cos t \cos s \sin s - \sin t \sin^2 s \, ds = -2\pi \sin t$$

and

$$T(\cos t) = \int_{-\pi}^{\pi} K(t, s) \cos s \, ds$$

$$= \int_{-\pi}^{\pi} \cos t \cos^2 s - \sin t \sin s \cos s \, ds = 2\pi \cos t.$$

For (ii) we write $K(t, s)$ in terms of Legendre polynomials from Example 10.8:

$$K(t, s) = 4 \left(\sqrt{\frac{3}{2}} t \right) \left(\sqrt{\frac{3}{2}} s \right) + \frac{8}{5} \left(\sqrt{\frac{5}{8}} (3t^2 - 1) \right) \left(\sqrt{\frac{5}{8}} (3s^2 - 1) \right).$$

Since $\left\{ \sqrt{\frac{3}{2}} t, \sqrt{\frac{5}{8}} (3t^2 - 1) \right\}$ are orthonormal, the integral operator T has

$$T(t) = 4t \qquad \text{and} \qquad T(3t^2 - 1) = \frac{8}{5}(3t^2 - 1).$$

16.8 (i) Given any n-dimensional subspace V of H, any $(n + 1)$-dimensional subspace W of H contains a vector orthogonal to V. Indeed, if $\{w_j\}_{j=1}^{n+1}$ is a basis of W, then $P_V w_j \in V$, where V is the orthogonal projection onto V. Since V is n-dimensional, the $n + 1$ vectors $P_V w_j$ are linearly dependent, so there exists k such that

$$P_V w_k = \sum_{j \neq k} \alpha_j P_V w_j.$$

It follows that

$$P_V [w_k - \sum_{j \neq k} \alpha_j w_j] = 0,$$

so $w_k - \sum_{j \neq k} \alpha_j w_j \in W$ is orthogonal to V (note that this is non-zero, otherwise $w_1, \ldots, w_{n+1}$ would be linearly dependent).

(ii) Suppose that $x \in \text{Span}\{e_1, \ldots, e_n\}$, i.e. $x = \sum_{j=1}^{n} \alpha_j e_j$; then

$$(Tx, x) = \frac{\left(\sum_{j=1}^{n} \alpha_j \lambda_j e_j, \sum_{k=1}^{n} \alpha_k e_k \right)}{\sum_{j=1}^{n} |\alpha_j|^2} = \frac{\sum_{j=1}^{n} \lambda_j |\alpha_j|^2}{\sum_{j=1}^{n} |\alpha_j|^2} \geq \lambda_n.$$

It follows from (i) that if V_n is any n-dimensional subspace, then there is a vector in $\text{Span}(e_1, \ldots, e_{n+1})$ contained in $V_n^\perp$, and so

$$\max_{V_n^\perp} \frac{(Tx, x)}{\|x\|^2} \geq \lambda_{n+1},$$

using (ii). If we take V to be the n-dimensional space spanned by the first n eigenvectors, then $V^\perp$ is the space spanned by $\{e_j\}_{j=n+1}^{\infty}$, and then

$$\max_{V^\perp} \frac{(Tx, x)}{\|x\|^2} = \lambda_{n+1};$$

this gives the required equality.

16.9 Suppose that $x_n + iy_n$ is a bounded sequence in $H_{\mathbb{C}}$; then (x_n) and (y_n) are bounded sequences in H. Since T is compact, there is a subsequence x_{n_j} such that $T x_{n_j}$ has a convergent subsequence. Since (y_{n_j}) is bounded in H, we can find a further subsequence such that both $T x_{n_j'}$ and $T y_{n_j'}$ are convergent. It follows that

$$T_{\mathbb{C}}(x_{n_j'} + iy_{n_j'}) = T x_{n_j'} + iT y_{n_j'}$$

converges, so $T_\mathbb{C}$ is compact. To show that $T_\mathbb{C}$ is self-adjoint, we write

$$(T_\mathbb{C}(x+iy), u+iv)_{H_\mathbb{C}} = (Tx+iTy, u+iv)$$

$$= (Tx, u) - i(Tx, v) + i(Ty, u) + (Ty, v)$$

$$= (x, Tu) - i(x, Tv) + i(y, Tu) + (y, Tv)$$

$$= (x+iy, Tu+iTv)$$

$$= (x+iy, T_\mathbb{C}(u+iv)).$$

Chapter 18

18.1 We have $f'(t) = t^{p-1} - 1$, so $f'(t) \geq 0$ for all $t \geq 1$, so $f(1) \leq f(t)$ for every $t \in \mathbb{R}$. With $1/p + 1/q = 1$ we obtain $f(1) = 0$ and so

$$\frac{t^p}{p} + \frac{1}{q} - t \geq 0.$$

Choosing $t = ab^{-q/p}$ now yields

$$\frac{a^p b^{-q}}{p} + \frac{1}{q} \geq ab^{-q/p},$$

and so

$$\frac{a^p}{p} + \frac{b^q}{q} \geq ab^{q(1-1/p)} = ab,$$

as required.

18.2 Note that $p/(p-1)$ and p are conjugate indices, so we have

$$\sum_{j=1}^{n} |x_j + y_j|^p \leq \sum_{j=1}^{n} |x_j + y_j|^{p-1} |x_j| + \sum_{j=1}^{n} |x_j + y_j|^{p-1} |y_j|$$

$$\leq \left(\sum_{j=1}^{n} |x_j + y_j|^p \right)^{(p-1)/p} \left(\sum_{j=1}^{n} |x_j|^p \right)^{1/p}$$

$$+ \left(\sum_{j=1}^{n} |x_j + y_j|^p \right)^{(p-1)/p} \left(\sum_{j=1}^{n} |y_j|^p \right)^{1/p}$$

$$= \left(\sum_{j=1}^{n} |x_j + y_j|^p \right)^{(p-1)/p} [\|x\|_{\ell^p} + \|y\|_{\ell^p}];$$

dividing both sides by $(\sum_j |x_y + y_j|^p)^{(p-1)/p}$ yields

$$\|x + y\|_{\ell^p} = \left(\sum_{j=1}^{n} |x_j + y_j|^p \right)^{1/p} \leq \|x\|_{\ell^p} + \|y\|_{\ell^p}.$$

18.3 We apply Hölder's inequality with exponents q/p and $q/(q-p)$ to obtain

$$\|f\|_{L^p}^p = \int_\Omega |f(x)|^p \, dx \le \left(\int_\Omega |f(x)|^q \, dx \right)^{p/q} \left(\int_\Omega 1 \, dx \right)^{(q-p)/q}$$

$$= |\Omega|^{(q-p)/q} \|f\|_{L^q}^p,$$

which yields (18.9).

If $1 \le p \le q < \infty$ the function $f(x) = (1 + |x|)^{-n/q}$ provides an example for which $f \in L^q(\mathbb{R}^n)$ but $f \notin L^p(\mathbb{R}^n)$. For $1 \le p < q = \infty$ simply take $f(x) = 1$.

18.4 We show that the map $x \mapsto L_x$, defined for $y \in c_0$ by setting

$$L_x(y) = \sum_{j=1}^\infty x_j y_j$$

is an isometric isomorphism. Note that L_x is linear in y.

We first show that it is a linear isometry. Take $x \in \ell^1$ and $y \in c_0$; then

$$|L_x(y)| = \left| \sum_{j=1}^\infty x_j y_j \right| \le \|x\|_{\ell^1} \|y\|_{\ell^\infty}$$

by Hölder's inequality, which shows that $L_x \in (c_0)^*$ with

$$\|L_x\|_{(c_0)^*} \le \|x\|_{\ell^1}. \qquad (S.23)$$

The linearity of the map $x \mapsto L_x$ is clear. We now show that it is an isometry, i.e. that we have equality in (S.23).

This is clear if $x = 0$, so we take $x \neq 0$, and we choose $y^{(n)} \in c_0$ with

$$y_j^{(n)} = \begin{cases} x_j/|x_j| & x_j \neq 0 \text{ and } j \le n \\ 0 & x_j = 0 \text{ or } j > n. \end{cases} \qquad (S.24)$$

Now $\|y^{(n)}\|_{\ell^\infty} = 1$, and

$$|L_x(y^{(n)})| = \sum_{j=1}^n |x_j| = \left(\sum_{j=1}^n |x_j| \right) \|y^{(n)}\|_{\ell^\infty}.$$

It follows that for every n we have

$$\|L_x\|_{(c_0)^*} \ge \sum_{j=1}^n |x_j|$$

and hence $\|L_x\|_{(c_0)^*} \ge \|x\|_{\ell^1}$. Combined with (S.23) this gives the desired equality of norms.

We now show that the map $x \mapsto L_x$ is onto. Given $L \in (c_0)^*$, arguing as in the proof of Theorem 18.5, if we can find an $x \in \ell^1$ such that $L = L_x$, then, by applying this operator to $e_j \in c_0$, we must have

$$L(e_j) = L_x(e_j) = x_j.$$

If x is an element of ℓ^1, then since we can write any $y \in c_0$ as $y = \sum_{j=1}^{\infty} y_j e_j$ (with the sum convergent in c_0), we will then have

$$L(y) = L\left(\sum_{j=1}^{\infty} y_j e_j\right) = \sum_{j=1}^{\infty} y_j L(e_j) = \sum_{j=1}^{\infty} x_j y_j$$

as required. So we need only show that $x \in \ell^1$. If we define $y^{(n)}$ as in (S.24), then

$$\sum_{j=1}^{n} |x_j| = |L(y^{(n)})| \leq \|L\|_{(c_0)^*} \|u y^{(n)}\|_{c_0} = \|L\|_{(c_0)^*};$$

this bound does not depend on n so $x \in \ell^1$ as required.

18.5 The linear functionals δ_x on $X = C([-1, 1])$ defined as $\delta_x(f) := f(x)$ for each $x \in [-1, 1]$ clearly have norm 1. However, $\|\delta_x - \delta_y\|_{X^*} = 2$; the upper bound is clear and we can find $f \in X$ such that $f(x) = -f(y) = \|f\|_\infty$. This gives an uncountable collection in X^* that are all a distance 2 apart, from which it follows that X^* cannot be separable.

18.6 (i) We have

$$\|f\|_{L^p}^p = \int_\Omega |f|^p \, dx \leq \left(\int_\Omega 1^{2/(2-p)} \, dx\right)^{(2-p)/2} \left(\int_\Omega |f|^2 \, dx\right)^{p/2}$$

$$= |\Omega|^{1-p/2} \|f\|_{L^2}^p.$$

(ii) Part (i) shows that if $f \in L^2$, then $f \in L^p$, so any linear functional $\ell \in (L^p)^*$ also acts on any $f \in L^2$, with

$$|\ell(f)| \leq \|\ell\|_{(L^p)^*} \|f\|_{L^p} \leq \|\ell\|_{(L^p)^*} |\Omega|^{1/p - 1/2} \|f\|_{L^2};$$

so $\ell \in (L^2)^*$.

(iii) The function f_k is an element of L^2, since $\|f_k\|_{L^\infty} \leq k$, so we can use f_k in (18.10) to obtain

$$\ell(f_k) = \int_\Omega f_k g \, dx = \int_\Omega |g_k|^{q-2} g_k g \, dx \geq \int_\Omega |g_k|^q \, dx.$$

However, we also have

$$\|f_k\|_{L^p}^p = \int_\Omega |g_k|^{(q-1)p} \, dx = \int_\Omega |g_k|^q,$$

and so, since $|\ell(f_k)| \leq \|\ell\|_{(L^p)^*} \|f_k\|_{L^p}$, it follows that

$$\|g_k\|_{L^q}^q \leq \|\ell\|_{(L^p)^*} \|g_k\|_{L^q}^{q/p},$$

i.e. $\|g_k\|_{L^q} \leq \|\ell\|_{(L^p)^*}$ uniformly in k. The Monotone Convergence Theorem (Theorem B.7) now ensures that $g \in L^q$.

Chapter 19

19.1 First note that Lemma 8.11 implies that U is itself a Hilbert space with the same norm and inner product as H. Since $\phi \in U^*$, we can use the Riesz Representation Theorem (Theorem 12.4) to find $v \in U$ with $\|v\| = \|\phi\|_{U^*}$ such that

$$\phi(u) = (u, v) \qquad \text{for every} \qquad u \in U.$$

Now define $f(u) := (u, v)$ for every $u \in H$; then $f \in H^*$ extends ϕ (i.e. $f(x) = \phi(x)$ for every $x \in U$) and

$$\|f\|_{H^*} = \|v\| = \|\phi\|_{U^*}$$

(see the proof of Theorem 12.4) as required.

19.2 We have one extension f with $f(x) = (x, v)$ with $v \in U$. Now suppose that $f' \in H^*$ is another extension of $\hat{f}$ to H with the same norm. Then there exists (by the Riesz Representation Theorem) $v' \in H$ such that $f'(x) = (x, v')$ for every $x \in H$, and $\|f'\|_{H^*} = \|v'\|_H$; but then, since these two functionals must agree on U, we have

$$(u, v') = (u, v) \qquad u \in U,$$

so $v' = v + q$ with $q \in U^\perp$. But then $\|v'\|^2 = \|v\|^2 + \|q\|^2$, and since $\|v'\| = \|v\|$ it follows that $q = 0$, so $v' = v$.

19.3 We extend f 'by continuity': if $y \in \overline{U}$, then there exist $(x_n) \in U$ such that $x_n \to y$, and we define

$$F(y) = \lim_{n \to \infty} f(x_n).$$

This is well defined, since if we also have $(y_n) \in U$ with $y_n \to x$, then

$$|f(x_n) - f(y_n)| = |f(x_n - y_n)| \le M\|x_n - y_n\| \to 0 \qquad \text{and} \qquad n \to \infty.$$

Furthermore, F is linear, since for $x, y \in \overline{U}$, $\alpha, \beta \in \mathbb{K}$, we can find sequences $(x_n), (y_n) \in U$ such that $x_n \to x$ and $y_n \to y$, and then, as $\alpha x_n + \beta y_n \in U$ and $\alpha x_n + \beta y_n \to \alpha x + \beta y$,

$$F(\alpha x + \beta y) = \lim_{n \to \infty} f(\alpha x_n + \beta y_n) = \lim_{n \to \infty} \alpha f(x_n) + \beta f(y_n) = \alpha F(x) + \beta F(y).$$

Finally, we have $|f(x_n)| \le M\|x_n\|$ for all n, and since $\|x_n\| \to \|x\|$ (as $x_n \to x$) we have $|F(x)| \le M\|x\|$ as required.

19.4 For (i) we use $p(\alpha x) = \alpha x$ with $\alpha = 0$. For (b) note that we have

$$p(x) + p(y - x) \le p(y) \qquad \text{and} \qquad p(y) + p(x - y) \le p(x),$$

and since $p(y - x) = p(x - y)$ the inequality follows. For (c) put $y = 0$ in (b). Finally, for part (d), if $p(x) = p(y) = 0$ and $\alpha, \beta \in \mathbb{K}$, then

$$0 \le p(\alpha x + \beta y) \le |\alpha|p(x) + |\beta|p(y) = 0,$$

so $p(\alpha x + \beta y) = 0$.

19.5 Suppose that $f, g \in X^*$ with $f(x) = g(x) = \phi(x)$ for all $x \in U$ and that $\|f\|_{X^*} = \|g\|_{X^*} = \|\phi\|_{U^*}$. Then, for any $\lambda \in (0, 1)$ and $x \in U$, we have

$$[\lambda f + (1 - \lambda)g](x) = (1 - \lambda)f(x) + \lambda g(x) = \phi(x),$$

and furthermore,

$$\|\lambda f + (1 - \lambda)g\|_{X^*} \le (1 - \lambda)\|f\|_{X^*} + \lambda\|g\|_{X^*} = \|\phi\|_{U^*}.$$

Since $\lambda f + (1 - \lambda g) = \phi$ on U, we also have $\|\lambda f + (1 - \lambda)g\|_{X^*} \ge \|\phi\|_{U^*}$, and so $\|\lambda f + (1 - \lambda)g\|_{X^*} = \|\phi\|_{U^*}$.

19.6 Let $\{\zeta_j\}_{j=1}^{\infty}$ be the countable sequence of unit vectors whose linear span is dense in X. Let $W_0 = W$ and $i_0 = 0$. Given W_n and i_n, choose i_{n+1} to be the smallest index $> i_n$ so that $\zeta_{i_{n+1}} \notin W_n$ and let $z_{n+1} = \zeta_{i_{n+1}}$. The resulting collection (z_j) has the required properties.

19.7 First, if W is closed use Exercise 19.3 to extend ϕ to a bounded linear functional ϕ on $\overline{W}$ with the same norm. It therefore suffices to prove the result assuming from the outset that W is closed.

Consider the sequence of linear subspaces and elements (z_j) from Exercise 19.6. Given a linear functional $f_n \in W_n^*$ with $\|f_n\|_* = \|\phi\|_{W^*}$ we can use the extension argument from the proof of the Hahn–Banach Theorem in Section 19.1 to extend f_n to $f_{n+1} \in W_{n+1}^*$ with $\|f_{n+1}\|_* = \|f_n\|_* = \|\phi\|_*$. In this way we can define a linear functional f_∞ on W_∞ by letting $f_\infty(x) = f_n(x)$ for any $x \in W_n$ (this is well defined, since if $x \in W_n \cap W_m$ with $m > n$ we know that f_m extends f_n). Finally, we use Exercise 19.3 again to extend f_∞ to an element $f \in X^*$ (since $X = \overline{W_\infty}$) that satisfies the same bound.

Chapter 20

20.1 If $x = 0$ we take $f(y) = (y, z)$ for any $z \in H$ with $\|z\| = 1$. For $x \neq 0$ we set $f(y) = (y, x/\|x\|)$; then $\|f\|_{H^*} = 1$ and $f(x) = \|x\|$.

20.2 Let $U = \mathrm{Span}(x_j)_{j=1}^n$, and define a linear functional $\phi : U \to \mathbb{K}$ by setting

$$\phi\left(\sum_{j=1}^n \alpha_j e_j\right) = \sum_{j=1}^n \alpha_j a_j.$$

Since U is finite-dimensional, ϕ is bounded, i.e. $\phi \in U^*$. Now extend ϕ to $f \in X^*$ using the Hahn–Banach Theorem.

20.3 If $\|x\| \leq M$, then $|f(x)| \leq \|f\|_* \|x\| \leq M$ for all $f \in X^*$ with $\|f\|_* \leq 1$. Conversely, if $|f(x)| \leq M$ for all $f \in X^*$ with $\|f\|_* = 1$, we can take f to be the support functional for x, which gives $|f(x)| = \|x\| \leq M$. The equality for $\|x\|$ follows from these two facts.

20.4 Since Y is a proper subspace of H, we can decompose $x = u + v$, where $u \in Y$ and $v \in Y^\perp$; note that $d = \|v\|$. Set $z = v/\|v\|$ and define $f \in H^*$ by setting $f(a) = (a, z)$. Then $\|f\| = 1$,

$$f(y) = (y, z) = 0 \text{ for every } y \in Y,$$

and

$$f(x) = f(u + v, z) = \left(v, \frac{v}{\|v\|}\right) = \|v\| = d$$

as required.

20.5 Take $x \in X$ with $\|x\| \neq 0$ and suppose that there are distinct $f, g \in X^*$ with $\|f\|_{X^*} = \|g\|_{X^*} = 1$ and

$$f(x) = g(x) = \|x\|_X.$$

Then, since X^* is strictly convex, $\|\frac{1}{2}(f+g)\|_{X^*} < 1$, and so

$$2\|x\|_X = |f(x) + g(x)| = 2|[\tfrac{1}{2}(f+g)(x)]| < 2\|x\|_X,$$

a contradiction.

20.6 A Hilbert space is uniformly convex (Exercise 8.10) and hence strictly convex (Exercise 10.4), so by Exercise 20.5 there is a unique linear functional $f \in X^*$ such that $\|f\|_{X^*} = \|x\|_X$ such that $f(x) = 1$, and this is just $(\cdot, x)$. For this functional we have

$$f(Tx) = (Tx, x)$$

and so $V(T) = \{(Tx, x) : x \in H\}$ as in Exercise 16.3.

20.7 Take $Y = \{0\}$ and $x \neq 0$. Then $\mathrm{dist}(x, Y) = \|x\|$, and so we have $f \in X^*$ with $\|f\|_{X^*} = 1$ and $f(x) = \|x\|$.

20.8 For any $y \in \mathrm{Ker} f$ we have

$$|f(x)| = |f(x-y)| \le \|f\|_{X^*}\|x-y\|;$$

so that $|f(x)| \le \|f\|_{X^*}\mathrm{dist}(x, \mathrm{Ker}(f))$. For the reverse inequality, we use the fact that for every $\varepsilon > 0$ there exists $u \in X$ with $\|u\| = 1$ such that $f(u) > \|f\|_{X^*} - \varepsilon$ and set

$$y = x - \frac{f(x)}{f(u)}u.$$

Then $f(y) = 0$, i.e. $y \in \mathrm{Ker}(f)$, and so

$$\mathrm{dist}(x, \mathrm{Ker}(f)) \le \|x - y\| = \frac{|f(x)|}{|f(u)|} < \frac{|f(x)|}{\|f\|_{X^*} - \varepsilon}.$$

Since this holds for all $\varepsilon > 0$, it follows that $\mathrm{dist}(x, \mathrm{Ker}(f)) \le |f(x)|/\|f\|_{X^*}$. Combining these two inequalities yields the result.

20.9 T is linear and since each ϕ has norm 1 we have

$$\|Tx\|_\infty \le \sup_n |\phi_n(x)| \le \sup_n \|x\| = \|x\|,$$

and so $\|T\| \le 1$. We now have to show that $\|Tx\|_\infty \ge \|x\|$ for every $x \in X$; since T is linear it is sufficient to do this for x with $\|x\| = 1$.

Given such an x and $\epsilon > 0$ choose x_n from the dense sequence with $\|x - x_n\| < \epsilon$, and then

$$\phi_n(x) = \phi_n(x_n) + \phi(x - x_n) = 1 + \phi(x - x_n) \ge 1 - \|\phi_n\|\|x - x_n\| > 1 - \epsilon.$$

So $\|Tx\|_\infty > 1 - \epsilon$ and since ϵ was arbitrary it follows that $\|Tx\|_\infty \ge 1$ as required.

20.10 If $z \notin \mathrm{clin}(E)$, then Proposition 20.4 furnishes an $f \in X^*$ so that $f|_{\mathrm{clin}(E)} = 0$ but $f(z) \neq 0$, which is disallowed by our assumption.

If $f \in X^*$ vanishes on E, then it vanishes on $\mathrm{Span}(E)$ (finite linear combinations of elements of E), and so vanishes on $\mathrm{clin}(E)$ since it is continuous: if $x \in \mathrm{clin}(E)$, then there exist $x_n \in \mathrm{Span}(E)$ such that $x_n \to x$, and then $f(x) = \lim_{n\to\infty} f(x_n) = 0$.

So if there exists $f \in X^*$ that vanishes on E but for which $f(z) \neq 0$, then we have

$$|f(z)| = |f(z-y)| \le \|f\|_{X^*}\|z - y\|$$

for any $y \in \mathrm{clin}(E)$. It follows that

$$\mathrm{dist}(z, \mathrm{clin}(E)) = \inf\{\|z - y\| : y \in \mathrm{clin}(E)\} \geq \frac{|f(z)|}{\|f\|_{X^*}} > 0.$$

20.11 If $T \in B(H, K)$, then we have $T^\times \in B(K^*, H^*)$ defined by $T^\times g = g \circ T$. The Hilbert adjoint is defined by setting

$$(Tu, v)_K = (u, T^*v)_H,$$

or

$$R_K v(Tu) = R_H(T^*v)(u).$$

We can rewrite this as $(R_K v) \circ T = (R_H \circ T^*)v$, or, using the Banach adjoint,

$$(T^\times \circ R_K)v = (R_H \circ T^*)v.$$

This shows that $T^\times \circ R_K = R_H \circ T^*$, and so, applying R_H^{-1} to both sides, we obtain

$$T^* = R_H^{-1} \circ T^\times \circ R_K.$$

20.12 Let B be a bounded subset of Y^*; we need to show that $T^\times(B)$ is precompact, i.e. has compact closure. If we let $K = \overline{T(\mathbb{B}_X)}$, then K is a compact subset of Y, since T is compact. Since any $f \in Y^*$ is a continuous map from Y into $\mathbb{K}$, we can consider its restriction to K, which gives an element of $C(K)$. So we can think of B as a subset of $C(K)$, and for any $f \in B$ and any $y_1, y_2 \in K$ we have

$$|f(y_1) - f(y_2)| \leq \|f\|_{Y^*}\|y_1 - y_2\|_Y \qquad \text{and} \qquad |f(y)| \leq \|f\|_{Y^*}\|y\|,$$

the second inequality since $0 \in K$. So B is a bounded equicontinuous family in $C(K)$, so precompact by the Arzelà–Ascoli Theorem.

Now take any sequence $(f_n) \in B$. By the above, (f_n) has a subsequence $(f_{n_j})_j$ that converges uniformly on K. Therefore $(f_{n_j} \circ T) \in X^*$ converges uniformly on $\mathbb{B}_X$; since X^* is complete, $f_{n_j} \circ T \to g$ in X^*, for some $g \in X^*$. Since $T^\times f_{n_j} = f_{n_j} \circ T$, it follows that $T^\times f_{n_j} \to g$ in X^*, which shows that $T^\times$ is compact.

20.13 Since (x_n) converges, it is bounded, with $|x_n| \leq M$, say. Since $x_n \to \alpha$, given any $\varepsilon > 0$ there exists N such that $|x_n - \alpha| < \varepsilon/2$ for all $n \geq N$. Now choose $N' \geq N$ sufficiently large that $N(M + |\alpha|)/N' < \varepsilon/2$, and then

$$\left|\frac{x_1 + \cdots + x_n}{n} - \alpha\right| = \left|\frac{(x_1 - \alpha) + \cdots + (x_n - \alpha)}{n}\right|$$

$$\leq \left|\frac{(x_1 - \alpha) + \cdots + (x_N - \alpha)}{N'}\right| + \left|\frac{(x_{N+1} - \alpha) + \cdots + (x_n - \alpha)}{N'}\right|$$

$$\leq \frac{\varepsilon}{2} + \frac{(N' - N)\varepsilon/2}{N'} < \varepsilon,$$

Chapter 21

21.1 We have

$$p(x) = \inf\{\lambda > 0 : \lambda^{-1}x \in B_X\} = \inf\{\lambda > 0 : \|\lambda^{-1}x\| < 1\} = \|x\|.$$

21.2 If $a, b \in \cap_{\alpha \in \mathbb{A}} K_\alpha$ and $\lambda \in (0, 1)$ then $\lambda a + (1 - \lambda)b \in K_\alpha$ for each $\alpha \in \mathbb{A}$, since $a, b \in K_\alpha$ and K_α is convex. Since this holds for every $\alpha \in \mathbb{A}$, it follows that $\lambda a + (1 - \lambda)b \in \cap_\alpha K_\alpha$ and hence $\cap_{\alpha \in \mathbb{A}} K_\alpha$ is convex.

21.3 Suppose that $a, b \in \overline{K}$; then there exist sequences $(a_n), (b_n) \in K$ such that $a_n \to a$ and $b_n \to b$. If we take $\lambda \in (0, 1)$, then $\lambda a_n + (1 - \lambda b_n) \in K$ and

$$\lambda a_n + (1 - \lambda)b_n \to \lambda a + (1 - \lambda)b,$$

so $\lambda a + (1 - \lambda b) \in \overline{K}$ which shows that $\overline{K}$ is convex.

21.4 We will show that conv(U) is totally bounded, i.e. that for any $\varepsilon > 0$ we can find a cover of U by a finite number of open balls of radius ε. It then follows from Exercise 6.10 that $\overline{\text{conv}(U)}$ is compact.

Given $\varepsilon > 0$ find a cover of U by a finite number of balls of radius $\varepsilon/2$ with centres in $F := \{x_1, \ldots, x_k\}$. The line segments $L_i = \{\lambda x_i : \lambda \in [0, 1]\}$ are all convex sets, so their sum $L := L_1 + \cdots + L_k$ is also convex, and contains all the x_i. Since L is compact, there exists a finite set G such that $L \subset G + \frac{\varepsilon}{2}B_X$. Then

$$U \subset F + \frac{\varepsilon}{2}B_X \subset L + \frac{\varepsilon}{2}B_X \subset G + \frac{\varepsilon}{2}B_X + \frac{\varepsilon}{2}B_X = G + \varepsilon B_X.$$

Since the set $L + \frac{\varepsilon}{2}B_X$ is convex and contains U, it must also contain conv(U), so conv$(U) \subset G + \varepsilon B_X$. In other words, conv$(U)$ has a finite cover by balls of radius ε for every $\varepsilon > 0$. It follows that $\overline{\text{conv}(U)}$ is compact.

21.5 Take any $x \in \text{conv}(U)$ with

$$x = \sum_{j=1}^{k} \lambda_j x_j, \qquad \text{where} \qquad x_j \in U \text{ and } \sum_{j=1}^{k} \lambda_j = 1,$$

and suppose that $k > n + 1$. Then $k - 1 > n$ and so the $k - 1$ vectors

$$\{x_2 - x_1, x_3 - x_1, \ldots, x_k - x_1\}$$

are linearly dependent: there exist $\{\alpha_j\}_{j=2}^{k}$, not all zero, such that

$$\sum_{j=2}^{k} \alpha_j (x_{j+1} - x_1) = 0.$$

If we set $\alpha_1 = -\sum_{j=2}^{k} \alpha_j$, then

$$\sum_{j=1}^{k} \alpha_j x_j = 0 \qquad \text{and} \qquad \sum_{j=1}^{k} \alpha_j = 0.$$

Since not all the α_j are zero and $\sum \alpha_j = 0$, it follows that at least one of them, α_i, say, is positive. For any choice of $\gamma \in \mathbb{R}$ we have

$$x = \sum_{j=1}^{k} \lambda_j x_j - \gamma \sum_{j=1}^{k} \alpha_j x_j = \sum_{j=1}^{k} [\lambda_j - \gamma \alpha_j] x_j. \qquad (S.25)$$

We choose

$$\gamma = \min\left\{\frac{\lambda_j}{\alpha_j} : \alpha_j > 0, \ j = 1, \ldots, k\right\};$$

then $\lambda_j - \gamma\alpha_j \geq 0$ for each $j = 1, \ldots, k$ and $\lambda_i - \gamma\alpha_i = 0$ for some $i \in \{1, \ldots, k\}$.

The sum in (S.25) is therefore a convex combination of the (x_j), since we have $\lambda_j - \gamma\alpha_j \geq 0$ and $\sum_{j=1}^k \lambda_j - \gamma\alpha_j = 1$. However, we can remove the term from the sum in which $\lambda_i - \gamma\alpha_i = 0$. We continue in this way until we have no more than $n + 1$ terms in the sum.

21.6 If K is convex, then for any x, y and $\lambda \in (0, 1)$ we have

$$
\begin{aligned}
\lambda d(x) + (1 - \lambda)d(y) &= \lambda \inf_{a \in K} \|x - a\| + (1 - \lambda) \inf_{b \in K} \|y - b\| \\
&= \inf_{a \in K} \|\lambda x - \lambda a\| + \inf_{b \in K} \|(1 - \lambda)y + (1 - \lambda)b\| \\
&= \inf_{a,b \in K} \left[\|\lambda x - \lambda a\| + \|(1 - \lambda)y + (1 - \lambda)b\|\right] \\
&\geq \inf_{a,b \in K} \|[\lambda x + (1 - \lambda)y)] - [\lambda a + (1 - \lambda)b]\|\| \\
&\geq \inf_{k \in K} \|[\lambda x + (1 - \lambda)y)] - k\| \\
&= d(\lambda x + (1 - \lambda)y),
\end{aligned}
$$

since $\lambda a + (1 - \lambda b) \in K$.

For the converse simply take $a, b \in K$: then

$$0 \leq d(\lambda a + (1 - \lambda b)) \leq \lambda d(a) + (1 - \lambda)d(b) = 0,$$

so $d(\lambda a + (1 - \lambda b)) = 0$, i.e. $\lambda a + (1 - \lambda b) \in K$ (since K is closed).

Chapter 22

22.1 First note that finite-dimensional subspaces are complete so they must be closed, and they contain no open balls (otherwise they would be infinite-dimensional) so they are nowhere dense. If we let $Y_n = \mathrm{Span}(x_1, \ldots, x_n)$, then Y_n is nowhere dense, and since X cannot be given as the countable union of nowhere dense subsets, it follows that $X \neq \cup_n Y_n$.

22.2 Consider again the sets

$$F_n = \{x \in X : \sup_{T \in \mathcal{S}} \|Tx\| \leq n\}$$

from the proof of the Principle of Uniform Boundedness. These sets are closed and must all have empty interior (so be nowhere dense), since if one set contained a non-empty open ball we could follow the proof of the Principle of Uniform Boundedness to show that $\sup_{T \in \mathcal{S}} \|T\| < \infty$. The complements of these sets F_n,

$$G_n = \{x \in X : \sup_{T \in \mathcal{S}} \|Tx\| > n\},$$

are therefore open and dense, and so their intersection

$$\bigcap_{n=1}^{\infty} G_n = \{x \in X : \sup_{T \in \mathcal{S}} \|Tx\| = \infty\}$$

is residual.

22.3 For any particular polynomial p we have $p(x) = \sum_{j=0}^{m} a_j x^j$, i.e. p can be written in the form in (22.5), with $a_j = 0$ for all $j \geq N$. So

$$|T_n p(x)| \leq \min(n, N) \max_j |a_j| \qquad \sup_n |T_n p(x)| \leq N \max_j |a_j| < \infty.$$

However, if we take $p_n(x) = \sum_{j=0}^{n} x^j$, then $\|p_n\| = 1$, but $T_n p_n = n$, so we have $\|T_n\|_{X^*} \geq n$. In order not to contradict the Principle of Uniform Boundedness, it must be the case that $(X, \|\cdot\|)$ is not a Banach space, i.e. not complete.

22.4 We treat the case when Y is a Banach space.

Define a map $T: X \to B(Y, Z)$ by setting $(Tx)(y) = b(x, y)$; then since $b(x, y)$ is linear and continuous in Y we have $Tx \in B(Y, Z)$ with

$$\|(Tx)(y)\|_Z = \|b(x, y)\|_Z \leq C(x)\|y\|_Y.$$

Now consider the collection $\{Tx : x \in \mathbb{B}_X\}$, which is a subset of $B(Y, Z)$. Arguing as above, for each $y \in Y$ we have

$$\|Tx(y)\|_Z = \|b(x, y)\|_Z \leq C(y)\|x\|$$

since $x \mapsto b(x, y)$ is linear and continuous for each $y \in Y$. This shows, in particular, that

$$\sup_{x \in \mathbb{B}_X} \|Tx(y)\|_Z \leq C(y) < \infty$$

for each $y \in Y$. The Principle of Uniform Boundedness therefore guarantees that

$$\sup_{x \in \mathbb{B}_X} \|Tx\|_{B(Y,Z)} \leq M.$$

It follows, since T is linear, that $\|Tx\|_{B(Y,Z)} \leq M\|x\|_X$, and therefore

$$\|b(x, y)\|_Z = \|(Tx)(y)\|_Z \leq \|Tx\|_{B(Y,Z)}\|y\|_Y \leq M\|x\|_X\|y\|_Y.$$

22.5 First recall (see Lemma 3.10) that $\ell^p \subset \ell^q$. We let

$$S_n := \{x \in \ell^p : \|x\|_{\ell^p} \leq n\}.$$

Now suppose that $(x^{(k)}) \in \ell^p$ with

$$\sum_{j=1}^{\infty} |x_j^{(k)}|^p \leq n^p$$

and $x^{(k)} \to x$ in ℓ^q. Then in particular, for any N we have

$$\sum_{j=1}^{N} |x_j^{(k)} - x_j|^q \to 0,$$

which implies that $x_j^{(k)} \to x_j$ for every k. So we have

$$\sum_{j=1}^{N} |x^{(k)}|^p \leq \lim_{n \to \infty} \sum_{j=1}^{N} |x_j^{(k)}|^p \leq n^p.$$

This shows that S_n is closed in ℓ^q.

However, S_n contains no open sets (so its interior is empty). To see this, take some $y \in \ell^q \setminus \ell^p$ (see Exercise 3.7): then $x + \varepsilon y \subset B_{\ell^q}(x, \varepsilon)$ for any $x \in S_n$ and $\varepsilon > 0$, but $x \notin \ell^p$.

Since S_n is closed and nowhere dense in ℓ^q, it follows that $\ell^p = \bigcup_{n=1}^{\infty} S_n$ is meagre in ℓ^q.

22.6 Consider the maps $T_n : \ell^p \to \mathbb{K}$ given by

$$T_n(y) := \sum_{j=1}^{n} x_j y_j.$$

Then $T_n \in B(\ell^p; \mathbb{K})$ for each n, since by Hölder's inequality

$$|T_n(y)| = \left| \sum_{j=1}^{n} x_j y_j \right| \leq \left(\sum_{j=1}^{n} |x_j|^q \right)^{1/q} \left(\sum_{j=1}^{n} |y_j|^p \right)^{1/p} = \left(\sum_{j=1}^{n} |x_j|^q \right)^{1/q} \|y\|_{\ell^p}. \tag{S.26}$$

If we choose

$$y_j = \begin{cases} |x_j|^q / x_j & 1 \leq j \leq n \text{ and } x_j \neq 0 \\ 0 & \text{otherwise} \end{cases}$$

then

$$|T_n(y)| = \sum_{j=1}^{n} |x_j|^q \leq \|T_n\| \|y\|_{\ell^p} = \|T_n\| \left(\sum_{j=1}^{n} |x_j|^{(q-1)p} \right)^{1/p}$$

$$= \|T_n\| \left(\sum_{j=1}^{n} |x_j|^q \right)^{1/p},$$

which shows, since $1 - (1/p) = (1/q)$, that

$$\|T_n\| \geq \left(\sum_{j=1}^{n} |x_j|^q \right)^{1/q},$$

which combined with (S.26) yields

$$\|T_n\| = \left(\sum_{j=1}^{n} |x_j|^q \right)^{1/q}. \tag{S.27}$$

We also know that $\sup_n |T_n(y)| < \infty$, since $\sum_{j=1}^{\infty} x_j y_j$ converges. The Principle of Uniform Boundedness now guarantees that $\sup_n \|T_n\| < \infty$; given (S.27), this implies that $x \in \ell^q$.

Chapter 23

23.1 If no (y_n) exists such that $\sum \alpha_n y_n < \infty$, then the map $T : \ell^\infty \to \ell^1$ is not only one-to-one but also onto. Since T is bounded, the Inverse Mapping Theorem would then guarantee that T is an isomorphism. However, ℓ^∞ is not isomorphic to ℓ^1 (ℓ^∞ is not separable but ℓ^1 is, and separability is preserved under isomorphisms); so in fact there must exist such a sequence (y_n).

23.2 If the $\{e_j\}$ are a basis, then their linear span is dense, and, using Corollary 23.5, we have

$$\left\| \sum_{i=1}^n a_i e_i \right\| = \left\| P_n \left(\sum_{j=1}^m a_j e_j \right) \right\| \le \left[\sup_n \| P_n \| \right] \left\| \sum_{j=1}^m a_j e_j \right\|.$$

Let $Y = \mathrm{Span}(\{e_j\})$, which is dense in X by (i). Not also that it follows from (ii), using induction, that the $\{e_j\}$ are linearly independent and

$$|a_n| = |a_n| \|e_n\| \le 2K \left\| \sum_{j=1}^m a_j e_j \right\| \qquad n < m.$$

Therefore the maps $\Pi_n : Y \to \mathrm{Span}(e_1, \dots, e_n)$ defined by setting

$$\Pi_n \left(\sum_{j=1}^m a_j e_j \right) = \sum_{j=1}^{\min(m,n)} a_j e_j$$

are bounded linear projections on Y with norm at most K. Since Y is dense in X, it follows using Exercise 19.3 that each map Π_n can be extended uniquely to a map $P_n : X \to \mathrm{Span}(e_1, \dots, e_n)$ that satisfies $\|P_n\|_{B(X)} \le K$.

Now, given any $x \in X$ and $\varepsilon > 0$, find $y = \sum_{j=1}^m a_j e_j$ such that $\|y - x\| < \varepsilon$; then, for all $n > m$, we have

$$\|x - P_n x\| \le \|x - y\| + \|y - P_n y\| + \|P_n y - P_n x\|$$
$$\le \varepsilon + 0 + \|P_n\| \varepsilon \le (1 + K)\varepsilon,$$

and so $P_n x \to x$ as $n \to \infty$.

23.3 We will show that if $x_n \to x$ and $T x_n \to y$, then $y = Tx$, and then the boundedness of T follows from the Closed Graph Theorem. Take any $z \in H$; then

$$(T x_n, z) = (x_n, T z).$$

Letting $n \to \infty$ on both sides we obtain

$$(y, z) = (x, Tz) = (Tx, z) \qquad \text{for every} \quad z \in H,$$

and so $y = Tz$ as required.

23.4 Clearly T is not bounded: if we set $f_n(x) = x^n$, then $f_n'(x) = nx^{n-1}$, so

$$\|f_n\|_\infty = 1 \qquad \text{and} \qquad \|f_n'\|_\infty = n.$$

However, suppose that $(f_n) \in X$, with $f_n \to f$ and $f'_n \to g$ in the supremum norm; then f_n is a Cauchy sequence in C^1, and so $f_n \to g$ with $g \in C^1$ and $f'_n = f'$ (see Theorem 4.12).

23.5 Suppose that $(x_n) \in X$ with $x_n \to x$ and $Tx_n \to f$, $f \in X^*$. Since

$$(Tx_n - Ty)(x_n - y) \geq 0,$$

we can take $n \to \infty$ to obtain

$$(f - Ty)(x - y) \geq 0.$$

Set $y = x + tz$ with $z \neq 0$ then

$$0 \leq (f - Tx - tTz)(tz) = t[f(z) - Tx(z)] - t^2(Tz)(z).$$

If this is to hold for all $t \in \mathbb{R}$, then we must have $f(z) = Tx(z)$, i.e. $Tx = f$. The Closed Graph Theorem now implies that T is bounded.

Chapter 26

26.1 Take $F \in X^{**}$. We need to find $x \in X$ such that, for every $f \in X^*$ we have

$$F(f) = f(x).$$

Note that $F \circ T_Y : Y \to \mathbb{K}$, and is both linear and bounded, since

$$|F \circ T_Y(y)| \leq \|F\|_{X^{**}} \|T_Y(y)\|_{X^*} \leq \|F\|_{X^{**}} \|T_Y\|_{B(Y,X^*)} \|y\|_Y.$$

It follows that $F \circ T_Y \in Y^*$, and so we can find an $x \in X$ such that

$$F \circ T_Y = T_X x.$$

Now, given $f \in X^*$, apply both sides to $T_Y^{-1} f$ to obtain

$$F(f) = [T_X x](T_Y^{-1} f) = (T_Y T_Y^{-1} f)(x) = f(x)$$

as required.

26.2 If U is bounded, then $\|u\| \leq M$ for all $u \in U$ and some $M > 0$, and then $|f(u)| \leq \|f\|\|u\| \leq M\|f\|$. To show the opposite implication, consider the family of maps

$$\mathcal{S} = \{u^{**} : u \in U\}$$

that all lie in X^{**}. Then for every $f \in X^*$ we have

$$\sup_{u \in U} |u^{**}(f)| = \sup_{u \in U} |f(u)| < \infty,$$

so the Principle of Uniform Boundedness guarantees that

$$\sup_{u \in U} \|u^{**}\| < \infty.$$

But we know that $\|u^{**}\| = \|u\|$, and so $\sup_{u \in U} \|u\| < \infty$, i.e. U is bounded.

26.3 If T is not bounded, then there exist $x_n \in X$ such that $\|x_n\| = 1$ and $\|Tx_n\| \geq n$. The (Tx_n) also form an unbounded sequence when considered as elements of Y^{**}, so there is a functional $\phi \in Y^*$ for which $\phi(Tx_n)$ is unbounded. (If this were not the case, then for each $\phi \in Y^*$, the scalars $(Tx_n)^{**}(\phi)$ would be bounded, and then the

Principle of Uniform Boundedness would show that $\|Tx_n\|_{Y^{**}}$ is bounded uniformly in n.) It follows that $\phi \circ T$ is not a bounded functional on X.

26.4 Define an element $\Phi \in X^{**}$ by setting

$$\Phi(f) = \int_0^T f(\xi(t))\,dt \qquad \text{for every } f \in X^*. \tag{S.28}$$

This map Φ is clearly linear, and it is bounded since

$$|\Phi(f)| \le \int_0^T \|f\|_{X^*}\|\xi(t)\|_X\,dt \le \left(\int_0^T \|\xi(t)\|_X\,dt\right)\|f\|_{X^*}$$

and $\int_0^T \|\xi(t)\|_X\,dt < \infty$. Since X is reflexive, it follows that there exists an element $y \in X$ such that $\Phi = y^{**}$, i.e.

$$\Phi(f) = f(y) \qquad \text{for all } f \in X^*.$$

Using (S.28) this yields

$$f(y) = \int_0^T f(\xi(t))\,dt. \tag{S.29}$$

To prove the inequality, let $f \in X^*$ be the support functional for y, i.e. $\|f\|_{X^*} = 1$ and $f(y) = \|y\|_X$; then from (S.29) we obtain

$$\left\| \int_0^T \xi(t)\,dt \right\|_X = \|y\|_X = f(y) = \int_0^T f(\xi(t))\,dt$$

$$\le \int_0^T \|f\|_{X^*}\|\xi(t)\|_X\,dt = \int_0^t \|\xi(t)\|_X\,dt.$$

26.5 Suppose that T does not map L^q onto $(L^p)^*$, i.e. that there exists $\ell \in (L^p)^*$ that cannot be realised as $T(g)$ for any $g \in L^q$. Since $T(L^q)$ is a closed subspace of $(L^p)^*$, it follows using the distance functional from Proposition 20.4 that there exists some non-zero $F \in (L^p)^{**}$ such that $F(\ell) = 0$ for every $\ell \in T(L^q)$. Since L^p is reflexive it follows that there exists $f \in L^p$ such that

$$F(\ell) = \ell(f) \qquad \text{for every } \ell \in (L^p)^*.$$

It follows that $\ell(f) = 0$ for every $\ell \in (L^p)^*$, and so $f = 0$, which in turn implies that $F = 0$. But this contradicts the fact that F is non-zero, so T must be onto.

Chapter 27

27.1 We know that, given any linear subspace M and $y \ne M$, there exists an $f \in X^*$ such that $f|_M = 0$ and $f(y) \ne 0$. So if $f(x) = 0$ for every $f \in X^*$ such that $f|_M = 0$ we must have $x \in M$. So if $x_n \in M$ and $x_n \rightharpoonup x$, then for all such f we have $f(x) = \lim_{n \to \infty} f(x_n) = 0$ and so $x \in M$. (Or, using Theorem 27.7, M is convex so closed implies weakly closed.)

27.2 Consider the bounded sequence $(e^{(j)})$ in ℓ^1, and suppose that $(e^{(n_k)})$ is a weakly convergent subsequence. Choose $f \in \ell^\infty$ such that $f_{n_k} = (-1)^k$ and is zero otherwise. Then $L_f(e^{(n_k)}) = (-1)^k$ which does not converge as $k \to \infty$. It follows that ℓ^1 is not reflexive, since the unit ball is weakly (sequentially) compact in any reflexive Banach space.

27.3 The closed convex hull H of (x_n) is closed and convex, so it is also weakly closed. It follows that x is contained in H. So x can be approximated by finite linear convex combinations of the x_n.

Now, denote by $\text{conv}(A)$ the convex hull of A, and notice that

$$\text{conv}(\{x_n\}_{n=1}^{\infty}) = \bigcup_{k=1}^{\infty} \text{conv}\{x_1, \ldots, x_k\};$$

it follows that their closures are equal. We know that

$$x \in \overline{\bigcup_{k=1}^{\infty} \text{conv}\{x_1 \ldots, x_k\}},$$

and so

$$0 = \lim_{n \to \infty} \text{dist}\left(x, \bigcup_{k=1}^{n} \text{conv}\{x_1, \ldots, x_k\}\right) = \lim_{n \to \infty} \text{dist}(x, \text{conv}\{x_1, \ldots, x_n\}).$$

It follows that we can choose a sequence (y_n) with $y_n \in \text{conv}\{x_1, \ldots, x_n\}$ such that $y_n \to x$.

27.4 If $x = 0$, then the result is immediate. So assume that $x \neq 0$. For n sufficiently large that $\|x_n\| \neq 0$, set

$$y_n = \frac{x_n}{\|x_n\|} \quad \text{and} \quad y = \frac{x}{\|x\|}.$$

Then $y_n \rightharpoonup y$ and $\|y_n\| = 1$. It follows $y_n + y \rightharpoonup 2y$ and so

$$\|y\| \leq \liminf_{n \to \infty} \|(y_n + y)/2\|.$$

But also $\|y\| = 1$ and $\|y_n\| = 1$, so

$$\limsup_{n \to \infty} \|\tfrac{1}{2}(y_n + y)\| \leq 1, \tag{S.30}$$

whence $\|\tfrac{1}{2}(y_n + y)\| \to 1$. The uniform convexity of X implies that $\|y_n - y\| \to 0$ as $n \to \infty$: suppose not; then there is an $\varepsilon > 0$ and a subsequence y_{n_k} such that

$$\|y_{n_k} - y\| > \varepsilon.$$

But then it follows from the uniform convexity of X that

$$\left\| \frac{y_{n_k} + y}{2} \right\| < 1 - \delta$$

for every k, contradicting (S.30).

27.5 Consider the sequence $f_n(x) = 1 + \sin nx$. Then $f_n \rightharpoonup 1$, $\|f_n\|_{L^1} = 2\pi$, but $\|f_n - 1\| = 4$ for every n.

27.6 For a counterexample in ℓ^1 take $(e^{(n)})$. Then, for every k, we have $e_k^{(n)} \to 0$ as $n \to \infty$ and $\|e^{(n)}\|_{\ell^1} = 1$ for every n, so if the lemma held in this case we would have to have $x^{(n)} \rightharpoonup \mathbf{0}$, where $\mathbf{0}$ is the sequence consisting entirely of zeros. However, if we consider the bounded linear functional on ℓ^1 given by

$$\ell(x) := \sum_{j=1}^{\infty} x_j$$

then $\ell(e^{(n)}) = 1$ for every n, but $\ell(\mathbf{0}) = 0$, so $e^{(n)} \not\rightharpoonup \mathbf{0}$.

For a counterexample in ℓ^∞, define ℓ on the subspace of ℓ^∞ consisting of all convergent sequences by setting

$$\ell(x) = \lim_{n \to \infty} x_n,$$

and then extend this to a bounded linear functional L on ℓ^∞ using the Hahn–Banach Theorem. Now let

$$x_j^{(n)} = \begin{cases} 1 & 1 \le j \le n \\ 0 & j > n, \end{cases}$$

so that $x_j^{(n)} \to 1$ as $n \to \infty$ for every j, and $\|x^{(n)}\|_{\ell^\infty} = 1$. So if the lemma held in ℓ^∞ we would have $x^{(n)} \rightharpoonup \mathbf{1}$, where $\mathbf{1}$ is the sequence consisting entirely of 1s. However, $L(x^{(n)}) = 0$ for every n, but $L(\mathbf{1}) = 1$, so $x^{(n)} \not\rightharpoonup \mathbf{1}$.

27.7 We have $e^{(j)} \rightharpoonup 0$ as $j \to \infty$ (see the beginning of Chapter 27). Since T is compact, we then have $Te^{(j)} \to 0$ using Lemma 27.4.

27.8 We first prove the 'if' part. Let $\hat{f}(x) = f(x) - f(0)$; then, given $\varepsilon > 0$, there exists $\delta(\varepsilon) > 0$ such that $|\hat{f}(x)| < \varepsilon$ whenever $|x| < 2\delta(\varepsilon)$. Define

$$\hat{f}_\varepsilon(x) = \begin{cases} \hat{f}(x) & 2\delta(\varepsilon) \le |x| \le 1 \\ \hat{f}(x)\left[\dfrac{|x|}{\delta(\varepsilon)} - 1\right] & \delta(\varepsilon) < |x| < 2\delta(\varepsilon) \\ 0 & |x| \le \delta(\varepsilon). \end{cases}$$

Then $\|\hat{f}_\varepsilon - \hat{f}\|_\infty < \varepsilon$ and $\hat{f}_\varepsilon = 0$ on $(-\delta/2, \delta/2)$. It follows using (27.11) that

$$\int_{-1}^{1} \hat{f}_\varepsilon(t)\phi_n(t)\,dt \to 0 \qquad \text{as } n \to \infty$$

and, using (27.12),

$$\left| \int_{-1}^{1} \hat{f}(t)\phi_n(t)\,dt - \int_{-1}^{1} \hat{f}_\varepsilon(t)\phi_n(t) \right| \le \int_{-1}^{1} |\hat{f}_\varepsilon(t) - \hat{f}(t)|\phi_n(t)\,dt \le \varepsilon.$$

Therefore, using (27.13) again,

$$\left| \int_{-1}^{1} f(t)\phi_n(t)\,dt - f(0) \right| = \left| \int_{-1}^{1} \hat{f}(t)\phi_n(t)\,dt \right|$$

$$\leq \left| \int_{-1}^{1} \hat{f}_{\varepsilon}(t)\phi_n(t)\,dt \right| + \varepsilon,$$

which proves (27.10).

For 'only if', (27.11) is clearly required since it is a consequence of (27.10) with $f \equiv 1$ on $[-1, 1]$, and (27.12) is immediate from (27.10) when $g(0) = 0$. The only thing left to prove is the boundedness in (27.13); to see this, we regard

$$f \mapsto \int_{-1}^{1} k_n(t)f(t)\,dt$$

as an element F_n of $C([-1, 1])^*$, and (27.10) says that $F_n \overset{*}{\rightharpoonup} \delta_0$ (where $\delta_0(f) := f(0)$ for $f \in C([-1, 1])$), so by $\|F_n\|_{C^*}$ must be bounded. We showed in Example 11.9 that

$$\|F_n\|_{C^*} = \int_{-1}^{1} |k_n(t)|\,dt,$$

and so (27.13) now follows.

27.9 Each C_n is clearly closed, convex, and non-empty; since C_n is a subset of $\mathbb{B}_X$ it is also bounded. Suppose that $x \in C_k$ for some k; then, for each $\varepsilon > 0$, there is an $m(\varepsilon)$ such $\|x - y\| < \varepsilon$ for some $y \in \text{conv}\{x_k, \ldots, x_{k+m(\varepsilon)}\}$ such that $\|x - y\| < \varepsilon$. It follows that

$$|f_n(x)| = |f_n(x - y)| < \varepsilon \qquad n > k + m(\varepsilon).$$

This shows that $f_n(x) \to 0$ as $n \to \infty$.

However, since $f_k(x_j) \geq \theta$ for all $j \geq k$, it follows that we have $f_k(z) \geq \theta$ for all $z \in \text{conv}\{x_k, x_{k+1}, x_{k+2}, \ldots\}$. In particular, if $x \in \cap_n C_n$, then we must have $f_n(x) \geq \theta$ for each n. But this contradicts the fact that $f_n(x) \to 0$, and so $\cap_n C_n = \varnothing$.

27.10 For each n choose some $x_n \in C_n$; then, since (x_n) is bounded, there is a subsequence (x_{n_j}) that converges weakly to some $x \in X$. If $x \notin C_k$ for some k then by Corollary 21.8 there exists $f \in X^*$ such that

$$f(x) < \inf_{y \in C_k} f(y) \leq f(x_{n_j})$$

for all j, since $x_{n_j} \in C_{n_j} \subset C_k$. It follows that

$$f(x) \leq \lim_{j \to \infty} f(x_{n_j}) = f(x),$$

a contradiction.

References

Adams, R.A. (1975) *Sobolev spaces*. Academic Press, New York.

Banach, S. ([1932] 1978) *Théorie des opérations linéaires*. Repr. Chelsea, New York.

Bergman, G. M. (1997) *The Axiom of Choice, Zorn's Lemma, and all that*. https://
math.berkeley.edu/~gbergman/grad.hndts/AC+Zorn+.ps.

Bollobás, B. (1990) *Linear analysis*. Cambridge University Press, Cambridge.

Brezis, H. (2011) *Functional analysis, Sobolev spaces and partial differential equations*. Springer, New York.

Brown, A. L. and Page, A. (1970) *Elements of functional analysis*. Van Nostrand Reinhold, London.

Carothers, N. L. (2005) *A short course on Banach space theory*. Cambridge University Press, Cambridge.

Clarkson, J. A. (1936) Uniformly convex spaces. *Trans. AMS* **40**, 396–414.

Costara, C. and Popa, D. (2003) *Exercises in functional analysis*. Kluwer Academic, Dordrecht, Netherlands.

Eberlein, W. F. (1947) Weak compactness in Banach spaces. I. *Proc. Natl. Acad. Sci. U.S.A.* **33**, 51–3.

Enflo, P. (1973) A counterexample to the approximation problem in Banach spaces. *Acta Math.* **130**, 309–17.

Evans, L. C. (1998) *Partial differential equations*. American Mathematical Society, Providence, RI.

Friedberg, S., Insel, A. and Spence, L. (2014) *Linear algebra*. Pearson, Harlow.

Giles, J. R. (2000) *Introduction to the analysis of normed linear spaces*. Cambridge University Press, Cambridge.

Goffman, C. and Pedrick, G. (1983) *First course in functional analysis*. Chelsea, New York.

Hartman, P. (1973) *Ordinary differential equations*. John Wiley, Baltimore.

Heinonen, J. (2003) *Geometric embeddings of metric spaces*, Report 90, Department of Mathematics and Statistics, University of Jyväskylä.

Helly, E. (1912) Über lineare Funkionaloperationen. *S.-B. K. Akad. Wiss. Wien Math.-Naturwiss. Kl.* **121**, 265–97.

Holland, F. (2016) A leisurely elementary treatment of Stirling's formula. *Irish Math. Soc. Bull.* **77**, 35–43.

394

James, R. C. (1951) A non-reflexive Banach space isometric with its second conjugate space. *Proc. Natl. Acad. Sci. U.S.A.* **37**, 174–17.

James, R. C. (1964) Weak compactness and reflexivity. *Israel J. Math.* **2**, 101–19.

Jordan, P. and von Neumann, J. (1935) On inner products in linear metric spaces. *Ann. Math.* **36**, 719–23.

Körner, T. W. (1989) *Fourier analysis.* Cambridge University Press, Cambridge.

Kreyszig, E. (1978) *Introductory functional analysis with applications.* John Wiley, New York.

Lax, P. D. (2002) *Functional analysis.* John Wiley, New York.

Megginson, R. E. (1998) *An introduction to Banach space theory.* Graduate Texts in Mathematics 183. Springer, New York.

Meise, R. and Vogt, D. (1997) *Introduction to functional analysis.* Oxford University Press, Oxford.

Munkres, J. R. (2000) *Topology,* second edition. Prentice Hall, Upper Saddle River, NJ.

Naylor, A. W. and Sell, G. R. (1982) *Linear operator theory in engineering and science.* Springer Applied Mathematical Sciences 40. Springer, New York.

Priestley, H. (1997) *Introduction to integration.* Oxford University Press, Oxford.

Pryce, J. D. (1973) *Basic methods of linear functional analysis.* Hutchinson, London.

Renardy, M. and Rogers, R. C. (1993) *An introduction to partial differential equations.* Texts in Applied Mathematics 13. Springer, New York.

Rudin, W. (1966) *Real and complex analysis.* McGraw-Hill, New York.

Rudin, W. (1991) *Functional analysis.* McGraw-Hill, New York.

Rynne, B. P. and Youngson, M. A. (2008) *Linear functional analysis.* 2nd edn. Springer, London.

Stein, E. M. and Shakarchi, R. (2005) *Real analysis: measure theory, integration, and Hilbert spaces.* Princeton University Press, Princeton, NJ.

Sutherland, W. A. (1975) *Introduction to metric and topological spaces.* Oxford University Press, Oxford.

Yosida, K. (1980) *Functional analysis.* Springer Classics in Mathematics. Springer, Berlin.

Young, N. (1988) *An introduction to Hilbert space.* Cambridge University Press, Cambridge.

Zeidler, E. (1995) *Applied functional analysis.* Springer Applied Mathematical Sciences 108. Springer, New York.

Index